国外优秀数学著作
原版系列

Number Theory(Dreaming in Dreams)
——Proceedings of the 5th China-Japan Seminar

数论（梦幻之旅）

——第五届中日数论研讨会演讲集

● ● ● ［日］青木崇 (Takashi Aoki)
● ● ● ［日］金光茂 (Shigeru Kanemitsu)
● ● ● 刘建亚 (Jianya Liu)

著

哈尔滨工业大学出版社
HARBIN INSTITUTE OF TECHNOLOGY PRESS

黑版贸审字 08—2017—078 号

图书在版编目(CIP)数据

数论：梦幻之旅：第五届中日数论研讨会演讲集：英文/(日)青木崇(Takashi Aoki)，(日)金光茂(Shigeru Kanemitsu)，刘建亚(Jianya Liu)著. —哈尔滨：哈尔滨工业大学出版社,2018.2

书名原文：Number Theory (Dreaming in Dreams)：Proceedings of the 5th China—Japan Seminar

ISBN 978 - 7 - 5603 - 7154 - 2

Ⅰ.①数⋯ Ⅱ.①青⋯ ②金⋯ ③刘⋯ Ⅲ.①数论-文集-英文 Ⅳ.①O156—53

中国版本图书馆 CIP 数据核字(2017)第 303269 号

策划编辑   刘培杰
责任编辑   张永芹   杜莹雪
封面设计   孙茵艾
出版发行   哈尔滨工业大学出版社
社   址   哈尔滨市南岗区复华四道街 10 号   邮编 150006
传   真   0451 - 86414749
网   址   http://hitpress. hit. edu. cn
印   刷   哈尔滨市工大节能印刷厂
开   本   787mm×1092mm   1/16   印张 19.25   字数 391 千字
版   次   2018 年 2 月第 1 版   2018 年 2 月第 1 次印刷
书   号   ISBN 978 - 7 - 5603 - 7154 - 2
定   价   68.00 元

Dedicated to Schinzel on the occasion of his 70th birthday.
Sto lat, sto lat. Niech zyje nam.

PREFACE

The present volume is the proceedings of the 5th Japan-China Seminar on Number Theory "Dreaming in dreams" held during August 27-31, 2008 at Kinki University, Higashi-osaka, Japan, the organizers being Shigeru Kanemitsu and Jianya Liu with Professor Takashi Aoki as the local organizer.

The title sounded somewhat romantic or exotic and one of the participants, Professor Tim. D. Browning, a relative of the world famous poet R. Browning, expressed a poetic view that the title suggested that one could dream of proving the RH or whatsoever of the hardest nuts to crack in dreams. But we chose this title in view of the following due reason. Osaka is most well-known for its world famous Osaka Castle. The builder of Osaka Castle, Hideyoshi Toyotomi, a hero in the 16th century made a poem at his deathbed. "Like a dew drop was I born and into a dew drop am I fading, all that prevails in Naniwa (the present world) is like dreams in a dream." Here Naniwa sounds the same as the old name of Osaka. Also many of us went to the center of the city (for empty orchestra, perhaps), Namba, which is the modern name of Naniwa. This gives a good reason to entitle the seminar. This may not sound poetic but associatively logical. Indeed, at the end we have a poem composed by Professor Chaohua Jia. Thus we now have at least four poets among participants, including Professors Tianxin Cai (a professional), Chaohua Jia, Jianya Liu (who composed a poem in the proceedings of the 4th China-Japan Seminar), and Tim D. Browning.

The atmosphere was enjoyable as usual and we believe everyone enjoyed the 5 rich days. We organized various social activities including reception party at Sheraton-Miyako Hotel to which we thank for their great hospitality and generosity of providing us with champagne bottles. We made a tour to Kyoto (the bus was arranged through Sheraton-Miyako) and visited a few musts there. Some of the foreign participants paid multiple visits to Kinkakuji Temple and Kiyomizu Temple. Evening events were also entertaining which included empty orchestra activities at various places. We found that not only Chinese participants who were known to be good and enthusiastic singers, but most of the Western participants shared the same spirit. We found Professors Trevor Wooley, Winfried Kohnen, Katsuya Miyake and Yumiko Hironaka most entertaining. Professor Jörg Brüdern, though didn't sing, promised to play the guitar at the next occasion. S. Kanemitsu, to his regret, missed the chance of attending the ever-night show including Professor Tim. D. Browning and Professor Koichi Kawada.

Now about the contents of the seminar and the present proceedings. The talks ranged over a wide spectrum of contemporary number theory. As can be seen from the papers themselves as well as in the following brief descriptions in this volume, we succeeded in assembling topics from Analytic Number Theory (Classical and Modern with emphasis on additive number theory), Theory of Modular Forms, Algebraic Groups and Algebraic Number Theory.

In the proceedings we collected not only papers from the participants but from those invitees who could not attend the seminar, including Professors Andrzej Schinzel (who was about to come), Igor Shparlinski and Ken Yamamura.

In [Browning] a new direction of research in analytic number theory is exhibited, i. e. a quantitative study on the distribution of rational points of some variety—specifically, a del Pezzo surface of degree 4, $V \subset \mathbb{P}^4$ which is defined over the rationals and is assumed to have a conic bundle structure. The main result is the asymptotic formula for the counting function

$$N_{U_0,H}(B) = \#\{x \in U_0(\mathbb{Q}) : H(x) \leq B\},$$

where U_0 is a certain Zariski open subset and $H(x)$ is a certain norm function. The formula reads

$$N_{U_0,H}(B) = c_{V_0,H} B(\log B)^4 + \text{error term},$$

establishing the Manin conjecture in this case, where $c_{V_0,H}$ is the constant conjectured by Peyre.

The proof involves various ingredients, the geometric Picard group $\text{Pic}(V_0) \simeq \mathbb{Z}^5$, analysis of conic sections and classical techniques including lattice points counting and divisor problems for binary forms.

[Brüdern-Kawada-Wooley] is the 8th of their series of papers "Additive representation in thin sequences" I-VII and is a timely summary which looks over their recent results in an enlightening way.

Their main concern is the diagonal form

$$\lambda_1 x_1^k + \cdots + \lambda_s x_s^k,$$

as $x_1, \cdots, x_s$ ranges over $\mathbb{Z}$ or a subset thereof, where $s \geq 2$, $k \geq 1$ and $\lambda_1, \cdots, \lambda_s$ are non-zero real numbers.

The purpose of the paper is two-fold; on one hand, it centers around Diophantine inequalities and on the other on the potential of the methods developed in the series, including the Davenport-Heilbronn Fourier transform method, a counterpart of the Hardy-Littlewood circle method for Diophantine inequalities. Starting from the case of additive cubic forms, the authors give a very clear survey on their hitherto contributions, giving proofs of some of the important theorems, which makes the paper more instructive and readable.

In the paper [Hoshi-Miyake] the authors are concerned with the FIP (Field Isomorphism Problem) on k-generic polynomial $f_{\mathbf{t}}^G(X) \in k(\mathbf{t})[X]$, where k is a field of arbitrary characteristic, G is one of finite groups D_3, D_4, D_5 (dihedral groups) and C_3, C_4 (cyclic groups), and $k(\mathbf{t})$ is the rational function field over k with n indeterminates $\mathbf{t} = (t_1, t_2, \cdots, t_n)$. A monic separable polynomial $f_{\mathbf{t}}^G(X)$ is called a k-generic polynomial for G if

- (G_1) the Galois group of $f_{\mathbf{t}}^G(X)$ over $k(\mathbf{t})$ is isomorphic to G and

- (G_2) every G-extension $L/K, K \supset k$ may be obtained as $L = \mathrm{Spl}_K f_{\mathbf{a}}^G$, the splitting field of $f_{\mathbf{a}}^G$ over K, for some $\mathbf{a} = (a_1, a_2, \cdots, a_n) \in K^n$.

The FIP reads: For a field $K \supset k$ and $\mathbf{a}, \mathbf{b} \in K^n$, determine whether $L_{\mathbf{a}} = \mathrm{Spl}_K f_{\mathbf{a}}^G$ and $L_{\mathbf{b}} = \mathrm{Spl}_K f_{\mathbf{b}}^G$ are isomorphic over K or not. Numerical examples are given in the cases $G = C_4$ and $G = D_5$.

Two more related problems are stated without details: Subfield Problem and Field Intersection Problem for generic polynomials.

In his paper [Jia], C. -H. Jia gives some developments over the results on dynamics of the w-function introduced by W. -S. Goldring in 2006, where dynamics refers to the orbit of successive iterates of the function. For $n = p_1 p_2 p_3 \in S := C_3 \cup B_3$ (p_i's are prime, not all equal), the w-function is defined by

$$w(n) = P(p_1 + p_2) P(p_2 + p_3) P(p_3 + p_1),$$

where $P(n)$ signifies the largest prime factor of n. The objective is to classify the elements of S and there are many results obtained by Goldring, Y.-G. Chen et al.

The inverse problem of finding the inverse image of the w-function is also of interest. One of the conjectures of Goldring states that every element of C_3, which is the set of all $n = p_1 p_2 p_3$ with p_i all distinct primes, has infinitely many C_3-parents, i.e. there are infinitely many $m \in C_3$ such that $w(m) = n$. Jia proves, by making a novel use of the large sieve method, some quantitative results including Theorem 9 saying that there is an element of $r_2 r_2 q \in C_3$, where $r_1, r_2 \sim \sqrt{x} \log x$, $q \leq 4x$ which has many distinct C_3-parents (where the data is quantitative).

In the paper [Kohnen], Kohnen surveys on the recent results his paper jointly with Mason on the generalized modular functions (GMF) (of weight zero) on Γ, where $\Gamma \subset SL_2(\mathbb{Z})$ is a subgroup of finite index. A GMF is a holomorphic function $f : \mathcal{H} \to \mathbf{C}$ satisfying the following two conditions.

i) $f(\gamma \circ z) = \chi(\gamma) f(z)$ $(\forall \gamma \in \Gamma)$, where $\chi : \Gamma \to \mathbf{C}^*$ is a (not necessarily unitary) character of Γ,

ii) f is meromorphic at the cusps of Γ.

In the paper [Komori-Matsumoto-Tsumura], the authors report on some recent developments on the second author's
Problem: Is it possible to find some functional relations for multiple zeta-functions which include some value-distribution for MZV's (multiple zeta values)? Here the multiple zeta-function of complex variables $\mathbf{s} = (s_1, s_2, \cdots . s_r)$ is defined by

$$\zeta(s_1, s_2, \ldots, s_r) = \sum_{m_1 > m_2 > \cdots > m_r \geq 1} \frac{1}{m_1^{s_1} m_2^{s_2} \cdots m_r^{s_r}}.$$

and the MZV of depth r is $\zeta(k_1, k_2, \ldots, k_r)$ with $k_1, k_2, \ldots, k_r \in \mathbb{N}, k_1 > 1$.

The authors are concerned with the multi-variable (version of the) Witten zeta-function defined by

$$\zeta_r(\mathbf{s}; \mathfrak{g}) = \sum_{m_1=1}^{\infty} \cdots \sum_{m_r=1}^{\infty} \prod_{\alpha \in \Delta_+} \langle \alpha^\vee, m_1 \lambda_1 + \cdots + m_r \lambda_r \rangle^{-s_\alpha},$$

where $\mathfrak{g}$ is a complex semi-simple Lie algebra with rank r, $\mathbf{s} = (s_\alpha)_{\alpha \in \Delta_+} \in \mathbb{C}^n$ and the data appearing on the right-hand side are certain quantities associated with $\mathfrak{g}$.

Theorem 5.1 seems to be the culmination of the results which gives a general form of the functional relations for the multiple zeta-function $\zeta_r(\mathbf{s}, \mathbf{y}; \Delta)$ with an additive character of a root system, i.e. for

$$\sum_{w \in W^I} \left(\prod_{\alpha \in \Delta_{w^{-1}}} (-1)^{-s_\alpha} \right) \zeta_r(w^{-1}\mathbf{s}, w^{-1}\mathbf{y}; \Delta).$$

However, to deduce explicit functional relations for specific root systems from Theorem 5.1 seems rather cumbersome and the authors give some handy formulas in §6 which they apply in later sections to deduce explicit relations for $\mathfrak{g} = A_3$, B_2, B_3, C_3. There are given many concrete examples which make the paper readable.

[Liu] is a short introduction to Maass wave forms, which originated from his lectures at Postech in 2007. Restricting to the simplest setting of the Maass forms of weight 0 on the full modular group, he manages to give a quick introduction to its formidable theory and renders the situation more familiar to analytic number-theorists. From the contents one can see the flowchart of the paper. He starts from Fourier expansion of Maass forms, proving thereby the Chowla-Selberg formula for the Eisenstein series, goes on to the spectral decomposition of the Hilbert space of square-integrable automorphic functions with respect to the non-Euclidean Laplacian, establishing the facts about the Laplace eigenvalues. After introducing Hecke theory, he introduces the automorphic L-functions and develop analytic methods to study them. Towards the end of the paper, a Linnik-type problem for Maass forms is studied, exhibiting how analytic number theory can develop on such exotic stages.

There are two nice collections of problems by Andrzej and Igor'.

In [Schinzel] there is a collection of problems concerning the number $N(f)$ of non-zero coefficients of a polynomial $f \in K$, K being a field. f is called an $N(f)$-nomial, e. g. $x^n - a$ is a binomial while $4x^{20} + 7x^{18} + 64$ is a trinomial. Problems are about the estimate from above or below on $N(f)$. E. g. Problem 2 asks about the existence (and boundedness if it exists) of a constant $C(K)$ such that every trinomial over K has an irreducible factor f with $N(f) < C(K)$.

In [Shparlinski] many open problems are given on the estimate of exponential and character sums according to the author's taste and interest, with enlightening annotation. Problems fall into two categories; in the first category new improvements of the estimates are asked for, while the second is concerned with applications of these sums. We shall talk about the second category.

Problem 3.6 has a natural interpretation in the study of polynomially growing sequences on orbits of the dynamical system generated by the map $u \to gu$ in $\mathbb{Z}/g\mathbb{Z}$. Problem 3.11 gives the estimate for complexity of factorization algorithm for polynomials over $\mathbb{F}_p$. Problem 3.13 has applications to the study of distribution of Selmer groups of a certain family of elliptic curves. Problem 3.13 is concerned with the estimate of the exponential sum of a sparse polynomial and has applications to number theory, computer science and cryptography. Problem 3.43 is about the exponential sum of a non-linear recurrence sequence and has applications to pseudo-random number generation.

We invited Professor Yamamura to contribute his list of determinantal expressions for the class number of Abelian number fields. The theory started from the paper of Carlitz and Olson in 1955 and has been pursued constantly. The list is in the spirit of Dilcher-Skula-Slavutskii volume on Bernoulli numbers, is complete and will be useful for researchers in the relevant problems.

And we also need to add one sentence, Professor Haruo Tsukada added corrigendum to his paper published in the last proceedings.

Finally, vote of thanks is due. We would like to thank Kinki University for its generous permission of using its excellent facilities. The conference room was equipped with modern conveniences and was very useful in conducting the seminar. We would like to thank Professor Kohji Chinen for his help in preparing posters of the seminar as well as constant support in keeping the working conditions in the conference room pleasant. We would like to thank Dr. Hiromitsu Tanaka for technical help in manipulating modern devices and for his thoughtful arrangement of things. We would like to thank Sheraton-Miyako Hotel for its excellent service and hospitality; especially thanks are due to Messers H. Fujihara and K. Morimoto for their thoughtful support throughout.

As in the case of the last proceedings, Professor Jing Ma from Jilin University made a devoted help in editing and we record here our hearty thanks to her for her excellent and beautiful preparation of the manuscript of the proceedings. It was a pity that she could not attend the seminar, but she came to Japan in July, 2009 to complete the editing work.

Finally, we would like to express our hearty thanks to S. Kanemitsu's students, Mr. N. -L. Wang and Ms. X. -H. Wang for their devoted support in making the stay of foreign participants more comfortable and pleasant.

As usual we complete the preface by a poem. This time Professor Chaohua Jia composed it.

大阪行

賈朝華

迷離難波夜
清澄澱河川
論道花叢裡
何須上深山

CONTENTS

x

RESENT PROGRESS ON THE QUANTITATIVE ARITHMETIC OF DEL PEZZO SURFACES

TIM D. BROWNING

School of Mathematics, University of Bristol,
University Walk, Clifton, Bristol BS8 1TW, United Kingdom
t.d.browning@bristol.ac.uk

We survey the state of affairs for the distribution of $\mathbb{Q}$-rational points on non-singular del Pezzo surfaces of low degree, highlighting the recent resolution of Manin's conjecture for a non-singular del Pezzo surface of degree 4 by la Bretèche and Browning [3].

1. Introduction

Let $V \subset \mathbb{P}^{n-1}$ be a non-singular projective variety defined over $\mathbb{Q}$. A fundamental theme in mathematics is to understand when the set of rational points $V(\mathbb{Q}) = V \cap \mathbb{P}^{n-1}(\mathbb{Q})$ is non-empty, and if it is non-empty, to provide a reasonable description of it. Our aim here is to survey recent joint work of the author with la Bretèche [3], which addresses the second question from a quantitative standpoint for a certain variety of dimension 2.

Before turning to the specifics of this work it is perhaps worth entertaining ourselves with some preliminary observations. In the classification of algebraic varieties the Fano varieties, namely those varieties V whose anticanonical divisor $-K_V$ is ample, are those on which we expect to stand the highest chance of finding rational points. The situation for curves is relatively well-understood and so we will consider here the case of Fano surfaces, beginning with some simple-minded numerics. Imagine that we are given a Fano variety V of dimension 2 and degree d, which is a non-singular complete intersection in $\mathbb{P}^{n-1}$. Quadrics are easy to analyse and so we assume that $d > 2$. Suppose that $V = W_1 \cap \cdots \cap W_t$ for hypersurfaces $W_i \subset \mathbb{P}^{n-1}$ of degree d_i and assume, furthermore, that the intersection is generically transversal. We are not interested in hyperplane sections of V, and so we will assume without loss of generality that $d_i \geq 2$. Then the following inequalities must be satisfied:

(1) $d_1 + \cdots + d_t < n$, [Fano]

(2) $n - 1 - t = 2$, [complete intersection of dimension 2]

(3) $d = d_1 \cdots d_t \geq 3$, [Bézout]

(4) $d_t \geq \cdots \geq d_1 \geq 2$.

It follows that the only possibilities are

$$(d; d_1, \ldots, d_t; n; t) \in \{(3; 3; 4; 1), (4; 2, 2; 5; 2)\}.$$

These surfaces correspond to a cubic surface in $\mathbb{P}^3$ and an intersection of 2 quadrics in $\mathbb{P}^4$, respectively. In fact these are the most famous examples of "del Pezzo surfaces". This article will focus its attention on the latter type of surfaces, which comprise the non-singular del Pezzo surfaces of degree 4.

Henceforth, we let $V \subset \mathbb{P}^4$ be such a del Pezzo surface, cut out by the intersection of two quadrics

$$\begin{cases} \Phi_1(x_0, \ldots, x_4) = 0, \\ \Phi_2(x_0, \ldots, x_4) = 0, \end{cases}$$

with Φ_1, Φ_2 quadratic forms defined over $\mathbb{Z}$. We will refer to such a surface as a "dP$_4$" for ease of notation. The geometry of dP$_4$s is classical and is discussed in Hartshorne [11, Section V.4], for example. The question of finding conditions under which $V(\mathbb{Q})$ is non-empty is a difficult problem in number theory that will not concern us here. We refer the interested reader to the thesis of Wittenberg [25].

Assuming that $V(\mathbb{Q}) \neq \emptyset$, it follows from work of Segre [22] that $V(\mathbb{Q})$ is dense in V under the Zariski topology. Thus it is natural to give some measure of the density of rational points on V. It is at this point that the Manin conjecture [8] enters the picture. This predicts an asymptotic formula for the growth rate of the counting function

$$N_{U,H}(B) = \#\{x \in U(\mathbb{Q}) : H(x) \leq B\},$$

as $B \to \infty$, where $U \subset V$ is the Zariski open subset formed by deleting the 16 lines from V and $H : \mathbb{P}^4(\mathbb{Q}) \to \mathbb{R}_{\geq 0}$ is defined via $H(x) = \|\mathbf{x}\|$, if $x = [\mathbf{x}]$ with $\mathbf{x} = (x_0, \ldots, x_4) \in \mathbb{Z}^5$ such that $\gcd(x_0, \ldots, x_4) = 1$. Here $\| \cdot \|$ is an arbitrary norm on $\mathbb{R}^5$. Let $\mathrm{Pic}_{\overline{\mathbb{Q}}}(V)$ be the geometric Picard group of V, which is just the group of Weil divisors $\mathrm{Div}(V)$ modulo linear equivalence. Suppose that K is a splitting field for the 16 lines on V. Assuming that $V(\mathbb{Q})$ is non-empty, the Manin conjecture predicts the existence of a positive constant $c_{V,H}$ such that

$$N_{U,H}(B) = c_{V,H} B (\log B)^{\mathrm{rankPic}(V)-1} \bigl(1 + o(1)\bigr), \qquad (1.1)$$

as $B \to \infty$, where $\mathrm{Pic}(V) = \mathrm{Pic}_{\overline{\mathbb{Q}}}(V)^{\mathrm{Gal}(K/\mathbb{Q})}$. Furthermore, there is a conjecture due to Peyre [20] concerning the value of the constant $c_{V,H}$.

It should be stressed that the Manin conjecture has received a great deal of attention in the context of singular del Pezzo surfaces, an account of which can be found in the author's companion survey [4]. The situation for non-singular dP$_4$s is rather less satisfactory. For an arbitrary such surface the best result we have is the upper bound

$$N_{U,H}(B) = O_{\varepsilon}(B^{\frac{3}{2}+\varepsilon}),$$

for any $\varepsilon > 0$. This is due to Salberger, but has yet to appear in print. It is based on a far-reaching refinement of Heath-Brown's "determinant method" developed in [14]. The implied constant in Salberger's estimate is allowed to depend on ε but is uniform in the coefficients of Φ_1 and Φ_2. It seems likely that by adapting further work of Heath-Brown [13] one could hope to replace the exponent $\frac{3}{2}$ by $\frac{5}{4}$, under a certain unproven hypothesis concerning the growth rate of the rank of elliptic curves over $\mathbb{Q}$.

One can do better when the dP$_4$ is assumed to have a "conic bundle structure". This simply boils down to Φ_1 taking the shape

$$\Phi_1(x_0, \ldots, x_4) = x_0 x_1 - x_2 x_3, \tag{1.2}$$

which we now suppose to be the case. In work communicated at the conference *Higher dimensional varieties and rational points* at Budapest in 2001, Salberger has established the upper bound

$$N_{U,H}(B) = O_{\varepsilon,V}(B^{1+\varepsilon}) \tag{1.3}$$

for any $\varepsilon > 0$, where now the implied constant is allowed to depend on ε and V. Leung [17] has subsequently shown how to replace B^{ε} in Salberger's work by a small power of $\log B$. His work implies that

$$N_{U,H}(B) = O_{\varepsilon,V}(B(\log B)^5). \tag{1.4}$$

This is best possible whenever the 16 lines are all defined over $\mathbb{Q}$, since the geometric Picard group of a dP$_4$ has rank 6.

As highlighted by Swinnerton-Dyer [24, Question 15], it has become something of milestone to establish the Manin conjecture for a single example of a dP$_4$. The gradual improvements recorded above all point the way towards the possibility of actually establishing an asymptotic formula for the counting function associated to a well-chosen dP$_4$. We will see that this is indeed the case for the surface defined by the forms (1.2) and

$$\Phi_2(x_0, \ldots, x_4) = x_0^2 + x_1^2 + x_2^2 - x_3^2 - 2x_4^2. \tag{1.5}$$

Let $V_0 \subset \mathbb{P}^4$ denote the corresponding conic bundle dP_4 and let U_0 be the Zariski open subset formed by deleting the lines from V_0. It is easy to check that V_0 is non-singular and we will prove shortly that $\mathrm{Pic}(V_0) \cong \mathbb{Z}^5$, with $K = \mathbb{Q}(i)$. We are now ready to reveal the main result that is surveyed in this note, and which confirms the conjectured estimate (1.1) for the surface under consideration.

Theorem. *We have*

$$N_{U_0,H}(B) = c_{V_0,H}B(\log B)^4 + O\Big(\frac{B(\log B)^4}{\log\log B}\Big),$$

where $c_{V_0,H} > 0$ is the constant predicted by Peyre.

For comparison, Fok [17] also studies the surface V_0, obtaining the upper bound

$$N_{U_0,H}(B) = O\big(B(\log B)^4\big). \tag{1.6}$$

We will see in due course how our argument leads to this estimate rather easily. In the following section we will show how to compute the Picard group associated to the surface V_0. While these calculations are routine they are perhaps not entirely familiar to analytic number theorists and so we will present fuller details than appear in [3]. Section 3 comprises the bulk of this survey and contains a discussion of the proof of the theorem. A number of innocent simplifications will be made along the way and so any serious reader would do better to consult [3] directly. Finally, in Section 4 we will suggest some avenues for future exploration.

Acknowledgements

Some of this paper was prepared while the author was enjoying the hospitality of Professor Jianya Liu at Shandong University in Weihei and Professor Shigeru Kanemitsu at the *5th China–Japan Conference* at Kinki University in Osaka. It is a great pleasure to thank these two gentlemen for the convivial atmosphere that they provided. While working on this paper the author was supported by EPSRC grant number EP/E053262/1.

2. Geometry of V_0

Our main task in this section is to calculate an explicit basis for the Picard group of the surface V_0 that is defined by (1.2) and (1.5). We begin by recalling some basic facts about the geometry of general dP_4s $V \subset \mathbb{P}^4$. All of the facts that we will need can be found in the book by Manin [18]. Such

V are isomorphic to the blow-up of $\mathbb{P}^2$ along a union of 5 points $p_1, \ldots, p_5$, no 3 of which are collinear. Let E_i denote the exceptional divisor above p_i, for $1 \le i \le 5$, let $L_{i,j}$ denote the strict transform of the line going through p_i and p_j, for $1 \le i < j \le 5$, and let Q denote the strict transform of the unique conic going through all 5 points. These 16 divisors constitute the famous 16 lines on V.

If Λ is the strict transform of a line in $\mathbb{P}^2$ not passing through any of the 5 distinguished points, then a free basis for the geometric Picard group $\mathrm{Pic}_{\overline{\mathbb{Q}}}(V)$ is given by $\{\Lambda, E_1, \ldots, E_5\}$. Here, as in all that follows, we have found it convenient to identify divisors with their classes in $\mathrm{Pic}_{\overline{\mathbb{Q}}}(V)$. In terms of this basis the anticanonical divisor class can be written $-K_V = 3\Lambda - \sum_{i=1}^{5} E_i$, and furthermore, we have $L_{i,j} = \Lambda - E_i - E_j$ and $Q = 2\Lambda - \sum_{i=1}^{5} E_i$. There is an intersection form $(\cdot, \cdot)$ on $\mathrm{Pic}_{\overline{\mathbb{Q}}}(V)$, which is non-degenerate, symmetric and bilinear. In terms of this intersection form the divisors $\Lambda, E_1, \ldots, E_5$ satisfy the relations

$$(\Lambda, \Lambda) = 1, \quad (\Lambda, E_i) = 0, \quad (E_i, E_j) = \begin{cases} -1, & \text{if } i = j, \\ 0, & \text{if } i \ne j. \end{cases}$$

One can now check that $-K_V$ is very ample, with self-intersection number $4 = \deg V$. We are now ready to produce an explicit basis for $\mathrm{Pic}(V_0)$.

Let $\varepsilon_1, \varepsilon_2 \in \{-1, +1\}$ and let $i = \sqrt{-1}$. Then a straightforward calculation reveals that the 16 lines on V_0 are given by

$$M_1(\varepsilon_1, \varepsilon_2) : \begin{cases} x_0 = \varepsilon_1 x_2 = \varepsilon_2 x_4, \\ x_1 = \varepsilon_1 x_3, \end{cases} \qquad M_2(\varepsilon_1, \varepsilon_2) : \begin{cases} x_1 = \varepsilon_1 x_2 = \varepsilon_2 x_4, \\ x_0 = \varepsilon_1 x_3, \end{cases}$$

$$M_3(\varepsilon_1, \varepsilon_2) : \begin{cases} x_3 = \varepsilon_1 i x_0 = \varepsilon_2 i x_4, \\ x_1 = \varepsilon_1 i x_2, \end{cases} \qquad M_4(\varepsilon_1, \varepsilon_2) : \begin{cases} x_3 = \varepsilon_1 i x_1 = \varepsilon_2 i x_4, \\ x_0 = \varepsilon_1 i x_2. \end{cases}$$

In particular all of the lines split over the Gaussian field $\mathbb{Q}(i)$. Let us write $\mathcal{G} = \mathrm{Gal}(\mathbb{Q}(i)/\mathbb{Q}) \cong \mathbb{Z}/2\mathbb{Z}$ for the Galois group of the splitting field. Among the lines above we will need to identify 5 mutually skew lines that are globally invariant under the action of $\mathcal{G}$. Let us set

$$E_1 = M_1(-1, 1), \quad E_2 = M_1(1, -1), \ E_3 = M_2(-1, -1),$$
$$E_4 = M_4(-1, -1), \ E_5 = M_4(1, 1).$$

These lines satisfy $(E_i, E_j) = 0$ for each $1 \le i < j \le 5$. Furthermore, E_1, E_2, E_3 are defined over $\mathbb{Q}$, and E_4, E_5 are defined over $\mathbb{Q}(i)$, but are conjugate under the action of $\mathcal{G}$. We may therefore take $\Lambda, E_1, \ldots, E_5$ as a basis for $\mathrm{Pic}_{\overline{\mathbb{Q}}}(V_0)$, whence $\mathrm{Pic}(V_0) = \{\Lambda, E_1, E_2, E_3, E_4 + E_5\}$. This establishes the fact that $\mathrm{Pic}(V_0) \cong \mathbb{Z}^5$, as recorded in the introduction. Using the intersection form it is now routine to show that we can choose the projection of V_0 to $\mathbb{P}^2$ in such a way that we have the equalities

$$L_{1,2} = M_2(1,-1), \ L_{1,3} = M_1(-1,-1), \ L_{1,4} = M_3(-1,-1),$$
$$L_{1,5} = M_3(1,1), \quad L_{2,3} = M_1(1,1), \qquad L_{2,4} = M_3(1,-1),$$
$$L_{2,5} = M_3(-1,1), \ L_{3,4} = M_4(-1,1), \quad L_{3,5} = M_4(1,-1),$$
$$L_{4,5} = M_2(1,1), \quad Q = M_2(-1,1),$$

in the description of the lines.

Let us identify $\mathrm{Pic}(V_0) \otimes_{\mathbb{Z}} \mathbb{R}$ with its dual using the intersection form. Let $\Lambda_{\mathrm{eff}}(V_0)$ denote the convex cone in $\mathrm{Pic}(V_0) \otimes_{\mathbb{Z}} \mathbb{R}$ that is generated by the classes of effective divisors, and let $\Lambda_{\mathrm{eff}}^{\vee}(V_0)$ denote its dual cone, with respect to the intersection form. As is well-known, $\Lambda_{\mathrm{eff}}(V_0)$ is generated by the classes $[O] = \sum_{L \in O} L$ associated to each orbit O of the action of $\mathcal{G}$ on the 16 lines. In our setting we have 12 orbits overall, given by

$$O_1(\varepsilon_1, \varepsilon_2) = \{M_1(\varepsilon_1, \varepsilon_2)\}, \ O_3(\varepsilon_1, \varepsilon_2) = \{M_3(\varepsilon_1, \varepsilon_2), M_3(-\varepsilon_1, -\varepsilon_2)\},$$
$$O_2(\varepsilon_1, \varepsilon_2) = \{M_2(\varepsilon_1, \varepsilon_2)\}, \ O_4(\varepsilon_1, \varepsilon_2) = \{M_4(\varepsilon_1, \varepsilon_2), M_4(-\varepsilon_1, -\varepsilon_2)\},$$

for the various $\varepsilon_1, \varepsilon_2 \in \{-1, +1\}$. We may therefore conclude that $\Lambda_{\mathrm{eff}}(V_0)$ is generated by the classes

$$
\begin{aligned}
&E_1, E_2, E_3, E_4 + E_5, \\
&\Lambda - E_i - E_j \text{ for } (i,j) \in \{(1,2),(1,3),(2,3),(4,5)\}, \\
&2\Lambda - E_1 - E_2 - E_3 - E_4 - E_5, \\
&2\Lambda - 2E_i - E_4 - E_5 \text{ for } i \in \{1,2,3\},
\end{aligned}
\tag{2.7}
$$

in $\mathrm{Pic}(V_0)$. We will take advantage of this information in §3.5.

3. Overview of the proof

The proof of the theorem can be broken into a number of basic steps. The primary ingredients involved are the following:

(1) reduction to conics of low height;
(2) parametrisation of the conics;
(3) lattice point counting in the plane;
(4) divisor problem for binary forms; and
(5) comparison with Peyre's constant.

We proceed to discuss each of these steps in a little more detail.

3.1. *Reduction to conics of low height*

Recall the definition of the quadratic forms Φ_1, Φ_2 from (1.2) and (1.5). Although this is not crucial to the success of the proof, it simplifies matters to work with the norm on $\mathbb{R}^5$ given by $\|\mathbf{x}\| = \max\{|x_0|, |x_1|, |x_2|, |x_3|, \sqrt{\frac{2}{3}}|x_4|\}$. The starting point of the investigation is the expression

$$N_{U_0, H}(B) = 8N_1(B) + O(B),$$

where

$$N_1(B) = \#\left\{\mathbf{x} \in \mathbb{N}^5 : \begin{array}{l} \gcd(x_0, \ldots, x_3) = 1, \quad \max\{x_0, \ldots, x_3\} \leq B, \\ \Phi_1(\mathbf{x}) = \Phi_2(\mathbf{x}) = 0, \quad \{x_0, x_1\} \neq \{x_2, x_3\} \end{array}\right\}.$$

This amounts to showing that we can restrict the count to positive integer solutions, an argument that follows from elementary considerations. It is here that our choice of norm plays a useful role. As should be clear from Section 2, the condition $\{x_0, x_1\} \neq \{x_2, x_3\}$ ensures that we remain in U_0.

The next stage of the argument involves parametrising the solutions to the equation $\Phi_1(\mathbf{x}) = 0$ and then substituting this parametrisation into the second equation $\Phi_2(\mathbf{x}) = 0$. Without labouring the details, the outcome is that $N_1(B)$ is the number of $(a, b, x, y, z) \in \mathbb{N}^5$ such that

$$C_{a,b} : \quad (a^2 - b^2)x^2 + (a^2 + b^2)y^2 = 2z^2, \tag{3.8}$$

with

$$\gcd(a, b) = \gcd(x, y) = 1, \quad ab, xy \neq 1, \quad \max\{a, b\} \max\{x, y\} \leq B.$$

Fundamental to our argument is the observation

$$(x, y, z) \in C_{a,b} \cap \mathbb{N}^3 \Leftrightarrow (b, a, z) \in C_{y,x} \cap \mathbb{N}^3. \tag{3.9}$$

For fixed coprime a, b such that $ab \neq 1$ it is clear that $C_{a,b} \subset \mathbb{P}^2$ defines a non-singular plane conic with discriminant equal to $-2(a^4 - b^4)$. In Figure 1 we have sketched the affine part of the system of fibres $\{C_{x,1}\}_{x \in \mathbb{A}^1}$. We let $M_{a,b}(B)$ denote the number of $(x, y, z) \in C_{a,b} \cap \mathbb{N}^3$ for which $\gcd(x, y) = 1$, $\max\{a, b\} < \max\{x, y\}$ and $\max\{a, b\} \max\{x, y\} \leq B$. Note that we must automatically have $xy \neq 1$ in this definition. Using (3.9) we find that

$$N_1(B) = \sum_{\substack{a, b \leq B \\ \gcd(a,b)=1, ab \neq 1}} M_{a,b}(B) = 2 \sum_{\substack{a, b < \sqrt{B} \\ \gcd(a,b)=1, ab \neq 1}} M_{a,b}(B) + E(B),$$

where $E(B)$ denotes the contribution from a, b, x, y, z for which $\max\{a, b\}$ is equal to $\max\{x, y\}$.

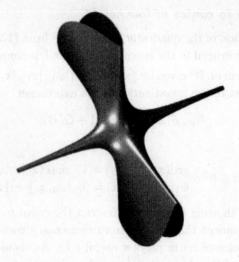

Fig. 1.   The fibres $\{C_{x,1}\}_{x \in \mathbb{A}^1}$

With a little work it is possible to show that $E(B) = O(B(\log B)^3)$, which is satisfactory. Thus we have succeeded in reducing the problem to one that involves a family of conics of relatively low height. In fact there are merely $O(B)$ conics to consider.

Were we charged instead with establishing an upper bound like (1.3) or (1.4) instead of our theorem, our analysis would now be relatively straight-forward, thanks to the control over the growth rate of rational points on conics found in the author's joint work with Heath-Brown [5, Corollary 2]. This implies that

$$M_{a,b}(B) \ll \tau(|a^4 - b^4|)\Big(1 + \frac{B}{\max\{a,b\}^{5/3}|a-b|^{1/3}}\Big),$$

where τ is the divisor function, since the corresponding conic has underlying matrix with determinant $-2(a^4 - b^4)$ and 2×2 minors with greatest common divisor $O(1)$. Here the fact that $\gcd(a,b) = 1$ and $ab \neq 1$ is crucial. Furthermore, we have employed the lower bound $|a^4 - b^4| \gg \max\{a,b\}^3|a-b|$ to simplify the second term in this upper bound.

The upper bound (1.6) is now an easy consequence of taking $x = \sqrt{B}$ in the pair of estimates

$$\sum_{a,b \leq x} \tau(|a^4 - b^4|) \ll x^2(\log x)^3, \qquad \sum_{a < b \leq x} \frac{\tau(|a^4 - b^4|)}{b^{5/3}|a-b|^{1/3}} \ll (\log x)^4.$$

Although we will not prove them here, both of these bounds make essential use of joint work of the author with la Bretèche [1] concerning uniform upper bounds for general sums of the shape

$$\sum_{n_1 \leq N_1} \sum_{n_2 \leq N_2} \varphi(|F(n_1, n_2)|),$$

with $\varphi : \mathbb{N} \to \mathbb{R}_{\geq 0}$ a suitable multiplicative arithmetic function and F a suitable binary form. This investigation extends work by Nair [19] on the corresponding situation for polynomials in only one variable.

3.2. *Parametrisation of the conics*

Turning the upper bound (1.6) into an asymptotic formula requires substantial further work. Recall the definition (3.8) of $C_{a,b}$. Fundamental to the proof is the observation that for each $a, b \in \mathbb{N}$ the conic $C_{a,b}$ always contains the rational point

$$\xi = [1, -1, a].$$

In the classical manner we can use this point to parametrise $C_{a,b}(\mathbb{Q})$, by considering the residual intersection with $C_{a,b}$ of an arbitrary line through ξ. Once carried out this eventually leads to the conclusion that

$$M_{a,b}(B) = \# \left\{ (s,t) \in \mathbb{Z}^2 : \begin{array}{l} \gcd(s,t) = 1, \ |s|, t > 0, \ 0 < Q_3(s,t), \\ 0 < Q_1(s,t), Q_2(s,t) \leq \frac{\lambda B}{\max\{a,b\}} \end{array} \right\} + O(1),$$

where

$$Q_1(s,t) = -2s^2 - (a^2 - b^2)t^2 + 4ast,$$
$$Q_2(s,t) = 2s^2 - (a^2 - b^2)t^2,$$
$$Q_3(s,t) = 2as^2 - 2(a^2 - b^2)st + a(a^2 - b^2)t^2,$$

and

$$\lambda = \gcd(Q_1, Q_2) = \gcd(Q_1 - Q_2, Q_2) = \gcd(4s(s - at), 2s^2 - (a^2 - b^2)t^2).$$

We have almost arrived at a lattice point counting problem, but there are still some annoying coprimality conditions to eliminate from the expression.

It is quite straightforward to discover that $\lambda = 2^\nu \lambda_1 \lambda_2$, with

$$\lambda_1 = \gcd(s, 2s^2 - (a^2 - b^2)t^2)_\flat = \gcd(s, a^2 - b^2)_\flat,$$
$$\lambda_2 = \gcd(s - at, 2s^2 - (a^2 - b^2)t^2)_\flat = \gcd(s - at, a^2 + b^2)_\flat,$$

and a certain explicit expression for ν. Here the symbol $\gcd(\cdot, \cdot)_\flat$ is used to denote the greatest odd common divisor of two integers. In order to simplify the exposition we will not discuss the value of ν here. This depends intimately on a, b, s, t and is calculated in [3, Lemma 13]. To rid ourselves of the coprimality relations we employ Möbius inversion. This gives

$$M_{a,b}(B) = \sum_{\substack{k_i \lambda_i | a^2 + (-1)^i b^2 \\ 2 \nmid k_1 k_2 \lambda_1 \lambda_2}} \mu(k_1)\mu(k_2) \sum_{\ell=1}^{\infty} \mu(\ell) \#(\Lambda \cap \mathcal{R}) + O(1) \qquad (3.10)$$

where $\mathcal{R} \subset \mathbb{R}^2$ is the set of $(s, t) \in \mathbb{R}^2$ such that

$$|s|, t > 0, \quad 0 < Q_3(s,t), \quad 0 < Q_1(s,t), Q_2(s,t) \leq \frac{2^\nu \lambda_1 \lambda_2 B}{\max\{a,b\}}$$

and

$$\Lambda = \left\{ (s,t) \in \mathbb{Z}^2 : k_1 \lambda_1 \mid s, \ k_2 \lambda_2 \mid s - at, \ \ell \mid (s,t) \right\}$$

Clearly Λ is a sublattice of $\mathbb{Z}^2$ and $\mathcal{R} \subset \mathbb{R}^2$ is a region defined by a piecewise continuous boundary. Thus we have succeeded in translating the problem into a lattice point counting problem in which special attention needs to be paid to the question of uniformity in all of the parameters a, b, λ_i, k_i and ℓ.

3.3. *Lattice point counting in the plane*

In an ideal world we would like to apply the classic estimate

$$\#(\Lambda \cap \mathcal{R}) = \frac{\text{vol}(\mathcal{R})}{\det \Lambda} + O(\partial \mathcal{R} + 1), \qquad (3.11)$$

where $\partial \mathcal{R}$ denotes the perimeter of $\mathcal{R}$, in order to handle the summand in (3.10). In our setting

$$\text{vol}(\mathcal{R}) = \frac{v(a,b)}{\sqrt{|a^2 - b^2|}} \cdot \frac{2^\nu \lambda_1 \lambda_2 B}{\max\{a,b\}}, \quad \partial \mathcal{R} \asymp \sqrt{\frac{2^\nu \lambda_1 \lambda_2 B}{\max\{a,b\}}},$$

with $v(a,b)$ a real-valued continuous function that is bounded above and below by absolute constants that do not depend on any of the parameters.

The overall contribution from the error term above is clearly

$$\gg \sum_{a,b} \sum_{\substack{k_i \lambda_i | a^2 + (-1)^i b^2 \\ 2 \nmid k_1 k_2 \lambda_1 \lambda_2}} |\mu(k_1)\mu(k_2)| \sum_{\ell=1}^{\infty} |\mu(\ell)| \left(1 + \sqrt{\frac{\lambda_1 \lambda_2 B}{\max\{a,b\}}}\right)$$

$$\geq \sum_{a,b} \left(1 + \sqrt{\frac{B}{\max\{a,b\}}}\right)$$

$$\gg B^{\frac{5}{4}},$$

which is disastrously big! Thus the classic lemma is too crude and we must work rather harder to make things succeed. Our approach involves using exponential sums to rewrite the divisibility information in the definition of Λ. In this way we can make the error terms more explicit and obtain cancellation in the summation over a, b and λ_1, λ_2. This part of the argument is long and technical and will be suppressed from the present exposition in order to maintain morale.

Ignoring the contribution from the error term, let us proceed to analyse the contribution from the obvious main term in the classic estimate above. Now an easy calculation reveals that $\det \Lambda = \ell^2 k_1 k_2 \lambda_1 \lambda_2 / \gcd(\ell, k_1 k_2 \lambda_1 \lambda_2)$. Hence our expected main term takes the shape

$$cB \sum_{\substack{a,b<\sqrt{B} \\ \gcd(a,b)=1, ab\neq 1}} \frac{v(a,b)}{\sqrt{|a^2-b^2|}\max\{a,b\}} g(|a^4-b^4|), \qquad (3.12)$$

for a certain constant $c > 0$ and a certain arithmetic function $g : \mathbb{N} \to \mathbb{Z}_{\geq 0}$, obtained by carrying out the summation over the k_i, λ_i and ℓ. In fact one finds that g is given multiplicatively by

$$g(p^\nu) = \begin{cases} \max\{1, \nu-1\}, & \text{if } p = 2, \\ 1 + \nu\left(\frac{1-\frac{1}{p}}{1+\frac{1}{p}}\right), & \text{if } p > 2. \end{cases} \qquad (3.13)$$

Recalling that $\tau(p^\nu) = 1 + \nu$ we see that g can be written as the Dirichlet convolution $g = h * \tau$ for a "small" arithmetic function h. For our purposes an arithmetic function h is said to be small if the infinite sum $\sum_{d=1}^{\infty} d^{-\frac{1}{4}} |h(d)|$ is convergent.

3.4. Divisor problem for binary forms

The next step is to analyse the main term (3.12). The factor $v(a,b)/(\sqrt{|a^2-b^2|}\max\{a,b\})$ is basically harmless and can be reintroduced later via an application of partial summation. We therefore need to investigate sums of the shape

$$S_*(X, \varphi) = \sum_{\substack{a,b\leq X \\ \gcd(a,b)=1}} \varphi(F(a,b)),$$

for suitable arithmetic functions $\varphi : \mathbb{N} \to \mathbb{R}_{\geq 0}$ and suitable integral binary forms F. Let $S(X, \varphi)$ denote the corresponding sum without the condition that $\gcd(a, b) = 1$ in the summation.

When $\varphi = \tau$ is the ordinary divisor function the problem of estimating $T(X) = S(X, \tau)$ has enjoyed considerable attention in the literature.

Dealing with forms of degree at most 2 is straightforward. Greaves [10] pioneered the study of higher degree forms by showing the existence of constants c_F, c'_F, with $c_F > 0$, such that

$$T(X) = c_F X^2 \log X + c'_F X^2 + O_{\varepsilon,F}(X^{\frac{27}{14}+\varepsilon}), \qquad (3.14)$$

for any $\varepsilon > 0$, when F is an irreducible cubic form with non-zero discriminant. Greaves' proof uses exponential sums. It remained a significant open problem to deal with forms of degree 4 until Daniel [6] was able to show that

$$T(X) = c_F X^2 \log X + O_F(X^2 \log \log X),$$

for a constant $c_F > 0$, when F is an irreducible quartic form with non-zero discriminant. Daniel also sharpens Greaves' asymptotic formula, showing that the error term in (3.14) may be replaced by $O_{\varepsilon,F}(X^{\frac{15}{8}+\varepsilon})$ for cubics. Daniel's argument avoids using exponential sums and is based on some ideas from the geometry of numbers instead.

In our work we need to extend the work of Daniel to an asymptotic formula for $T(X)$, but with

- $F = L_1 L_2 Q$ reducible, where L_1, L_2 are non-proportional integral binary linear forms and Q is an integral binary quadratic form which is irreducible over $\mathbb{Q}$;
- τ replaced by $\tau * h$ for small h; and
- a summation over coprime a, b.

In other words we need to investigate $S_*(X; \tau * h)$ for small arithmetic functions h. All of the necessary estimates are corralled into a separate investigation [2]. As a special case we are able to show in [2, Théorème 1] that

$$T(X) = c_{L_1,L_2,Q} X^2 (\log X)^3 + O_{\varepsilon,L_1,L_2,Q}(X^2 (\log X)^{2+\varepsilon}),$$

for any $\varepsilon > 0$, and furthermore, in [2, Corollaire 2] an asymptotic formula of similar strength for $S_*(X, \tau * h)$. We also investigate the asymptotic behaviour of the more general sum

$$S_*(X, h; Y, V) = \sum_{\substack{\mathbf{x}=(x_1,x_2)\in\mathbb{Z}^2\cap X\mathcal{B} \\ \gcd(x_1,x_2)=1}} h(L_1, L_2, Q; Y, V),$$

for $Y \geq 2$ and $\mathcal{B} \subseteq [-1,1]^2$ a convex region whose boundary is defined by a piecewise continuously differentiable function, with $L_i(\mathbf{x}) > 0$ and $Q(\mathbf{x}) > 0$

for every $\mathbf{x} \in \mathcal{B}$. Furthermore, $V \subseteq [0,1]^4$ is a region cut out by a finite number of hyperplanes with absolutely bounded coefficients, and the summand is defined to be

$$h(L_1, L_2, Q; Y; V) = \sum_{\substack{d|L_1(\mathbf{x})L_2(\mathbf{x})Q(\mathbf{x}) \\ d_i=\gcd(d,L_i(\mathbf{x})), \ d_3=\gcd(d,Q(\mathbf{x})) \\ (\frac{\log d_1}{\log Y}, \frac{\log d_2}{\log Y}, \frac{\log d_3}{2\log Y}, \frac{\log \max\{|x_1|,|x_2|\}}{\log Y}) \in V}} (1 * h)(d). \quad (3.15)$$

Suppose for simplicity that $L_i(\mathbf{x}) > 0$ and $Q(\mathbf{x}) > 0$ for every $\mathbf{x} \in [0,1]^2$. If one takes $V = [0,1]^4$ and $\mathcal{B} = [0,1]^2$, together with

$$h(n) = \begin{cases} 1, & \text{if } n = 1, \\ 0, & \text{if } n > 1, \end{cases}$$

and Y sufficiently large in terms of X, then $S_*(X, h; Y, V)$ simplifies to become the sum $S_*(X, \tau)$ that was introduced above. It might seem strange to study such a general function, but it turns out that this is what is required for the theorem, as we will discuss below.

The proofs in [2] are based on the approach using the geometry of numbers that was developed by Daniel in his study of $T(X)$ for irreducible forms of degree at most 4. The need to maintain complete uniformity in all of the relevant parameters causes considerable extra labour.

3.5. *Comparison with Peyre's constant*

According to the conjecture of Peyre [20], the constant $c_{V_0,H}$ in the theorem should be a product of two constants $\alpha(V_0)$ and $\tau_H(V_0)$.

Let $L(s, V_0) = \zeta(s)^5 L(s, \chi)$, where χ is the real non-principal character modulo 4. It turns out that this is the Hasse–Weil L-function associated to our non-singular dP$_4$. The constant $\tau_H(V_0)$ is basically a product of local densities and convergence factors. Specifically, if $\omega_{H,v}$ denotes the usual v-adic density of points on $V_0(\mathbb{Q}_v)$ for any place v of $\mathbb{Q}$, then

$$\tau_H(V_0) = \lim_{s \to 1} ((s-1)^5 L(s, V_0)) \omega_{H,\infty} \prod_p \frac{\omega_{H,p}}{L_p(1, V_0)}$$

$$= \omega_{H,\infty} \prod_p \left(1 - \frac{1}{p}\right)^5 \omega_{H,p}.$$

Checking that the predicted local factors $\omega_{H,v}$ really do match up with the factors that emerge in the theorem is quite involved.

The constant $\alpha(V_0)$ is a rational number defined as the volume of a certain polytope obtained by intersecting the cone of effective divisors of

V_0 with a hyperplane. Determining its value for non-singular del Pezzo surfaces of low degree can be a very challenging problem in its own right. For example, for the non-singular del Pezzo surface of degree 1 in which all of the 240 exceptional divisors are defined over $\mathbb{Q}$ one would need to calculate the volume of a polytope with 19440 generators! An alternative approach has been developed by Derenthal, Joyce and Teitler [7]. For V_0 we have

$$\alpha(V_0) = \frac{1}{36}, \tag{3.16}$$

as calculated by Derenthal in [3, Appendix].

For the sake of general edification, we present here an alternative proof of (3.16) using the method developed by Peyre and Tschinkel [21]. The constant $\alpha(V_0)$ is defined by Peyre [20] as the volume of the polytope

$$\{\mathbf{t} = (t_0, \dots, t_4) \in \Lambda_{\text{eff}}^{\vee}(V_0) : (\mathbf{t}, -K_{V_0}) = 1\}.$$

Here $\mathbf{t} \in \operatorname{Pic}(V_0) \otimes_{\mathbb{Z}} \mathbb{R}$ is understood to mean $t_0\Lambda + \cdots + t_3E_3 + t_4(E_4 + E_5)$, with $t_0, \dots, t_4 \in \mathbb{R}$. Recall that the effective cone $\Lambda_{\text{eff}}(V_0)$ is generated by the classes (2.7) in $\operatorname{Pic}(V_0)$. Then it follows that $\alpha(V_0)$ is the volume of the polytope

$$P = \left\{ \mathbf{t} \in \mathbb{R}^5 : \begin{array}{l} 3t_0 - t_1 - t_2 - t_3 - t_4 = 1, \\ t_0 - t_4, t_1, t_2, t_3, t_4 > 0, \\ t_0 - t_i - t_j > 0 \text{ for } (i,j) \in \{(1,2),(1,3),(2,3)\}, \\ 2t_0 - t_1 - t_2 - t_3 - t_4 > 0, \\ 2t_0 - 2t_i - t_4 > 0 \text{ for } i \in \{1,2,3\} \end{array} \right\}.$$

Eliminating t_4, we are left with the the non-simplicial polytope

$$P = \left\{ \mathbf{t} \in \mathbb{R}^4 : \begin{array}{l} -1 + 3t_0 - t_1 - t_2 - t_3 > 0, \\ 1 - t_0, t_1, t_2, t_3 > 0, \\ t_0 - t_i - t_j > 0 \text{ for } (i,j) \in \{(1,2),(1,3),(2,3)\}, \\ 1 - 2t_0 + t_1 + t_2 + t_3 > 0, \\ 1 - t_0 - t_i + t_j + t_k > 0 \text{ for } \{i,j,k\} = \{1,2,3\} \end{array} \right\}. \tag{3.17}$$

We can calculate the volume of this using the software package `polymake` [9], for example, the outcome being that (3.16) holds.

We now arrive at one of the most subtle aspects of the proof of the theorem. In Section 3.3 we considered the effect of working with the main term in the classic estimate (3.11), which in turn led us to estimate (3.12), as discussed in Section 3.4. However, if one were to follow this course of action exactly, one would ultimately be led to a final main term of the shape

$$\frac{1}{32} \cdot \tau_H(V_0) \cdot B(\log B)^4,$$

where $\tau_H(V_0)$ is the product of local densities introduced above. One notes immediately that $\frac{1}{32} > \frac{1}{36} = \alpha(V_0)$ and so somewhere along the way we have over-counted things!

What is at fault here is our initial idea of approximating $\#(\Lambda \cap \mathcal{R})$ in (3.10) by the volume of $\mathcal{R}$ divided by the determinant of Λ. We have failed to observe that for some ranges of the parameters λ_1, λ_2 that appear in (3.10) we will have $\#(\Lambda \cap \mathcal{R}) = 0$ even if the approximation $\text{vol}(\mathcal{R})/\det \Lambda$ is non-zero. Thus a further reduction on the set of allowable parameters λ_1, λ_2 is necessary. Once taken into account this will eventually lead to the expected main term that we see in the statement of the theorem.

For any $R > 0$ and coprime $a, b < \sqrt{B}$ with $ab \neq 1$, define the region

$$V_{a,b}(R) = \left\{ (t_1, t_2) \in \mathbb{R}^2_{\geq 1} : \begin{array}{l} \max\{a,b\}t_1 \leq Rt_2, \\ \max\{a,b\}t_2 \leq Rt_1, \\ \max\{a,b\}^3/R \leq t_1 t_2, \\ t_1 t_2 \leq \max\{a,b\}R \end{array} \right\}. \tag{3.18}$$

It can then be shown that the summation over λ_1, λ_2 is necessarily restricted to $(\lambda_1, \lambda_2) \in V_{a,b}(R)$ for a suitable choice of R depending on B. Doing so requires one to take into account all of the available symmetry in our counting problem. Since we have $\lambda_i \mid a^2 + (-1)^i b^2$, so it follows that there exist $\mu_1, \mu_2 \in \mathbb{Z}$ such that $\lambda_i \mu_i = a^2 + (-1)^i b^2$. Switching between the λ_i and the μ_i is a crucial ingredient in deriving some of the inequalities present in the definition of $V_{a,b}(R)$. To illustrate a simple case let us indicate how the first inequality $\max\{a,b\}\lambda_1 \leq R\lambda_2$ arises in (3.18). We have already recorded an expression for the volume of $\mathcal{R}$ in Section 3.3. With a little extra work one is able to show that

$$\mathcal{R} \subset [-c\sqrt{X}, c\sqrt{X}] \times (0, c\sqrt{X/|a^2 - b^2|}],$$

where $c > 0$ is an absolute constant and $X = 2^\nu \lambda_1 \lambda_2 B/\max\{a,b\}$. Given that $s \neq 0$ and $\lambda_1 \mid s$ for any $(s,t) \in \Lambda \cap \mathcal{R}$, it therefore follows that $\lambda_1 \ll \sqrt{X}$, whence the first inequality in (3.18) is satisfied by (λ_1, λ_2) for any $R \gg 2^\nu B$.

The outcome of this analysis is that the function g in (3.12) should be replaced by

$$\tilde{g}(a,b;R) = \sum_{\substack{d \mid a^4 - b^4 \\ d_i = \gcd(d, a + (-1)^i b) \\ d_3 = \gcd(d, a^2 + b^2)}} (1 * h)(d) \chi_{d_1 d_2, d_3}(R),$$

for a suitable value of $R \geq 1$, where $\chi_{t_1, t_2}(R)$ is the characteristic function of the set (3.18) and h is such that $g = \tau * h$ in (3.13). Define the region

$$W = \left\{ \mathbf{w} \in \mathbb{R}^4 : \begin{array}{c} w_i \geq 0 \text{ for } i \in \{1,2,3,4\}, \\ w_1 + w_2 + w_4 \leq 1 + 2w_3, \\ 2w_3 + w_4 \leq 1 + w_1 + w_2, \\ 3w_4 \leq 1 + w_1 + w_2 + 2w_3, \\ w_1 + w_2 + 2w_3 \leq 1 + w_4 \end{array} \right\} \subset [0,1]^4.$$

One easily checks that $\tilde{g}(a,b;R) = h(L_1, L_2, Q; R; W)$, in the notation of (3.15), with the appropriate choices of L_i, Q.

Once these changes are carried through one is ultimately led to a main term of the shape $\mathrm{vol}(W_0) \cdot \tau_H(V_0) \cdot B(\log B)^4$, where

$$W_0 = \left\{ (w_1, w_2, \frac{w_3}{2}, w_4) \in W : \begin{array}{c} \max\{w_1, w_2, \frac{1}{2}w_3\} \leq w_4, \\ 2w_4 \leq 1 \end{array} \right\}.$$

We can calculate the volume of W_0 using **polymake** [9], the outcome being that $\mathrm{vol}(W_0) = \frac{1}{36} = \alpha(W_0)$. The reader is invited to compare W_0 with P in (3.17). Both are polytopes in $\mathbb{R}^4$ defined by a minimal set of 12 inequalities.

4. Further exploration

It remains to discuss a few of the ways in which the ideas in [3] and [2] could be developed further. While it still seems hopeless to deal with the divisor problem for binary forms of degree 5 or more, one could try to handle the remaining binary quartic forms. In view of [2] and [6] it remains to handle reducible binary quartic forms which factor as a linear form and a cubic form, or as the product of two quadratic forms, or split completely as the product of four linear forms. All of this seems doable and would no doubt prove useful in related counting problems of the nature considered in our theorem.

Problem 4.1. *Establish analogues of [2] and [6] for other reducible binary quartic forms.*

Returning to the Manin conjecture, it is very natural to wonder how far the ideas behind the proof of the theorem can be extended to other surfaces.

Problem 4.2. *Establish the Manin conjecture for other conic bundle del Pezzo surfaces of degree* 4.

A good starting point would be to consider the family of surfaces $V \subset \mathbb{P}^4$ defined by (1.2) and

$$\Phi_2(x_0, \ldots, x_4) = c_0 x_0^2 + c_1 x_1^2 + c_2 x_2^2 + c_3 x_3^2 + c_4 x_4^2.$$

It is easy to check that V will be non-singular if and only if $c_0 \cdots c_4 \neq 0$ and $c_0 c_1 \neq c_2 c_3$. In attacking Problem 4.2 one might also begin by considering only those V for which the corresponding conic bundle morphisms have a section such as for the surface considered in the theorem. This is equivalent to the map $V(\mathbb{Q}) \to \mathbb{P}^1(\mathbb{Q})$ being surjective. Assuming this to be the case there will still be considerable work to be done, since the corresponding fibres take the shape

$$f_1(a, b)x^2 + f_2(a, b)y^2 + c_4 z^2 = 0,$$

for binary forms $f_1(a, b) = c_0 a^2 + c_3 b^2$ and $f_2(a, b) = c_2 a^2 + c_1 b^2$. Thus one is led to Problem 4.1 for the case in which $f_1 f_2$ factors over $\mathbb{Q}$ as the product of two quadratic forms or splits completely.

It would be even more interesting to handle a non-singular del Pezzo surface of degree 4 with a conic bundle structure, for which the morphism $V \to \mathbb{P}^1$ does not possess a section.

Problem 4.3. *Establish the Manin conjecture for a conic bundle del Pezzo surface of degree* 4 *whose morphisms to* $\mathbb{P}^1$ *do not have a section.*

Thus far we have dealt only with non-singular del Pezzo surfaces of degree 4. One might ask how far things can be pushed for the corresponding surfaces of degree 3: ie. non-singular cubic surfaces $V \subset \mathbb{P}^3$ defined over $\mathbb{Q}$. In this setting one should take $U = V \setminus \{27 \text{ lines}\}$. The following table summarises some of our progress to date:

who?	$N_{U,H}(B)$	conditions?
Slater & Swinnerton-Dyer [23]	$\gg B(\log B)^{\mathrm{rankPic}(V)-1}$	2 skew lines $/\mathbb{Q}$
Heath-Brown [12]	$\ll B^{\frac{4}{3}+\varepsilon}$	3 coplanar lines $/\mathbb{Q}$
Hooley [16]	$\ll B^{\frac{5}{3}+\varepsilon}$	V diagonal & 1 line $/\mathbb{Q}$
Salberger (in preparation)	$\ll B^{\frac{12}{7}+\varepsilon}$	—

It should be remarked that Heath-Brown's result covers the Fermat cubic surface $x_0^3 + x_1^3 + x_2^3 + x_3^3 = 0$. This particular surface has already been the object of much study. Previously, Hooley [15] had used exponential sums to show that $N_{U,H}(B) = O_\varepsilon(B^{\frac{5}{3}+\varepsilon})$ for any $\varepsilon > 0$. Later, a much shorter proof of this estimate was provided by Wooley [26], and it is the ideas in this argument that provided the inspiration behind Heath-Brown's sharper bound in [12]. With the latter result in mind we end with the following challenging problem.

Problem 4.4. *Prove an upper bound of the shape* $N_{U,H}(B) = O_V(B^\theta)$, *with* $\theta < 4/3$, *for a single non-singular cubic surface* $V \subset \mathbb{P}^3$.

In attacking Problem 4.4 it makes sense to start by analysing a cubic surface with a conic bundle structure. The example

$$x_0 x_1 (x_0 + x_1) = x_2 x_3 (x_2 + x_3)$$

seems worthy of attention.

References

1. R. de la Bretèche and T.D. Browning, Sums of arithmetic functions over values of binary forms. *Acta Arith.* **125** (2007), 291–304.
2. R. de la Bretèche and T.D. Browning, Le problème des diviseurs pour des formes binaires de degré 4. *Submitted*, 2008.
3. R. de la Bretèche and T.D. Browning, Manin's conjecture for quartic del Pezzo surfaces with a conic fibration. *Submitted*, 2008.
4. T.D. Browning, An overview of Manin's conjecture for del Pezzo surfaces. *Analytic number theory — A tribute to Gauss and Dirichlet (Göttingen, 20th June – 24th June, 2005)*, 39–56, Clay Mathematics Proceedings **7**, AMS, 2007.
5. T.D. Browning and D.R. Heath-Brown, Counting rational points on hypersurfaces. *J. reine angew. Math.* **584** (2005), 83–115.

6. S. Daniel, On the divisor-sum problem for binary forms. *J. reine angew. Math.* **507** (1999), 107–129.

7. U. Derenthal, M. Joyce and Z. Teitler, The nef cone volume of generalized Del Pezzo surfaces. *Algebra & Number Theory* **2** (2008), 157–182.

8. J. Franke, Y.I. Manin and Y. Tschinkel, Rational points of bounded height on Fano varieties. *Invent. Math.* **95** (1989), 421–435.

9. E. Gawrilow and M. Joswig, `polymake`: a framework for analyzing convex polytopes. *Polytopes—combinatorics and computation (Oberwolfach, 1997)*, 43–73, DMV Sem. **29**, Birkhäuser, 2000.

10. G. Greaves, On the divisor-sum problem for binary cubic forms. *Acta Arith.* **17** (1970), 1–28.

11. R. Hartshorne, *Algebraic Geometry*, Springer-Verlag, 1977.

12. D.R. Heath-Brown, The density of rational points on cubic surfaces. *Acta Arith.* **79** (1997), 17–30.

13. D.R. Heath-Brown, Counting rational points on cubic surfaces. *Astérisque* **251** (1998), 13–29.

14. D.R. Heath-Brown, The density of rational points on curves and surfaces. *Annals of Math.* **155** (2002), 553–595.

15. C. Hooley, On the numbers that are representable as the sum of two cubes. *J. reine angew. Math.* **314** (1980), 146–173.

16. C. Hooley, The density of rational points on cubic surfaces. *Glasgow Math. J.* **42** (2000), 225–237.

17. F.-S. Leung, *Manin's conjecture on a non-singular quartic del Pezzo surface*, D.Phil thesis, Oxford 2008.

18. Y.I. Manin, *Cubic forms*, 2nd ed., North-Holland Math. Library **4**, North-Holland, Amsterdam, New York and Oxford, 1986.

19. M. Nair, Multiplicative functions of polynomial values in short intervals. *Acta Arith.* **62** (1992), 257–269.

20. E. Peyre, Hauteurs et nombres de Tamagawa sur les variétés de Fano. *Duke Math. J.* **79** (1995), 101–218.

21. E. Peyre and Y. Tschinkel, Tamagawa numbers of diagonal cubic surfaces of higher rank. *Rational points on algebraic varieties*, 275–305, Progr. Math., **199**, Birkhäuser, 2001.

22. B. Segre, A note on arithmetical properties of cubic surfaces. *J. London Math. Soc.* **18** (1943), 24–31.

23. J.B. Slater and P. Swinnerton-Dyer, Counting points on cubic surfaces. I. *Astérisque* **251** (1998), 1–12.

24. P. Swinnerton-Dyer, Diophantine equations: progress and problems. *Arithmetic of higher-dimensional algebraic varieties (Palo Alto, CA, 2002)*, 3–35, Progr. Math. **226**, Birkhäuser, 2004.

25. O. Wittenberg, *Intersections de deux quadriques et pinceaux de courbes de genre 1*. Lecture Notes in Math. **1901**, Springer-Verlag, 2007, viii+218 pp.

26. T.D. Wooley, Sums of two cubes. *Internat. Math. Res. Notices* **4** (1995), 181–185.

ADDITIVE REPRESENTATION IN THIN SEQUENCES, VIII: DIOPHANTINE INEQUALITIES IN REVIEW

JÖRG BRÜDERN, KOICHI KAWADA AND TREVOR D. WOOLEY

Jörg Brüdern

Institut für Algebra und Zahlentheorie, Universität Stuttgart,
70511 Stuttgart, Germany
E-mail: bruedern@mathematik.uni-stuttgart.de

Koichi Kawada

Department of Mathematics, Faculty of Education, Iwate University,
Morioka, 020–8550 Japan
E-mail: kawada@iwate-u.ac.jp

Trevor D. Wooley

School of Mathematics, University of Bristol,
University Walk, Clifton, Bristol BS8 1TW, United Kingdom
E-mail: matdw@bristol.ac.uk

Recent developments in the theory of diophantine inequalities and the Davenport-Heilbronn method are discussed and then directed toward specific inequalities of definite character. Special emphasis is on the value-distribution of diagonal forms near thin test sequences.

1. Theme and results

1.1. *Diophantine inequalities*

The focus of our attention in this survey article is the distribution of the values of the diagonal form

$$\lambda_1 x_1^k + \lambda_2 x_2^k + \cdots + \lambda_s x_s^k, \tag{1.1}$$

as $x_1, \ldots, x_s$ range over $\mathbb{Z}$ or an interesting subset thereof; here $s \geq 2$ and $k \geq 1$ are given integers and $\lambda_1, \ldots, \lambda_s$ are non-zero real numbers. Our goal is twofold. In one direction we wish to emphasise recent developments in the analytic theory of diophantine inequalities, in another discuss the potential of methods developed in earlier papers of this series pertaining to this circle of ideas. Some basic principles are also readdressed herein, so that the paper should introduce the uninitiated reader to the subject.

When the polynomial (1.1) is a real multiple of a form with integer coefficients, its values are discrete and can be studied by classical methods as well as the techniques developed in this series. The complementary case, in which (1.1) is not a multiple of a rational form, arises when at least one of the ratios λ_1/λ_j $(2 \leq j \leq s)$ is irrational, and by renumbering the variables, we shall from now on suppose that λ_1/λ_2 is irrational. In this situation, and when $s \geq k+1$, one expects that the values of (1.1) are dense on the real line unless k is even and all λ_j have the same sign. In the latter case, one might hope that the gaps between these values shrink to zero as they approach infinity. When $k = 1$, this is classical territory as long as the variables x_j are allowed to vary over the integers, but, for example, much less is known about the distribution of $\lambda_1 p_1 + \lambda_2 p_2$ when p_1, p_2 denote primes. When $k = 2$, an affirmative theorem is available for indefinite quadratic forms from the work of Margulis [28] (see also [19]), and also for the definite case when $s \geq 5$. This was shown by Götze [25], following the pivotal contribution by Bentkus and Götze [2]. Their work may be viewed as a formidable refinement of a Fourier transform method developed by Davenport and Heilbronn [17]. This very general method is a variant for diophantine inequalities of the famous circle method of Hardy and Littlewood that delivered enormous insight into the labyrinth of diophantine equations during the last century. Davenport and Heilbronn themselves studied the indefinite case $k = 2$ of (1.1) and showed that the values taken at integer points are dense on the real line. Their pioneering paper was the igniting spark for much work on related questions. It would take us too far afield to sketch the development of the subject at large, so instead we follow a fruitful tradition in additive number theory, where new ideas have often been tested on Waring's problem for cubes or on Goldbach's problem. Thus, we now concentrate on the case $k = 3$ in (1.1), and later move on to the value distribution of $\lambda_1 p_1 + \lambda_2 p_2$, although the methods that we develop apply in a much wider context.

1.2. *Additive cubic forms*

We discuss a slightly narrower problem from now on, by enforcing a kind of definiteness. Let $\lambda_1, \ldots, \lambda_s$ denote *positive* real numbers, to be considered as fixed once and for all. For $0 < \tau \leq 1$ and $\nu > 0$, let $\rho_s(\tau, \nu)$ be the number of solutions of the inequality

$$|\lambda_1 x_1^3 + \lambda_2 x_2^3 + \cdots + \lambda_s x_s^3 - \nu| < \tau \qquad (1.2)$$

in natural numbers x_j. Note that for any solution counted here one has $x_j \ll \nu^{1/3}$. In accordance with our earlier remarks, we still expect the gaps among large values of $\lambda_1 x_1^3 + \lambda_2 x_2^3 + \cdots + \lambda_s x_s^3$ to shrink to zero provided that $s \geq 4$ and λ_1/λ_2 is irrational. One anticipates that $\rho_s(\tau, \nu)$ should be large when τ is fixed and ν is large. The work of Freeman [21], [22] applies to this problem, and so does the refinement by Wooley [44], and it is implicit in the latter that for τ fixed, one has

$$\rho_7(\tau, \nu) \gg \nu^{4/3}, \quad \rho_8(\tau, \nu) = c\nu^{5/3}(1 + o(1)). \tag{1.3}$$

Here c is a certain positive constant depending only on τ and the coefficients λ_j. Freeman and Wooley consider in detail a different scenario: ν is fixed, one counts solutions of (1.2) in integers x_j with $|x_j| \leq X$, and examines the growth for $X \to \infty$. Little change is necessary to derive the "definite" versions (1.3) by the arguments of Wooley [44], but older routines based on the original work of Davenport and Heilbronn [17] do not apply. Instead one would find only that (1.2) has infinitely many solutions in *integers* x_j.

Thus far, the situation is in direct correspondence to what is known when the cubic form in (1.2) is a multiple of a rational form. Focusing on this case temporarily, when τ is sufficiently small, the inequality (1.2) reduces to an equation. Hence there is no loss in assuming here that all λ_j are natural numbers, and that the equation is

$$\lambda_1 x_1^3 + \lambda_2 x_2^3 + \cdots + \lambda_s x_s^3 = \nu. \tag{1.4}$$

Let $\rho_s(\nu)$ denote the number of its solutions with $x_j \in \mathbb{N}$. The methods of Vaughan [33], [34] provide the lower bound $\rho_7(\nu) \gg \mathfrak{S}_7(\nu)\nu^{4/3}$, and the asymptotic formula $\rho_8(\nu) = C\mathfrak{S}_8(\nu)\nu^{5/3}(1 + o(1))$, where $\mathfrak{S}_s(\nu)$ is the singular series associated with (1.4), and C is a positive constant depending only on the λ_j. Similar formulae are expected when $s \geq 4$, and are at least implicitly known *on average*. In fact, the envisaged formula for $\rho_4(\nu)$ holds for all but $O(N(\log N)^{\varepsilon-3})$ of the natural numbers ν not exceeding N. This much follows from the work of Vaughan [33] and Boklan [3]. For an analogue in the irrational case, one must first address the question of how one should average over the now real number ν. One could choose a discrete sequence of test points that are suitably spaced, and then count how often an asymptotic formula for $\rho_4(\tau, \nu)$ fails. Alternatively, one may estimate the measure of all such real $\nu \in [1, N]$. Only very recently Parsell and Wooley [31] proved that this measure is $o(N)$. As an illustration of the averaging process, we improve their estimate when λ_1/λ_2 is an algebraic irrational. The result, which we deduce as a consequence of Theorem 2.1 in §2.3 below, fully reflects the current state of knowledge for forms with integer coefficients, but it appears difficult to do equally well under the sole assumption that λ_1/λ_2 is irrational.

Theorem 1.1. *Let $\lambda_1, \lambda_2, \lambda_3, \lambda_4$ denote positive real numbers with λ_1/λ_2 irrational and algebraic. Then, whenever $(\log N)^{-3} < \tau \leq 1$, one has*

$$\int_0^N \left| \rho_4(\tau, \nu) - \frac{2\Gamma(\frac{4}{3})^3 \tau \nu^{1/3}}{(\lambda_1 \lambda_2 \lambda_3 \lambda_4)^{1/3}} \right|^2 d\nu \ll \tau^{3/2} N^{5/3} (\log N)^{\varepsilon - 3/2}.$$

Previous articles within this series exposed methods for testing the conjectural behaviour of counting functions such as $\rho_s(\nu)$ when ν varies over a thin sequence, for example the values of a polynomial. A typical result is contained in Theorem 1.1 of III* that we now recall. It will be convenient to describe a polynomial $\phi \in \mathbb{R}[t]$ of degree $d \geq 1$ as a *positive polynomial* if its leading coefficient is positive. If $\phi \in \mathbb{Q}[t]$ and $\phi(n)$ is integral for all $n \in \mathbb{Z}$, then ϕ is an *integral polynomial*. Let $r_s(n)$ be the number of positive solutions of $n = x_1^3 + x_2^3 + \cdots + x_s^3$. Then, for any $0 < \delta \leq \frac{1}{2}$, the inequality

$$|r_6(\nu) - \Gamma(\tfrac{4}{3})^6 \mathfrak{S}_6(\nu)\nu| > \nu(\log \nu)^{-\delta}$$

can hold for no more than $O(N(\log N)^{2\delta - 5/2 + \varepsilon})$ of the values $\nu = \phi(n)$ with $n \leq N$ assumed by a positive integral quadratic polynomial ϕ. There is no difficulty in extending this to forms with positive integral coefficients.

Our primary concern in the later chapters of this paper is to describe methods that allow one to derive similar results in the context of diophantine inequalities. Later we will comment on some of the difficulties that arise, and we shall find the desired generalisation not as straightforward as one might hope. A conclusion for $\rho_6(\tau, \nu)$ of strength comparable to the aforementioned theorem on $r_6(\nu)$ is contained in the next result, the proof of which may be found in §5.2.

Theorem 1.2. *Let ϕ denote a positive integral quadratic polynomial, and let $\lambda_1, \ldots, \lambda_6$ denote positive real numbers with λ_1/λ_2 irrational. Also, let $0 < \tau \leq 1$. Then there exists a function $\xi(\nu)$, with $\xi(\nu) = o(1)$ as $\nu \to \infty$, such that the inequality*

$$|\rho_6(\tau, \nu) - 2\Gamma(\tfrac{4}{3})^6 (\lambda_1 \cdots \lambda_6)^{-1/3} \tau \nu| > \nu \xi(\nu)$$

holds for at most $O(N(\log N)^{\varepsilon - 5/2})$ of the positive values $\nu = \phi(n)$ with $1 \leq n \leq N$.

---

*Here and later we refer to our papers "Additive representation in thin sequences" by their numeral within the series, I–VII. Hence, III refers to [9], for example.

One might object that although it is rather natural to average over the values of an integral polynomial in the case of diophantine equations, this is not adequate for inequalities, and one should take the values of a real polynomial as test points, or even a monotone sequence with a certain rate of growth. In principle, our methods still apply in this wider context, but some techniques such as certain divisor estimates do no longer have their full impact on the problem at hand, and the ensuing results are sometimes considerably weaker. We illustrate this in §§6.2 and 6.3 with an analysis of following example.

Theorem 1.3. *Let ϕ denote a positive quadratic polynomial, and let $\lambda_1, \ldots, \lambda_6$ denote positive real numbers for which λ_1/λ_2 is irrational. Fix $0 < \tau \leq 1$. Then, there exists a real number $c > 0$ such that, for all but $O(N^{23/27})$ of the integers $n \in [1, N]$, one has $\rho_6(\tau, \phi(n)) \geq c\phi(n)$.*

We have not been able to establish the expected asymptotic formula almost always when ϕ is not an integral polynomial. Theorem 1.3 should also be compared with Theorem 1.1 of I where the exceptional set is shown to be $O(N^{19/28})$ when $\rho_6(\tau, \nu)$ is replaced by $r_6(\nu)$.

Similar results for forms in five variables are not yet available. This applies even to the simplest examples in the rational case: it is not known whether almost all squares are the sum of five positive cubes. In such situations our methods can sometimes be turned toward a lower bound estimate. The idea is discussed in detail in IV, and then used to show that when ϕ is a positive integral quadratic polynomial, then amongst the integers n with $1 \leq n \leq N$, the equation $\phi(n) = x_1^3 + \cdots + x_5^3$ has solutions with $x_j \in \mathbb{N}$ for at least $N^{129/136}$ values of n (see Theorem 1.1 of IV). In §6.4 we prove a result of similar flavour.

Theorem 1.4. *Let ϕ denote a positive quadratic polynomial, and let $\lambda_1, \ldots, \lambda_5$ be positive real numbers with λ_1/λ_2 irrational. Fix $0 < \tau \leq 1$. Then $\rho_5(\tau, \phi(n)) \geq 1$ for at least $N^{3/4}$ natural numbers $n \in [1, N]$.*

Cubic forms in seven variables, in the rational case, have also been discussed in III and VII, although only in the context of sums of cubes. Moments of $r_s(n)$ over polynomial sequences are one of the objectives in VII, and in particular, Theorem 1 of VII contains an asymptotic formula for the sum

$$\sum_{n \leq N} r_7(\phi(n))^2$$

when ϕ is a positive integral quadratic polynomial. The result coincides with the formula that arises from summing the leading term in the anticipated asymptotic expansion of $r_7(\nu)$. Similarly, under the assumptions of Theorem 1.3 (suitably adapted to the current context with seven variables), one can derive an asymptotic formula for

$$\sum_{n \leq N} \rho_7(\tau, \phi(n))^2.$$

It suffices to follow the pattern laid out in VII, but the argument is simpler in the absence of a singular series, and we spare the reader any detail.

1.3. *Linear forms in primes*

We now turn to the value distribution of the binary form $\lambda_1 p_1 + \lambda_2 p_2$ with positive coefficients λ_1, λ_2 and large prime variables p_1, p_2. When λ_1/λ_2 is rational, we may as well suppose that $\lambda_1, \lambda_2 \in \mathbb{N}$, and one then wishes to solve $\lambda_1 p_1 + \lambda_2 p_2 = n$ for a given natural number n. A necessary condition is that $\lambda_1 x_1 + \lambda_2 x_2 \equiv n \bmod 2\lambda_1\lambda_2$ has a solution in integers x_1, x_2 both coprime to $2\lambda_1\lambda_2$ (the *congruence condition*). It is at least implicit in the work of Montgomery and Vaughan [29] and of Liu and Tsang [27] that the number of natural numbers $n \leq N$ which satisfy the congruence condition, but have no representation in the form $n = \lambda_1 p_1 + \lambda_2 p_2$, does not exceed $O(N^{1-\delta})$, for some $\delta > 0$. Pintz has recently announced that one may take any $\delta < \frac{1}{3}$ here, at least when $\lambda_1 = \lambda_2 = 1$. A result of comparable strength is available for the irrational case $\lambda_1/\lambda_2 \notin \mathbb{Q}$ when this ratio is algebraic. This was observed by Brüdern, Cook and Perelli [6]. We illustrate the underlying idea in the second half of §2.3 with a related result. For $0 < \tau \leq 1$ and $\nu > 0$, let $\sigma(\tau, \nu)$ denote the number of prime solutions to

$$|\lambda_1 p_1 + \lambda_2 p_2 - \nu| < \tau, \tag{1.5}$$

with each solution p_1, p_2 counted with weight $(\log p_1)(\log p_2)$.

Theorem 1.5. *Let λ_1, λ_2 denote positive real numbers such that λ_1/λ_2 is an algebraic irrational. Then, for any $A \geq 1$, the set of real numbers ν with $1 \leq \nu \leq N$, for which*

$$\left| \sigma(\tau, \nu) - \frac{2\tau\nu}{\lambda_1\lambda_2} \right| > \frac{\tau\nu}{(\log N)^A}, \tag{1.6}$$

has measure $O(\tau^{-1} N^{2/3+\varepsilon})$, uniformly in $0 < \tau \leq 1$.

For comparison, Parsell [30] works under the weaker hypothesis that λ_1/λ_2 is irrational, and obtains a result that is essentially equivalent to

$$\int_0^N \left| \sigma(\tau, \nu) - \frac{2\tau\nu}{\lambda_1\lambda_2} \right|^2 \mathrm{d}\nu = o(N^3). \tag{1.7}$$

Our method also gives a proof of (1.7), as well as an improvement when λ_1/λ_2 is algebraic, but not by a power of N. Limitations arise from our current knowledge concerning the zeros of the Riemann zeta function. Thus, when λ_1/λ_2 is algebraic, one may apply the methods described below to confirm that the integral on the left hand side of (1.7) is $O(N^3 \exp(-c\sqrt{\log N}))$ for some $c > 0$, and with only moderate extra effort one obtains a saving that corresponds to the sharpest one currently known in the error term for the prime number theorem.

Next, we investigate averages over polynomial sequences. Theorem 1 of II asserts that there is a constant $\delta > 0$ such that, for any positive integral polynomial ϕ of degree d, the number of even values of $\phi(n)$ with $1 \leq n \leq N$ that are not the sum of two primes does not exceed $O(N^{1-\delta/d})$. In the irrational case we require λ_1/λ_2 to be algebraic, and the conclusion is decidedly weaker. In §5.6 we sketch a proof of the following result.

Theorem 1.6. *Let λ_1, λ_2 denote positive real numbers such that λ_1/λ_2 is an algebraic irrational. Fix $0 < \tau \leq 1$ and $A \geq 1$. Let ϕ denote a positive polynomial of degree d, and let $E_\phi(N)$ denote the number of integers n with $1 \leq n \leq N$ for which the inequality (1.6) holds with $\nu = \phi(n)$. Then, there is an absolute constant $\delta > 0$ such that*

$$E_\phi(N) \ll N^{1-\delta/(d \log d)}.$$

In the absence of the hypothesis that λ_1/λ_2 be algebraic, it seems difficult to establish a quantitative bound for $E_\phi(N)$, but proving that $E_\phi(N) = o(N)$ is straightforward. In chapter 4 we average over even thinner sets. Brüdern and Perelli [15] have a corresponding result on Goldbach's problem.

Theorem 1.7. *Let λ_1, λ_2 denote positive real numbers such that λ_1/λ_2 is an algebraic irrational. Fix $0 < \tau \leq 1$ and $A \geq 1$. Let $1 < \gamma < \frac{3}{2}$, and write $\phi(t) = \exp((\log t)^\gamma)$. Let $E_\phi(N)$ be the number of natural numbers n with $1 \leq n \leq N$ for which the inequality (1.6) holds with $\nu = \phi(n)$. Then, there is a $\kappa > 0$ such that*

$$E_\phi(N) \ll N \exp(-\kappa(\log N)^{3-2\gamma}).$$

1.4. *Further applications*

Various other examples of averages over thin sequences can be found in I–VI, and one may extend most of them to diophantine inequalities along the lines indicated above. We single out two results from V and VI that concern the form (1.1) when the degree k is large. Theorem 1.5 of V shows that whenever $s \geq \frac{1}{2} k \log k + O(k \log \log k)$ and ϕ is a quadratic, positive and integral polynomial, then for almost all $n \in \mathbb{N}$ the number $\phi(n)$ is the sum of s positive k-th powers. This may be somewhat surprising, because one cannot do substantially better in the seemingly simpler problem in which the quadratic polynomial $\phi(n)$ is replaced by a linear one; the lower bound on s required here is again of the type $s \geq \frac{1}{2} k \log k + O(k \log \log k)$. This much is implicit in the work of Wooley [40].

Now suppose that $\lambda_1 x_1^k + \cdots + \lambda_s x_s^k$ is an irrational form with positive coefficients, and let $0 < \tau \leq 1$ and $\nu > 0$. Let $\rho_{k,s}(\tau, \nu)$ denote the number of solutions of the inequality

$$|\lambda_1 x_1^k + \lambda_2 x_2^k + \cdots + \lambda_s x_s^k - \nu| < \tau$$

in positive integers x_j (whence $\rho_{3,s} = \rho_s$ in the notation of §1.2). One can now combine the methods used to prove Theorem 1.2 in this paper with the strategy explained in V to confirm that whenever ϕ is a positive *integral* quadratic polynomial and τ is fixed, then for almost all n, one has $\rho_{k,s}(\tau, \phi(n)) > 0$, provided only that $s \geq \frac{1}{2} k \log k + O(k \log \log k)$. This is a proper analogue of the aforementioned result on Waring's problem. For a general real polynomial, as we discover in §6.7, the problem is more difficult.

Theorem 1.8. *Let $\lambda_1, \ldots, \lambda_s$ be positive real numbers, and suppose that λ_1 / λ_2 is irrational. Let $0 < \tau \leq 1$. Let ϕ be a positive quadratic polynomial. Then there is a number $s_0(k)$, with $s_0(k) = \frac{3}{4} k \log k + O(k \log \log k)$, such that whenever $s \geq s_0(k)$, then for almost all n one has $\rho_{k,s}(\tau, \phi(n)) > 0$.*

Similar conclusions can be obtained when ϕ is a polynomial of degree $d \geq 3$, by a method akin to that used to establish Theorem 1.6.

One may also ask whether the form (1.1) takes values near an arithmetic sequence, such as the primes. This theme was discussed in VI, and we derive in §6.8 an analogue of Theorem 1 from that paper.

Theorem 1.9. *Suppose that all Dirichlet L-functions satisfy the Riemann hypothesis. Let λ_j be as in Theorem 1.8, and suppose that $s \geq \frac{8}{3} k + 2$. Then, the integer parts of $\lambda_1 x_1^k + \lambda_2 x_2^k + \cdots + \lambda_s x_s^k$ are prime infinitely often for natural numbers x_j.*

1.5. *A related diophantine inequality*

A polynomial $\phi \in \mathbb{R}[t]$ is described as *irrational* if it is not a real multiple of
an integral polynomial. It is in this case that our results are usually rather
weaker than their equation counterparts in other papers of this series; for
integral polynomials we experienced little difficulty in extending our ideas
for averaging over polynomial values to diophantine inequalities. In part,
this is due to the available estimates for the number of solutions of such
inequalities as

$$\left| \sum_{j=1}^{t} (\phi(n_j) - \phi(m_j)) \right| < 1, \tag{1.8}$$

in natural numbers $n_j, m_j \leq N$. These differ substantially according to
whether ϕ is integral or irrational. Auxiliary bounds of this type are also
relevant for a class of diophantine inequalities recently discussed by Freeman
[24]. He considers a set of s non-zero positive polynomials

$$\phi_j(t) = \sum_{l=1}^{d} \lambda_{jl} t^l \quad (1 \leq j \leq s)$$

without constant terms, and of degree at most d. Suppose that at least one
of the ratios $\lambda_{jl}/\lambda_{km}$ is irrational. The number $\rho_\phi(\tau, \nu)$ of solutions of

$$|\phi_1(x_1) + \cdots + \phi_s(x_s) - \nu| < \tau, \tag{1.9}$$

in positive integers, generalises the counter $\rho_{k,s}(\tau, \nu)$ in a natural way. Freeman's result is that there is a number $s_0(d)$, with $s_0(d) \sim 4d \log d$, and such
that for fixed $\tau > 0$ and large ν one has $\rho_\phi(\tau, \nu) \geq 1$ (see Theorem 2 of [24]).
Our investigation of (1.8) in §5.5 implies the following result, which we es-
tablish in chapter 7.

Theorem 1.10. *For any $d \in \mathbb{N}$, there exists a number $s_1(d)$ with $s_1(d) \sim$
$2d^2 \log d$ and the following property. When $\phi_1, \ldots, \phi_s$ are polynomials of
respective degrees $d_1, \ldots, d_s$ at most d subject to the conditions described in
the preceding paragraph, then*

$$\rho_\phi(\tau, \nu) = c(\phi)\tau\nu^{D-1} + o(\nu^{D-1}),$$

in which $D = d_1^{-1} + \cdots + d_s^{-1}$ and

$$c(\phi) = \frac{2\Gamma(1 + d_1^{-1}) \ldots \Gamma(1 + d_s^{-1})}{\Gamma(D)\lambda_1^{1/d_1} \ldots \lambda_s^{1/d_s}}.$$

Under the current stipulations on ϕ, the inequality (1.9) implies an upper bound on x_j, whence the problem remains "definite". Freeman [24] also considers a cognate "indefinite" problem, and the corresponding analogue of Theorem 1.10 follows *mutatis mutandis*. These results are the first recorded instances of asymptotic formulae for $\rho_\phi(\tau, \nu)$.

The individual chapters of this memoir are equipped with sections that describe the methods used herein in greater detail than would be possible at this point. The next chapter is intended as an introduction to the Davenport-Heilbronn method. Emphasis is on newer techniques that yield asymptotic formulae. In particular, we discuss a central contribution from the work of Bentkus and Götze [2] and of Freeman [21]. Rather than following their line of thought, we describe their device as an interference estimate for certain major arcs (Theorems 2.3 and 2.4). This yields an independent approach to asymptotic formulae for diophantine inequalities. In chapter 3, the work of chapter 2 is then illustrated with proofs for Theorems 1.1 and 1.5. A classical use of Plancherel's identity suffices here. In chapter 4, a discrete mean square approach is explained within a proof of Theorem 1.7. In chapter 5 we discuss the averaging tools from earlier papers in this series, that are then used to establish Theorems 1.2 and 1.6. Along the way, new mean value estimates for certain Weyl sums are obtained in §§5.4–5.5. The next chapter combines the ideas from chapter 5 with mean values of smooth Weyl sums, and also describes a lower bound method that was introduced in IV. The final chapter is an appendix on the problem described in Theorem 1.10.

The notation used in this memoir is standard, or otherwise explained at the appropriate stage of the proceedings. We write $e(\alpha) = \exp(2\pi i \alpha)$. The distance of a real number α to the nearest integer is $\|\alpha\|$. The integer part of α is $[\alpha]$, and $\lceil \alpha \rceil$ is the smallest integer n with $n \geq \alpha$. We apply the following convention concerning the letter ε. Whenever ε occurs in a statement, it is asserted that this statement is true for all real $\varepsilon > 0$, but constants implicit in Landau or Vinogradov symbols may depend on the actual value of ε.

2. The Fourier transform method

2.1. *Some classical integrals*

We begin with a review of the Fourier transform method, as pioneered by Davenport and Heilbronn. The scene is set up to cover more recent developments which yield asymptotic formulae, not only accidental lower bounds. Before we can embark on details, a few classical integral formulae are required that we now collect.

The Fourier transform of an integrable function $w : \mathbb{R} \to \mathbb{C}$ is

$$\widehat{w}(x) = \int_{-\infty}^{\infty} w(y)e(-xy)\mathrm{d}y, \tag{2.1}$$

and for any positive real number η, the functions

$$w_\eta(x) = \eta\Big(\frac{\sin \pi\eta x}{\pi\eta x}\Big)^2, \quad \widehat{w}_\eta(x) = \max\Big(0, 1 - \frac{|x|}{\eta}\Big) \tag{2.2}$$

are Fourier transforms of each other. Note that w and $\widehat{w}$ are non-negative. One can use (2.3) to construct a continuous approximation to the indicator function of an interval. When $\tau > 0$ and $\delta > 0$, define $W_{\tau,\delta} : \mathbb{R} \to [0,1]$ by

$$W_{\tau,\delta}(x) = \begin{cases} 1, & \text{for } |x| \leq \tau, \\ 1 - (|x| - \tau)/\delta, & \text{for } \tau < |x| < \tau + \delta, \\ 0, & \text{for } |x| \geq \tau + \delta. \end{cases}$$

By (2.3), one has

$$W_{\tau,\delta}(x) = \Big(1 + \frac{\tau}{\delta}\Big)\widehat{w}_{\tau+\delta}(x) - \frac{\tau}{\delta}\widehat{w}_\tau(x). \tag{2.3}$$

Before we return to diophantine inequalities, we apply the function $w_\eta(x)$, given by (2.3), to define a measure $\mathrm{d}_\eta x$ on $\mathbb{R}$ with the property that for any bounded continuous function $\psi : \mathbb{R} \to \mathbb{C}$ and a measurable set $\mathcal{B}$, one has

$$\int_{\mathcal{B}} \psi(x)\mathrm{d}_\eta x = \int_{\mathcal{B}} \psi(x)w_\eta(x)\mathrm{d}x. \tag{2.4}$$

We omit explicit mention of the range of integration when $\mathcal{B} = \mathbb{R}$. This convenient notation avoids repeated occurences of the kernel w_η in most of the many integrals to follow.

2.2. *Counting solutions of diophantine inequalities*

The basic idea that underpins the strategy followed by Davenport and Heilbronn [17] is best explained in broad generality. Consider the polynomial $F \in \mathbb{Z}[x_1, \ldots, x_s]$, and suppose that $u : \mathbb{Z}^s \to [0, \infty)$ is a weight that vanishes outside a finite subset of $\mathbb{Z}^s$. In practice u will be supported in a box depending on a size parameter, say B. For a positive real number τ, we then wish to evaluate the sum

$$\mathrm{P}(\tau; u) = \sum_{\substack{\mathbf{x} \in \mathbb{Z}^s \\ |F(\mathbf{x})| < \tau}} u(\mathbf{x}), \tag{2.5}$$

at least asymptotically, as $B \to \infty$. This is approached through the cognate yet analytically simpler expression

$$\mathrm{P}^*(\tau; u) = \sum_{\mathbf{x} \in \mathbb{Z}^s} u(\mathbf{x}) \widehat{w}_\tau(F(\mathbf{x})). \qquad (2.6)$$

Occasionally, we shall use also the clumsier notation $\mathrm{P}_F(\tau; u)$, $\mathrm{P}_F^*(\tau; u)$ to point out the dependence on F more explicitly. If only a lower bound for $\mathrm{P}(\tau; u)$ is desired, then it will suffice to proceed through the inequality

$$\mathrm{P}(\tau; u) \geq \mathrm{P}^*(\tau; u) \qquad (2.7)$$

that is readily inferred from (2.3), (2.5) and (2.6). Moreover, it is an exercise in elementary analysis to show that an asymptotic formula for $\mathrm{P}^*(\tau; u)$ implies a related one for the unweighted counter $\mathrm{P}(\tau; u)$. If one has to take care of error terms, this needs some dexterity. There are several ways to perform the transition. Freeman [22] and Wooley [44] describe a sandwich technique. Another option is to choose δ in the range $0 < \delta < \frac{1}{2}\tau$. Then observe that

$$\mathrm{P}(\tau; u) \leq \sum_{\mathbf{x} \in \mathbb{Z}^s} u(\mathbf{x}) W_{\tau,\delta}(F(\mathbf{x})), \qquad (2.8)$$

an upper bound that uses only the definition of $W_{\tau,\delta}$ and the non-negativity of $u(\mathbf{x})$. However, when δ is much smaller than τ, the right hand side here should be approximately equal to $\mathrm{P}(\tau; u)$. In fact, the difference between the left and right hand sides of (2.8) arises only from solutions of $\tau < |F(\mathbf{x})| < \tau + \delta$, and by taking into account the actual definition of $W_{\tau,\delta}$, we find that

$$\mathrm{P}(\tau; u) \geq \sum_{\mathbf{x} \in \mathbb{Z}^s} u(\mathbf{x}) W_{\tau,\delta}(F(\mathbf{x})) - \mathrm{P}_{F-\tau}^*(\delta; u) - \mathrm{P}_{F+\tau}^*(\delta; u).$$

We may combine these inequalities and apply (2.4) to conclude as follows.

Lemma 2.1. *Let* $0 < \delta < \frac{1}{2}\tau$. *Then, in the notation introduced in this section, one has*

$$\mathrm{P}(\tau; u) = \left(1 + \frac{\tau}{\delta}\right) \mathrm{P}^*(\tau + \delta; u) - \frac{\tau}{\delta} \mathrm{P}^*(\tau; u) - E,$$

where E *satisfies the inequalities* $0 \leq E \leq \mathrm{P}_{F+\tau}^*(\delta; u) + \mathrm{P}_{F-\tau}^*(\delta; u)$.

2.3. Weighted counting

With the transition relations in (2.7) and Lemma 2.1 now in hand, we may concentrate on $P^*(\tau; u)$. By (2.1), (2.3), (2.6) and (2.6) we have

$$P^*(\tau; u) = \int \sum_{\mathbf{x} \in \mathbb{Z}^s} e(\alpha F(\mathbf{x})) u(\mathbf{x}) \mathrm{d}_\tau \alpha. \qquad (2.9)$$

It then suffices to establish an asymptotic formula for $P^*(\tau; u)$; a mild uniformity in τ comes with it at no cost so that Lemma 2.1 yields the desired formula for $P(\tau; u)$.

The integral (2.9) is the starting point for the Davenport-Heilbronn method. Before we enter this subject, we make the rather abstract discussion in §2.2 more concrete and show how this can be used to reduce the proofs of Theorems 1.1 and 1.5 to related weighted versions of these theorems.

We begin with additive cubic forms. The classical Weyl sum

$$f(\alpha) = \sum_{x \le X} e(\alpha x^3) \qquad (2.10)$$

will be prominently featured. When $\lambda_1, \ldots, \lambda_s$ are positive we choose

$$X = 2(\lambda_1^{-1/3} + \cdots + \lambda_s^{-1/3} + 1) N^{1/3}. \qquad (2.11)$$

Then, for any solution of $|\lambda_1 x_1^3 + \cdots + \lambda_s x_s^3 - \nu| < 1$ with $\nu \le N$, one has $x_j \le X$. Hence, we may take $F = \lambda_1 x_1^3 + \cdots + \lambda_s x_s^3 - \nu$ and $u(\mathbf{x})$ as the indicator function on the box $1 \le x_j \le X$ $(1 \le j \le s)$. Then $\rho_s(\tau, \nu) = P_F(\tau; u)$ in the notation of the previous section, and the weighted analogue $P_F^*(\tau; u)$ now becomes

$$\rho_s^*(\tau, \nu) = \int f(\lambda_1 \alpha) \cdots f(\lambda_s \alpha) e(-\nu \alpha) \mathrm{d}_\tau \alpha,$$

as a special case of (2.9). In this form, Lemma 2.1 shows that whenever $0 < \delta < \frac{1}{2}\tau$, one has

$$\rho_s(\tau, \nu) = \left(1 + \frac{\tau}{\delta}\right) \rho_s^*(\tau + \delta, \nu) - \frac{\tau}{\delta} \rho_s^*(\tau, \nu) + O\left(\rho_s^*(\delta, \nu + \tau) + \rho_s^*(\delta, \nu - \tau)\right). \qquad (2.12)$$

For Theorem 1.2, we will have to work directly from this formula; see the last part of §5.2. In order to establish Theorem 1.1, on the other hand, we take $s = 4$, and invoke the following weighted variant which involves the moment

$$\Upsilon(\tau, N) = \int_0^N \left| \rho_4^*(\tau, \nu) - \frac{\Gamma(\frac{4}{3})^3 \tau \nu^{1/3}}{(\lambda_1 \lambda_2 \lambda_3 \lambda_4)^{1/3}} \right|^2 \mathrm{d}\nu. \qquad (2.13)$$

Theorem 2.1. *Let* $\lambda_1, \lambda_2, \lambda_3, \lambda_4$ *be positive real numbers with* λ_1/λ_2 *irrational and algebraic. Then, uniformly in* $(\log N)^{-3} < \tau \le 1$, *one has*

$$\Upsilon(\tau, N) \ll \tau N^{5/3}(\log N)^{\varepsilon-3}.$$

This will be proved in §§3.1–3.3. For the time being, we take Theorem 2.1 for granted and deduce Theorem 1.1. One uses (2.12) together with the identity

$$2\tau = \left(1 + \frac{\tau}{\delta}\right)(\tau + \delta) - \frac{\tau^2}{\delta} - \delta \tag{2.14}$$

within the integral in Theorem 1.1. Then, by (2.13) and the trivial inequality $|\alpha + \beta|^2 \le 2(|\alpha|^2 + |\beta|^2)$, one finds that

$$\int_0^N \left| \rho_4(\tau, \nu) - \frac{2\Gamma(\frac{4}{3})^3 \tau \nu^{1/3}}{(\lambda_1\lambda_2\lambda_3\lambda_4)^{1/3}} \right|^2 d\nu$$

$$\ll \left(\frac{\tau}{\delta}\right)^2 \left(\Upsilon(\tau + \delta, N) + \Upsilon(\tau, N)\right) + \int_0^{2N} |\rho_4^*(\delta, \nu)|^2 d\nu + \delta^2 N^{5/3}$$

holds uniformly for $0 < \delta < \frac{1}{2}\tau \le \frac{1}{2}$. The term $N^{5/3}\delta^2$ arises from integrating the term $-\delta$ in the expansion of 2τ, and the integral on the right hand side stems from the error terms $\rho^*(\delta, \nu \pm \tau)$ in (2.12). A similar argument gives

$$\int_0^{2N} |\rho_4^*(\delta, \nu)|^2 d\nu \ll \Upsilon(\delta, 2N) + \int_0^{2N} \nu^{2/3}\delta^2 d\nu,$$

and thus we conclude that

$$\int_0^N \left| \rho_4(\tau, \nu) - \frac{2\Gamma(\frac{4}{3})^3 \tau \nu^{1/3}}{(\lambda_1\lambda_2\lambda_3\lambda_4)^{1/3}} \right|^2 d\nu$$

$$\ll \left(\frac{\tau}{\delta}\right)^2 \left(\Upsilon(\tau + \delta, N) + \Upsilon(\tau, N)\right) + \Upsilon(\delta, 2N) + \delta^2 N^{5/3}.$$

The conclusion of Theorem 1.1 now follows from Theorem 2.1 on taking $\delta = \frac{1}{2}\tau^{3/4}(\log N)^{-3/4}$.

Similar techniques yield Theorem 1.5, but the details are more involved. For $\lambda_1, \lambda_2 > 0$ we study the linear problem (1.5). In this new context, put

$$X = 2(\lambda_1^{-1} + \lambda_2^{-1} + 1)N, \tag{2.15}$$

and observe that for any $\nu \le N$ and any solution of (1.5), one has $p_j \le X$. In (2.5) we insert $F = \lambda_1 x_1 + \lambda_2 x_2 - \nu$, and set $u(x_1, x_2) = 0$ unless x_1, x_2 are primes p_1, p_2 not exceeding X, in which case we take $u(p_1, p_2) = (\log p_1)(\log p_2)$. In the notation of Theorem 1.5 and that used in §2.2, we see that $\sigma(\tau,\nu) = P_F(\tau;u)$. By (2.6) and (2.9), the formulae for $P^*(\tau;u)$ mutate into

$$\sigma^*(\tau,\nu) = \int h(\lambda_1\alpha)h(\lambda_2\alpha)e(-\nu\alpha)d_\tau\alpha,$$

where now

$$h(\alpha) = \sum_{p\leq X}(\log p)e(\alpha p)$$

is a Weyl sum over primes. The formula (2.12) remains valid, with ρ_s, ρ_s^* replaced by σ, σ^*, respectively. We conclude as follows.

Theorem 2.2. *With the hypotheses of Theorem 1.5, for any $A \geq 1$, there is a measurable function $E_\tau(\nu)$ such that, for $1 \leq \nu \leq N$, one has*

$$\sigma^*(\tau,\nu) = \frac{\tau\nu}{\lambda_1\lambda_2} + E_\tau(\nu) + O(\tau N(\log N)^{-A})$$

and

$$\int_0^N |E_\tau(\nu)|^2 d\nu \ll \tau N^{8/3+\varepsilon}.$$

We defer the proof of this result to §3.4, for now is the moment to deduce Theorem 1.5. Observing that the desired conclusion is trivial when τ is smaller than $N^{-1/3}$, we are entitled to assume the contrary. Equipped with the newly interpreted versions of (2.12) and (2.13), with σ in place of ρ_s, we deduce from (2.14) that for $0 < \delta < \frac{1}{2}\tau$, one has

$$\sigma(\tau,\nu) - \frac{2\tau\nu}{\lambda_1\lambda_2} = \left(1+\frac{\tau}{\delta}\right)E_{\tau+\delta}(\nu) - \frac{\tau}{\delta}E_\tau(\nu) + \frac{\delta\nu}{\lambda_1\lambda_2}$$

$$+ O\left(\sigma^*(\delta,\nu+\tau) + \sigma^*(\delta,\nu-\tau) + \frac{\tau^2}{\delta}N(\log N)^{-8A}\right).$$

Choosing $\delta = \tau(\log N)^{-4A}$, it is an easy exercise in the theory of uniform distribution to show that

$$(\log N)^{-2}\sigma^*(\delta,\nu) \leq \operatorname{card}\{n, m \leq X : |\lambda_1 n + \lambda_2 m - \nu| < \delta\} \ll \delta N.$$

Here it is worth recalling that $X \asymp N$ and λ_1/λ_2 is algebraic, and thus one even obtains an asymptotic formula for the counting problem in the middle term. The formula central to our discussion now reduces to

$$\left|\sigma(\tau,\nu) - \frac{2\tau\nu}{\lambda_1\lambda_2}\right| \ll (\log N)^{4A}(|E_{\tau+\delta}(\nu)| + |E_\tau(\nu)|) + \frac{\tau N}{(\log N)^{2A}}. \qquad (2.16)$$

Next, consider any real number $\nu \in [\frac{1}{2}N, N]$ for which the inequality (1.6) holds. Then, by (2.16), one of the two inequalities

$$|E_\tau(\nu)| \geq \frac{\tau N}{(\log N)^{6A}} \quad \text{or} \quad |E_{\tau+\delta}(\nu)| \geq \frac{\tau N}{(\log N)^{6A}}$$

must also hold. However, by Theorem 2.2, the measure of all $\nu \leq N$, for which this lower bound for $|E_\tau(\nu)|$ holds, cannot exceed

$$\frac{(\log N)^{12A}}{\tau^2 N^2} \int_0^N |E_\tau(\nu)|^2 d\nu \ll \tau^{-1} N^{2/3+\varepsilon}.$$

Since the same argument may be applied to $E_{\tau+\delta}(\nu)$, the conclusion of Theorem 1.5 follows via a dyadic dissection argument.

2.4. *The central interval*

As remarked earlier, the Davenport-Heilbronn method for diophantine inequalities embarks from (2.9). Whenever it succeeds, an asymptotic formula is produced where the main term arises from an interval centered at the origin, hereafter called the *central interval*. It has become common to refer to the latter interval as the major arc, by analogy with the circle method, but we reserve the term "major arcs" for classical major arcs.

We proceed in moderate detail and discuss the central interval for the integrals $\rho_s^*(\tau, \nu)$ and $\sigma^*(\tau, \nu)$. Our treatment of $\rho_s^*(\tau, \nu)$ can be taken as a model for any other application of the Fourier transform method to definite diophantine inequalities. There is a certain overlap with the exposition in Wooley [44], but the discussion there emphasises indefinite forms, and uses a different kernology. We shall conclude as follows.

Lemma 2.2. *Suppose that $s \geq 4$, that $\lambda_1, \ldots, \lambda_s$ are positive real numbers and that X is defined by (2.11). Let $C > 0$ denote a real number with $6C\lambda_j < 1$ for all $1 \leq j \leq s$, and let $\mathfrak{C} = [-CX^{-2}, CX^{-2}]$. In addition, put*

$$I = \int_{\mathfrak{C}} f(\lambda_1\alpha)\cdots f(\lambda_s\alpha)e(-\alpha\nu)d_\tau\alpha.$$

Then, uniformly in $0 < \tau \leq 1$ and $1 \leq \nu \leq N$, one has

$$I = \Gamma(\tfrac{4}{3})^s \Gamma(\tfrac{s}{3})^{-1}(\lambda_1 \cdots \lambda_s)^{-1/3}\tau\nu^{s/3-1} + O\big(1 + \tau X^{s-10/3}\big).$$

Proof. Let

$$v(\alpha) = \int_0^X e(\alpha\beta^3)d\beta. \tag{2.17}$$

Then, according to Theorem 4.1 of Vaughan [35], in the range $|\alpha| \leq \frac{1}{6}X^{-2}$ one has $f(\alpha) = v(\alpha) + O(1)$. Also, Theorem 7.3 of Vaughan [35] yields

$$v(\alpha) \ll X(1 + X^3|\alpha|)^{-1/3}. \tag{2.18}$$

For $\alpha \in \mathfrak{C}$, we therefore deduce that

$$f(\lambda_1\alpha) \cdots f(\lambda_s\alpha) = v(\lambda_1\alpha) \cdots v(\lambda_s\alpha) + O\big(1 + X^{s-1}(1 + X^3|\alpha|)^{-(s-1)/3}\big).$$

Then, on multiplying the previous display by $w_\tau(\alpha)e(-\alpha\nu)$ and integrating, we obtain

$$I = \int_\mathfrak{C} v(\lambda_1\alpha) \cdots v(\lambda_s\alpha)e(-\alpha\nu)\mathrm{d}_\tau\alpha + O(1 + \tau X^{s-4}\log X).$$

By (2.18), the *singular integral*

$$I_\infty = \int v(\lambda_1\alpha) \cdots v(\lambda_s\alpha)e(-\nu\alpha)\mathrm{d}_\tau\alpha$$

converges, and by (2.18) and (2.3), for any $Y > 0$, one has

$$\int_Y^\infty |v(\lambda_1\alpha) \cdots v(\lambda_s\alpha)|\mathrm{d}_\tau\alpha \ll \tau \int_Y^\infty \alpha^{-s/3}\mathrm{d}\alpha.$$

It is now immediate that

$$I = I_\infty + O(1 + \tau X^{s-10/3}). \tag{2.19}$$

Within the singular integral, we resubstitute (2.17) and apply (2.1) and (2.6) to arrive at the interim identity

$$I_\infty = \int_{[0,X]^s} \widehat{w}_\tau(\lambda_1\beta_1^3 + \cdots + \lambda_s\beta_s^3 - \nu)\mathrm{d}\boldsymbol{\beta}.$$

Since the λ_j are positive, it follows from (2.3) that we may extend the range of integration to $[0,\infty)^s$. A change of variable then yields the alternative formula

$$I_\infty = (\lambda_1 \cdots \lambda_s)^{-1/3} \int_{[0,\infty)^s} \widehat{w}_\tau(z_1^3 + \cdots + z_s^3 - \nu)\mathrm{d}\mathbf{z}.$$

Consider the equation $t = z_1^3 + \cdots + z_s^3$, which defines a surface in $\mathbb{R}^s$ of codimension 1. The area of this manifold in the quadrant with all z_j positive is $\Gamma(\frac{4}{3})^s\Gamma(\frac{s}{3})^{-1}t^{\frac{s}{3}-1}$. Hence, by the transformation formula and Fubini's theorem, one confirms that

$$I_\infty = (\lambda_1 \cdots \lambda_s)^{-1/3}\Gamma\Big(\frac{4}{3}\Big)^s\Gamma\Big(\frac{s}{3}\Big)^{-1} \int_0^\infty t^{s/3-1}\widehat{w}_\tau(t - \nu)\mathrm{d}t.$$

By (2.3), the remaining integral on the right hand side is equal to

$$\int_{-\tau}^{\tau} \left(1 - \frac{|\alpha|}{\tau}\right)(\nu + \alpha)^{s/3-1}\mathrm{d}\alpha = \tau\nu^{s/3-1} + O(\tau^2\nu^{s/3-2}),$$

and the lemma follows from (2.19). □

The analysis of the prime variables case is deeper because we need a rather wide central interval, for a reason that will become more transparent in due course. In such cases, the distribution of primes in short intervals comes into play. Fortunately, our discussion may be abbreviated by appealing to Brüdern, Cook and Perelli [6].

Lemma 2.3. *Let* $\lambda_1, \lambda_2 > 0$ *and* $0 < \tau < 1$, *and suppose that* X *is defined by means of* (2.15). *In addition, let* $\mathfrak{C} = [-X^{-1/2}, X^{-1/2}]$. *Then, for any* $A > 1$, *and uniformly in* $0 < \tau \leq 1$ *and* $1 \leq \nu \leq N$, *one has*

$$\int_{\mathfrak{C}} h(\lambda_1\alpha)h(\lambda_2\alpha)e(-\nu\alpha)\mathrm{d}_\tau\alpha = \frac{\tau\nu}{\lambda_1\lambda_2} + O(\tau X(\log X)^{-A}).$$

Proof. Let

$$h^*(\alpha) = \sum_{x \leq X} e(\alpha x).$$

Then one may apply the methods underlying the proof of Lemma 2 of [6] to establish the estimate

$$\int_{\mathfrak{C}} |h(\lambda_j\alpha) - h^*(\lambda_j\alpha)|^2\mathrm{d}\alpha \ll X(\log X)^{-2A}. \tag{2.20}$$

Here we note that although Lemma 2 of [6] states (2.20) only with $A = 1$, an inspection of the proof, which combines only Lemma 1 and estimate (5) of that paper, shows that any positive value of A is permissible. Next, let

$$v_1(\alpha) = \int_0^X e(\alpha\beta)\mathrm{d}\beta.$$

By Euler's summation formula, one finds that $h^*(\alpha) - v_1(\alpha) \ll 1 + X|\alpha|$, whence from (2.20) we obtain

$$\int_{\mathfrak{C}} |h(\lambda_j\alpha) - v_1(\lambda_j\alpha)|^2\mathrm{d}\alpha \ll X(\log X)^{-2A}.$$

Then, on noting the trivial estimate $w_\tau(\alpha) \ll \tau$ evident from (2.3), it follows from Schwarz's inequality that

$$\int_{\mathfrak{C}} h(\lambda_1\alpha)h(\lambda_2\alpha)e(-\nu\alpha)\mathrm{d}_\tau\alpha$$

$$= \int_{\mathfrak{C}} v_1(\lambda_1\alpha)v_1(\lambda_2\alpha)e(-\nu\alpha)\mathrm{d}_\tau\alpha + O(\tau X(\log X)^{-A}). \tag{2.21}$$

We can now proceed as we have explained in detail for sums of cubes. By partial integration, one has $v_1(\alpha) \ll X(1 + X|\alpha|)^{-1}$, and consequently, the integral on the right hand side of (2.21) can be extended to the whole real line, with the introduction of acceptable errors. Then, applying Fourier inversion as before we confirm that

$$\int_{-\infty}^{\infty} v_1(\lambda_1\alpha)v_1(\lambda_2\alpha)e(-\nu\alpha)\mathrm{d}_\tau\alpha = \frac{\tau\nu}{\lambda_1\lambda_2},$$

the details being considerably simpler. This proves the lemma.   □

2.5. *The interference principle*

In the previous section, we evaluated the contribution from the central interval $\mathfrak{C}$ to the Fourier transform that counts solutions of a diophantine inequality. The complementary set $\mathfrak{c} = \mathbb{R}\backslash\mathfrak{C}$ consists of two disjoint half-lines, and we therefore refer to it as the *complementary compositum*. For a successful analysis, its contribution to the count should be of a lower order of magnitude. The most important ingredient in any proof of this is an *interference principle* asserting that, when λ_1, λ_2 are non-zero real numbers, and λ_1/λ_2 is irrational, then two exponential sums such as $f(\lambda_1\alpha)$ and $f(\lambda_2\alpha)$, or $h(\lambda_1\alpha)$ and $h(\lambda_2\alpha)$, say, cannot be large simultaneously unless α lies in the central interval. However, loosely speaking, when $|f(\lambda_j\alpha)|$ is large, then as a consequence of Weyl's inequality, or a suitable variant thereof, one finds that $\lambda_j\alpha$ has a rational approximation a_j/q_j with small denominator. If this happens simultaneously for $j = 1, 2$, then $a_1q_2/(a_2q_1)$ is an approximation to λ_1/λ_2, and so the measure of the set of all α where $|f(\lambda_1\alpha)|$ and $|f(\lambda_2\alpha)|$ are simultaneously large, should be quite small. In this form, the interference principle becomes a statement about diophantine approximations alone, and references to exponential sums can be removed entirely. For another, rather different view of this phenomenon, see section 3 of Brüdern [5].

We shall present here a simple derivation of the interference principle along the lines indicated, based on ideas of Watson [38]. It appears to us that the potential of this approach has been overlooked in the past. As we shall demonstrate below, our method provides easy access to asymptotic formulae for diophantine inequalities, avoiding to a large extent the entangled interplay between diophantine approximations and major arc information for exponential sums, as in the celebrated works of Bentkus and Götze [2], and Freeman [21], who were the first to obtain such asymptotic formulae at all. The stronger bounds that are available when λ_1/λ_2 is not only irrational but also algebraic, moreover, follow from the same principles.

Before making these comments precise, we need to introduce some notation. Let $N \geq 1$ denote the main parameter. For $1 \leq Q \leq \frac{1}{2}\sqrt{N}$ the intervals $|q\alpha - a| \leq Q/N$ with $1 \leq q \leq Q$, $a \in \mathbb{Z}$ and $(a, q) = 1$, are pairwise disjoint. Their union, the *major arcs* $\mathfrak{M}(Q)$, forms a 1-periodic set. The subset $\mathfrak{N}(Q) = \mathfrak{M}(Q) \cap [0, 1]$ is the familiar set of major arcs in the classical circle method. We also define here the *minor arcs*

$$\mathfrak{m}(Q) = \mathbb{R} \backslash \mathfrak{M}(Q), \quad \mathfrak{n}(Q) = [0, 1] \backslash \mathfrak{N}(Q),$$

although these will not be needed until the next section. Watson's method is imported through Lemma 4 of Brüdern, Cook and Perelli [6] that we restate as Lemma 2.4. Temporarily, we suppose only that λ_1, λ_2 are *non-zero* real numbers, so that our main estimates Theorems 2.3 and 2.4 apply also to indefinite problems. With $\lambda_1, \lambda_2 \in \mathbb{R} \backslash \{0\}$ fixed, we define

$$\mathfrak{K}(Q_1, Q_2) = \{\alpha \in \mathbb{R} : \lambda_j \alpha \in \mathfrak{M}(Q_j) \ (j = 1, 2)\}.$$

In addition, when $y > 0$, we put

$$\mathfrak{K}_y(Q_1, Q_2) = \{\alpha \in \mathfrak{K}(Q_1, Q_2) : y < |\alpha| \leq 2y\}.$$

Lemma 2.4. *Let λ_1, λ_2 be non-zero real numbers such that λ_1/λ_2 is irrational. There exists a positive real number $\varepsilon_0 = \varepsilon_0(\lambda_1, \lambda_2)$ with the following property. Suppose that $1 \leq Q_j \leq \frac{1}{2}\sqrt{N}$ $(j = 1, 2)$, and $r \in \mathbb{N}$ satisfies $r \leq \varepsilon_0 N/(Q_1 Q_2)$ and $\|r\lambda_1/\lambda_2\| < 1/r$. Then, for any $y > 0$ with $|\lambda_j| y \geq 2Q_j/N$ $(j = 1, 2)$, one has mes $\mathfrak{K}_y(Q_1, Q_2) \ll yN^{-1}Q_1Q_2 r^{-1}$.*

Note that $\mathfrak{N}(Q)$ has measure about $Q^2 N^{-1}$. An application of Schwarz's inequality therefore reveals that mes $\mathfrak{K}_y(Q_1, Q_2) \ll yN^{-1}Q_1Q_2$. Lemma 2.4 improves on this estimate by a factor $1/r$, and it is this saving that implies that $\lambda_1\alpha$ and $\lambda_2\alpha$ simultaneously lie on major arcs only for a slim set of real numbers α.

For a non-zero real number λ, define

$$T_\lambda(R) = \max\{r \in \mathbb{N} : r \leq R, \|\lambda r\| \leq 1/r\}. \tag{2.22}$$

When λ is irrational, then $\|\lambda r\| \leq 1/r$ has infinitely many solutions, and consequently $T_\lambda(R) \to \infty$ as $R \to \infty$. If r_m is the sequence of solutions of $\|\lambda r\| \leq 1/r$, arranged in increasing order, then for algebraic irrational numbers λ, Roth's theorem gives $r_{m+1} \ll r_m^{1+\varepsilon}$. Therefore, in this case,

$$T_\lambda(R) \gg R^{1-\varepsilon}. \tag{2.23}$$

We can now put Lemma 2.4 into a form more readily applied. Subject to the conditions of this lemma, we deduce from (2.3) and (2.6) that

$$\int_{\mathfrak{K}_y(Q_1,Q_2)} \mathrm{d}_\tau\alpha \ll Q_1 Q_2 (NT)^{-1} \min(\tau y, (\tau y)^{-1}),$$

where $T = T_{\lambda_1/\lambda_2}(\varepsilon_0 N/(Q_1 Q_2))$. We choose a number Y with $|\lambda_j|Y \geq 2Q_j/N$ $(j = 1, 2)$ and sum the previous estimate over $y = 2^l Y$. This gives

$$\int_{\substack{|\alpha| \geq Y \\ \alpha \in \mathfrak{R}(Q_1,Q_2)}} \mathrm{d}_\tau\alpha \ll Q_1 Q_2 (NT)^{-1}. \tag{2.24}$$

Theorem 2.3. *Let λ_1, λ_2 be non-zero real numbers. If λ_1/λ_2 is irrational and algebraic, then uniformly in $0 < \tau \leq 1$, $Q_j \leq \frac{1}{2}\sqrt{N}$ and $Y \geq 2Q_j/(|\lambda_j|N)$ $(j = 1, 2)$, one has*

$$\int_{\substack{|\alpha| \geq Y \\ \alpha \in \mathfrak{R}(Q_1,Q_2)}} \mathrm{d}_\tau\alpha \ll N^{\varepsilon-2} Q_1^2 Q_2^2.$$

The proof is immediate from (2.23) and (2.24).

Theorem 2.4. *Let $Q = Q(N)$ be a function that is increasing, unbounded, and satisfies $Q(N)/\sqrt{N} \to 0$ as $N \to \infty$. Let λ_1, λ_2 be non-zero real numbers such that λ_1/λ_2 is irrational. Then, uniformly in $0 < \tau \leq 1$, $1 \leq Q_j \leq Q(N)$ and $Y \geq 2Q(N)/(|\lambda_j|N)$, one has*

$$\int_{\substack{|\alpha| \geq Y \\ \alpha \in \mathfrak{R}(Q_1,Q_2)}} \mathrm{d}_\tau\alpha \ll N^{-1} Q_1 Q_2 T_{\lambda_1/\lambda_2}\left(\frac{\varepsilon_0 N}{Q(N)^2}\right)^{-1}.$$

Again, this is merely a restatement of (2.24) if one observes that under the current hypotheses, one has $Q_1 Q_2 \leq Q(N)^2$, and $T_\lambda(R)$ is a non-decreasing function. We note that the condition $N/Q(N)^2 \to \infty$ ensures that the upper bound in Theorem 2.4 is $o(N^{-1}Q_1 Q_2)$ as $N \to \infty$.

3. Classical mean square methods

3.1. *Plancherel's identity*

In this chapter, we complete the proofs of Theorems 1.1 and 1.5 by demonstrating Theorems 2.1 and 2.2. However, our primary concern is to illustrate, in this and related enterprises, the use of Plancherel's identity for take-off and Wooley's amplifier [44] coupled with the interference principle for landing.

As in previous sections, we begin with additive cubic forms. Under the hypotheses of Theorem 2.1, for any measurable set $\mathfrak{A} \subset \mathbb{R}$ we write

$$\rho_{\mathfrak{A}}^*(\tau, \nu) = \int_{\mathfrak{A}} f(\lambda_1 \alpha) f(\lambda_2 \alpha) f(\lambda_3 \alpha) f(\lambda_4 \alpha) e(-\nu\alpha) \mathrm{d}_\tau \alpha. \tag{3.1}$$

We define the central interval $\mathfrak{C}$ as in Lemma 2.2, and the complementary compositum $\mathfrak{c}$ as in §2.5, and then have $\rho_4^*(\tau, \nu) = \rho_{\mathfrak{C}}^*(\tau, \nu) + \rho_{\mathfrak{c}}^*(\tau, \nu)$. Consequently, recalling (2.13), we find that

$$\Upsilon(\tau, N) \ll \int_0^N \left| \rho_{\mathfrak{C}}^*(\tau, \nu) - \frac{\Gamma(\frac{4}{3})^3 \tau \nu^{1/3}}{(\lambda_1 \lambda_2 \lambda_3 \lambda_4)^{1/3}} \right|^2 \mathrm{d}\nu + \int_0^N |\rho_{\mathfrak{c}}^*(\tau, \nu)|^2 \mathrm{d}\nu.$$

Next, applying Lemma 2.2 to estimate the term involving the central interval, we deduce that

$$\Upsilon(\tau, N) \ll N(1 + N^{4/9} \tau^2) + \int_0^N |\rho_{\mathfrak{c}}^*(\tau, \nu)|^2 \mathrm{d}\nu. \tag{3.2}$$

Now, by (3.1) and (2.6), the number $\rho_{\mathfrak{c}}(\tau, \nu)$ is the Fourier transform, at ν, of the function that is

$$f(\lambda_1 \alpha) f(\lambda_2 \alpha) f(\lambda_3 \alpha) f(\lambda_4 \alpha) w_\tau(\alpha)$$

for $\alpha \in \mathfrak{c}$, and 0 on $\mathfrak{C}$. By Plancherel's identity,

$$\int_{-\infty}^\infty |\rho_{\mathfrak{c}}^*(\tau, \nu)|^2 \mathrm{d}\nu = \int_{\mathfrak{c}} |f(\lambda_1 \alpha) \cdots f(\lambda_4 \alpha)|^2 w_\tau(\alpha)^2 \mathrm{d}\alpha,$$

from which we deduce, via (2.3) and (2.6), the important inequality

$$\int_0^N |\rho_{\mathfrak{c}}^*(\tau, \nu)|^2 \mathrm{d}\nu \leq \tau \int_{\mathfrak{c}} |f(\lambda_1 \alpha) \cdots f(\lambda_4 \alpha)|^2 \mathrm{d}_\tau \alpha \tag{3.3}$$

through which the estimation will proceed.

3.2. *Some mean values*

We summarize here a few standard bounds for Weyl sums $f(\alpha)$. It will be convenient, temporarily, to write $\mathfrak{M} = \mathfrak{M}(X^{3/4})$ and $\mathfrak{m} = \mathfrak{m}(X^{3/4})$, with $\mathfrak{N}$ and $\mathfrak{n}$ defined mutatis mutandis. Combining the methods used to prove Lemma 1 in Vaughan [33] with the bounds for Hooley's Δ-function in Hall and Tenenbaum [26], we obtain an enhanced form of Weyl's inequality which asserts that

$$\sup_{\alpha \in \mathfrak{m}} |f(\alpha)| \ll X^{3/4} (\log X)^{1/4 + \varepsilon}. \tag{3.4}$$

Boklan [3] supplies the bound

$$\int_{\mathfrak{n}} |f(\alpha)|^8 d\alpha \ll X^5 (\log X)^{\varepsilon-3} \tag{3.5}$$

as an improvement over Theorem B of Vaughan [33]. Standard applications of the circle method (see Chapter 4 of [35]) yield the complementary estimates

$$\int_{\mathfrak{N}} |f(\alpha)|^4 d\alpha \ll X^{1+\varepsilon}, \quad \int_{\mathfrak{N}} |f(\alpha)|^{4+\theta} d\alpha \ll X^{1+\theta}, \tag{3.6}$$

the latter being valid for any fixed $\theta > 0$. This last bound combines with (3.8) to deliver the estimate

$$\int_0^1 |f(\alpha)|^8 d\alpha \ll X^5, \tag{3.7}$$

a conclusion that is also implied by Theorem 2 of Vaughan [33]. We now transport these mean value bounds into integrals over 1-periodic sets, against the measure $d_\tau \alpha$.

Lemma 3.1. *Let $G : \mathbb{R} \to \mathbb{C}$ be a function of period 1 that is integrable on $[0, 1]$. Then, for any $\tau > 0$ and any $u \in \mathbb{R}$, one has*

$$\int G(\alpha)e(-\alpha u)d_\tau \alpha = \sum_{n=-\infty}^{\infty} \widehat{w}_\tau(n - u) \int_0^1 G(\alpha)e(-\alpha n)d\alpha.$$

Proof. This is a special case of formula (4) in Brüdern [4], with $w_\tau(\alpha)$ in the role of the kernel K employed in [4]. Note that the sum on the right is over $|n - u| \leq \tau$ only.   □

As an example, take $u = 0$, and put $G(\alpha) = |f(\alpha)|^8$ when $\alpha \in \mathfrak{m}$, and $G(\alpha) = 0$ otherwise. Then

$$\int_{\mathfrak{m}} |f(\alpha)|^8 d_\tau \alpha = \sum_{|n| \leq \tau} \widehat{w}_\tau(n) \int_{\mathfrak{n}} |f(\alpha)|^8 e(-\alpha n)d\alpha,$$

whence, in particular,

$$\int_{\mathfrak{m}} |f(\alpha)|^8 d_\tau \alpha \ll X^5 (\log X)^{\varepsilon-3}. \tag{3.8}$$

Similarly, when $\tau \ll 1$ and $\theta > 0$, one infers from (3.6) that

$$\int_{\mathfrak{M}} |f(\alpha)|^4 d_\tau \alpha \ll X^{1+\varepsilon}, \quad \int_{\mathfrak{M}} |f(\alpha)|^{4+\theta} d_\tau \alpha \ll X^{1+\theta}, \tag{3.9}$$

and from (3.7) that

$$\int |f(\alpha)|^8 \mathrm{d}_\tau \alpha \ll X^5. \tag{3.10}$$

3.3. *The amplification technique*

The first steps in the estimation of the right hand side of (3.2) follow Wooley [44]. We need to cover $\mathfrak{c}$ by the sets

$$\mathfrak{d} = \{\alpha \in \mathfrak{c} : \lambda_j \alpha \in \mathfrak{m} \ (1 \leq j \leq 4)\},$$
$$\mathfrak{D} = \{\alpha \in \mathfrak{c} : \lambda_j \alpha \in \mathfrak{M} \ (1 \leq j \leq 4)\},$$
$$\mathfrak{E}_{ij} = \{\alpha \in \mathfrak{c} : \lambda_i \alpha \in \mathfrak{M}, \lambda_j \alpha \in \mathfrak{m}\} \qquad (1 \leq i, j \leq 4, \ i \neq j).$$

Each of these sets $\mathfrak{a}$ requires a different argument. For convenience, we write

$$I(\mathfrak{a}) = \int_{\mathfrak{a}} |f(\lambda_1 \alpha) \cdots f(\lambda_4 \alpha)|^2 \mathrm{d}_\tau \alpha.$$

By Hölder's inequality, one infers that

$$I(\mathfrak{d}) \leq \prod_{j=1}^{4} \left(\int_{\lambda_j \alpha \in \mathfrak{m}} |f(\lambda_j \alpha)|^8 \mathrm{d}_\tau \alpha \right)^{1/4}.$$

Since a change of variable reveals that

$$\int_{\lambda \alpha \in \mathfrak{m}} |f(\lambda \alpha)|^8 \mathrm{d}_\tau \alpha = \int_{\mathfrak{m}} |f(\alpha)|^8 \mathrm{d}_{\tau/\lambda} \alpha,$$

one derives from (3.8) the important bound

$$I(\mathfrak{d}) \ll X^5 (\log X)^{\varepsilon - 3}. \tag{3.11}$$

Next, we use (3.4) and Hölder's inequality to conclude that

$$I(\mathfrak{E}_{12}) \ll X^{3/2+\varepsilon} \left(\int_{\lambda_1 \alpha \in \mathfrak{M}} |f(\lambda_1 \alpha)|^4 \mathrm{d}_\tau \alpha \right)^{1/2}$$
$$\times \left(\int |f(\lambda_3 \alpha)|^8 \mathrm{d}_\tau \alpha \right)^{1/4} \left(\int |f(\lambda_4 \alpha)|^8 \mathrm{d}_\tau \alpha \right)^{1/4}.$$

Thus, by changes of variable together with (3.9) and (3.10), it follows that $I(\mathfrak{E}_{12}) \ll X^{9/2+\varepsilon}$. By symmetry, this shows that

$$I(\mathfrak{E}_{ij}) \ll X^{9/2+\varepsilon} \quad (1 \leq i, j \leq 4, \ i \neq j). \tag{3.12}$$

The *amplification* is now complete: since $\mathfrak{c}$ is the union of $\mathfrak{d}, \mathfrak{D}$ and the $\mathfrak{E}_{ij}$, then in view of (3.11) and (3.14), it suffices to consider $\mathfrak{D}$. Here all $\lambda_j \alpha$ lie on a major arc. But by comparison with the analysis just undertaken, major arc moments are far easier to control (compare (3.8), (3.6)), and the interference principle can be brought into play.

For $|q\alpha - a| \leq QX^{-3}$ and $q \leq Q \leq X^{3/4}$ it follows from Theorem 4.1 and Lemma 4.6 of [35] that

$$f(\alpha) \ll X(q + X^3|q\alpha - a|)^{-1/3}, \qquad (3.13)$$

and consequently, when $\alpha \in \mathfrak{M}(2Q)\backslash\mathfrak{M}(Q)$, we find that $f(\alpha) \ll XQ^{-1/3}$. Therefore, slicing $\mathfrak{D}$ into sections

$$\mathfrak{D}(Q_1, Q_2) = \{\alpha \in \mathfrak{D} : \lambda_j\alpha \in \mathfrak{M}(2Q_j)\backslash\mathfrak{M}(Q_j) \text{ for } j = 1, 2\}, \qquad (3.14)$$

with $1 \leq Q_j \leq X^{3/4}$, we see from Theorem 2.3 that

$$\int_{\mathfrak{D}(Q_1, Q_2)} |f(\lambda_1\alpha)f(\lambda_2\alpha)|^2 \mathrm{d}_\tau\alpha \ll N^{\varepsilon-2}X^4(Q_1Q_2)^{4/3} \ll N^\varepsilon.$$

By a dyadic dissection argument and trivial bounds for $|f(\lambda_3\alpha)f(\lambda_4\alpha)|$, it therefore follows that $I(\mathfrak{D}) \ll N^\varepsilon X^4$. In combination with (3.11) and (3.14), we may infer that the integral on the left hand side of (3.2) is $O(\tau X^5(\log X)^{\varepsilon-3})$, and Theorem 2.1 follows from (3.6). In view of the discussion following the statement of Theorem 2.1, this also completes the proof of Theorem 1.1.

3.4. *Linear forms in primes*

We now establish Theorem 2.2 by an argument paralleling that of the previous two sections. By Lemma 2.3, if we put $\mathfrak{c} = \{\alpha : |\alpha| \geq X^{-1/2}\}$ and

$$E_\tau(\nu) = \int_\mathfrak{c} h(\lambda_1\alpha)h(\lambda_2\alpha)e(-\nu\alpha)\mathrm{d}_\tau\alpha, \qquad (3.15)$$

then just as in the discussion leading to (3.2), Plancherel's identity gives

$$\int_0^N |E_\tau(\nu)|^2\mathrm{d}\nu \ll \tau \int_\mathfrak{c} |h(\lambda_1\alpha)h(\lambda_2\alpha)|^2\mathrm{d}_\tau\alpha. \qquad (3.16)$$

The dissection of $\mathfrak{c}$ this time is much simpler. For $j = 1$ and 2, we consider

$$\mathfrak{m}_j = \{\alpha \in \mathfrak{c} : \lambda_j\alpha \in \mathfrak{m}(X^{1/3})\}.$$

Then $\mathfrak{c}$ is the union of $\mathfrak{m}_1, \mathfrak{m}_2$ and $\mathfrak{c} \cap \mathfrak{K}(X^{1/3}, X^{1/3})$.

Vinogradov's estimate for exponential sums (Vaughan [35], Theorem 3.1) shows both that

$$\sup_{\alpha \in \mathfrak{m}(X^{1/3})} |h(\alpha)| \ll X^{5/6}(\log X)^4,$$

and, whenever $Q \le X^{1/3}$, that

$$\sup_{\alpha \in \mathfrak{M}(2Q) \setminus \mathfrak{M}(Q)} |h(\alpha)| \ll XQ^{-1/2}(\log X)^4. \tag{3.17}$$

Paired with the mean square bound

$$\int |h(\lambda_1 \alpha)|^2 \mathrm{d}_\tau \alpha = \sum_{\substack{\lambda_1 |p_1 - p_2| < \tau \\ p_1, p_2 \le X}} (\log p_1)(\log p_2) \ll X \log X, \tag{3.18}$$

the first of these estimates yields

$$\int_{\mathfrak{m}_2} |h(\lambda_1 \alpha)h(\lambda_2 \alpha)|^2 \mathrm{d}_\tau \alpha \ll X^{8/3}(\log X)^9,$$

and the same is true for the contribution from $\mathfrak{m}_1$, by symmetry. For the set $\mathfrak{K}(X^{1/3}, X^{1/3})$, we apply the same slicing technique as we used for $\mathfrak{D}$ in §3.3. Then, importing (3.16) into Theorem 2.3, we find that

$$\int_{\mathfrak{c} \cap \mathfrak{K}(X^{1/3}, X^{1/3})} |h(\lambda_1 \alpha)h(\lambda_2 \alpha)|^2 \mathrm{d}_\tau \alpha \ll X^{8/3+\varepsilon},$$

an estimate that may also be found on p. 97 of [6]. Collecting the upper bounds obtained thus far, we conclude that

$$\int_{\mathfrak{c}} |h(\lambda_1 \alpha)h(\lambda_2 \alpha)|^2 \mathrm{d}_\tau \alpha \ll X^{8/3+\varepsilon}, \tag{3.19}$$

and Theorem 2.2 follows from (3.16) and Lemma 2.3.

3.5. *Bessel's inequality*

The continuous averages in Theorems 1.1, 2.1 and 2.2 could be replaced by discrete ones. In such a setting, one chooses an increasing sequence of test points ν_m that is roughly of linear growth; see [6] for one of the many possibilities to make this precise. One would then study, for example, a mean square of the type

$$\sum_{\nu_m \le N} \left| \rho_4(\tau, \nu_m) - \frac{2\Gamma(\frac{4}{3})^3 \tau \nu_m^{1/3}}{(\lambda_1 \lambda_2 \lambda_3 \lambda_4)^{1/3}} \right|^2$$

as the appropriate analogue of the integral in Theorem 1.1. This approach, which has been used successfully by Brüdern, Cook and Perelli [6], Parsell [30], and others, is roughly of the same strength as the methods described herein. Bessel's inequality performs the averaging in the discrete world, and replaces Plancherel's identity in the work of §3.1 and elsewhere. For thinner averages, it appears that it is almost mandatory to work with discrete test points, and this will be the theme in most of the following sections.

4. Semi-classical averaging

4.1. *Another mean square approach*

The main purpose of this chapter is to establish Theorem 1.7. Thus, we continue the discussion begun in §§3.4 and 3.5, and examine the distribution of the linear form $\lambda_1 p_1 + \lambda_2 p_2$ near the sequence

$$\nu_m = \exp((\log m)^\gamma), \tag{4.1}$$

where $1 < \gamma < \frac{3}{2}$ is fixed once and for all. As much as is possible, the notation from the proof of Theorem 1.5 is kept throughout this chapter. The parameters X and N are linked as in (2.15), and we apply the same notation as in §3.4 and the statement of Lemma 2.3. This defines the central interval and its complement $\mathfrak{c}$. With $E_\tau(\nu)$ as in (3.15), we recall that Lemma 2.3 asserts that for any $A > 1$, uniformly in $0 < \tau \le 1$ and $1 \le \nu \le N$, one has

$$\sigma^*(\tau, \nu) = \frac{\tau\nu}{\lambda_1\lambda_2} + E_\tau(\nu) + O(\tau N(\log N)^{-A}). \tag{4.2}$$

Now define M through the equation

$$\exp((\log 2M)^\gamma) = N.$$

Then $\log M \asymp (\log N)^{1/\gamma}$, and whenever $m \le 2M$ one has $\nu_m \le N$.

Lemma 4.1. *Let $1 < \gamma < \frac{3}{2}$. Also, let $\lambda_1, \lambda_2 > 0$, and suppose that λ_1/λ_2 is an algebraic irrational. Then, there exists $\kappa > 0$ such that*

$$\sum_{M < m \le 2M} |E_\tau(\nu_m)|^2 \ll MN^2 \exp\left(-2\kappa(\log M)^{3-2\gamma}\right).$$

This implies Theorem 1.7, for in view of (4.1) we infer that for any fixed $0 < \tau < 1$, the asymptotic formula

$$\sigma^*(\tau, \nu) = \frac{\tau\nu}{\lambda_1\lambda_2} + O(N(\log N)^{-A})$$

holds for all but $O(M \exp(-\kappa(\log M)^{3-2\gamma}))$ of the values $\nu = \nu_m$ with $M < m \le 2M$. By the transfer principle (Lemma 2.1) and an argument almost identical to the one given after the statement of Theorem 2.2, this can be reformulated as an asymptotic formula for $\sigma(\tau, \nu)$. Theorem 1.7 then follows by summing over dyadic intervals.

4.2. *Exponential sums over test sequences*

Let

$$\Phi(\zeta) = \sum_{M < m \le 2M} e(\zeta\nu_m), \quad H(\zeta) = h(\lambda_1\zeta)h(\lambda_2\zeta). \tag{4.3}$$

Then, by (3.15), one has

$$\sum_{M<m\leq 2M}|E_\tau(\nu_m)|^2 = \iint_{\mathfrak{c}\times\mathfrak{c}} H(\alpha)H(-\beta)\Phi(\beta-\alpha)\mathrm{d}_\tau\alpha\,\mathrm{d}_\tau\beta. \qquad (4.4)$$

The next lemma is a special case of Theorem 2 of Brüdern and Perelli [15].

Lemma 4.2. *Let $1<\gamma<\frac{3}{2}$ and $0<\delta<\min(\frac{1}{100},\gamma-1)$. Then, there is a real number $\kappa>0$ such that uniformly in $N^{\delta-1}\leq|\zeta|\leq N^\delta$, one has*

$$\Phi(\zeta)\ll M\exp(-2\kappa(\log M)^{3-2\gamma}).$$

In order to establish Lemma 4.1, we split the integration in (4.4) into the three regions

$$\mathfrak{T}=\{(\alpha,\beta)\in\mathfrak{c}\times\mathfrak{c}:|\alpha-\beta|>2N^\delta\},$$
$$\mathfrak{U}=\{(\alpha,\beta)\in\mathfrak{c}\times\mathfrak{c}:N^{\delta-1}<|\alpha-\beta|\leq 2N^\delta\},$$
$$\mathfrak{V}=\{(\alpha,\beta)\in\mathfrak{c}\times\mathfrak{c}:|\alpha-\beta|\leq N^{\delta-1}\}.$$

The corresponding contributions to (4.4) are denoted by $J(\mathfrak{T})$, $J(\mathfrak{U})$, $J(\mathfrak{V})$ respectively, and we examine each in turn.

The set $\mathfrak{T}$ presents little difficulty. When $|\alpha-\beta|>2N^\delta$, then $\max(|\alpha|,|\beta|)>N^\delta$. By symmetry and a trivial bound for $\Phi(\zeta)$, this implies that

$$|J(\mathfrak{T})|\leq 4M\int|H(\beta)|\mathrm{d}_\tau\beta\int_{N^\delta}^\infty|H(\alpha)|\mathrm{d}_\tau\alpha.$$

By Schwarz's inequality and (3.18), the β-integral here is $O(X\log X)$. For the α-integral, note that whenever $\lambda>0$, then

$$\lambda\int_y^{y+1/\lambda}|h(\lambda\alpha)|^2\mathrm{d}\alpha=\int_0^1|h(\alpha)|^2\mathrm{d}\alpha\ll X\log X,$$

irrespective of $y\in\mathbb{R}$. Hence, we may split the range $[N^\delta,\infty)$ into intervals of length $1/\lambda$ to deduce via (2.3) and (2.6) that

$$\int_{N^\delta}^\infty|h(\lambda\alpha)|^2\mathrm{d}_\tau\alpha\ll\tau^{-1}N^{-\delta}X\log X.$$

We apply this bound with $\lambda=\lambda_1$ and $\lambda=\lambda_2$. Recalling that $X\asymp N$ in the current context, another application of Schwarz's inequality reveals that

$$J(\mathfrak{T})\ll MN^{2-\delta}(\log N)^2, \qquad (4.5)$$

which is more than is required.

For the set $\mathfrak{U}$, we argue similarly, but this time one estimates $\Phi(\alpha - \beta)$ using Lemma 4.2. The remaining double integral can be taken over $\mathbb{R}^2$, and is then readily reduced to four integrals of the type (3.18). This yields

$$J(\mathfrak{U}) \ll MN^2(\log N)^2 \exp(-2\kappa(\log M)^{3-2\gamma}). \qquad (4.6)$$

Rather more care is required for $\mathfrak{V}$. The treatment begins with (4.4) and the substitution $\beta = \alpha + \zeta$. With a trivial bound for $\Phi(\zeta)$ and (2.6), we infer that

$$J(\mathfrak{V}) \ll M \int_{\mathfrak{c}} |H(\alpha)| \int_{-N^{\delta-1}}^{N^{\delta-1}} |H(\alpha + \zeta)| w_\tau(\alpha + \zeta) d\zeta \, d_\tau \alpha.$$

Reverse the order of integrations, and then apply Schwarz's inequality to the integral over $\mathfrak{c}$ to bring in the integral (3.19). This implicitly applies the interference estimates from the work of §3.4, and reveals that

$$J(\mathfrak{V}) \ll MX^{4/3+\varepsilon} \int_{-N^{\delta-1}}^{N^{\delta-1}} \left(\int |H(\alpha + \zeta) w_\tau(\alpha + \zeta)|^2 d_\tau \alpha \right)^{1/2} d\zeta.$$

A trivial estimate for $h(\lambda_2(\alpha + \zeta))$, in combination with (2.3) and (3.18), bounds the inner integral by

$$\tau^2 X^2 \int_{-\infty}^{\infty} |h(\lambda_1(\alpha + \zeta))|^2 w_\tau(\alpha + \zeta) d\alpha \ll \tau^2 X^3 \log X,$$

which then implies the bound

$$J(\mathfrak{V}) \ll \tau MN^{11/6+\delta+\varepsilon}. \qquad (4.7)$$

The conclusion of Lemma 4.1 follows from (4.4), (4.5), (4.6) and (4.7).

4.3. *Potential applications*

The method exposed here is particularly useful if the test sequence ν_m is uniformly distributed modulo 1, as is the case with the example (4.1). The Fourier series $\Phi(\zeta)$ then peaks only at $\zeta = 0$. The mean square method produces a double integral that Φ collapses to an expression reminiscent of a one-dimensional situation, but with two sets of generating functions. This far, there is a strong resemblence to the analysis via Plancherel's identity. The success of the method then depends on the savings that one can obtain for $\Phi(\zeta)$, and in the case (4.1) this is our Lemma 4.2.

At least in principle, the method is also applicable when ν_m runs through an arithmetic sequence, such as the values of a polynomial, but then $\Phi(\zeta)$ may have large values when ζ is in some set of major arcs $\mathfrak{M}(Q)$. Yet, at the cost of extra complication in detail, the method can still be pressed home when the initial estimations stemming from a suitable analogue of Lemma 4.2 turn out to be successful. Perelli [32] is an example where these ideas were used, and one could obtain a weaker version of Theorem 1.6 along these lines as well. However, for polynomial sequences in particular, the methods developed in I - VI are more promising, and we now turn to their initiation in the context of diophantine inequalities.

5. Fourier analysis of exceptional sets

5.1. *An illustrative example*

The work of chapter 4 depends on an analysis of the Fourier series of the test sequence. From now on we take a different point of view, and examine the Fourier series of the *exceptional set* to a representation problem.

Most of the results stated in chapter 1 that we have not yet established have a common flavour. One investigates a diophantine inequality involving a (large) parameter ν, and it is expected that there are many solutions. This expectation is tested on average over the values ν_m of a positive polynomial (Theorems 1.2, 1.3, 1.6, 1.8). There may be exceptions to the anticipated behaviour, but these are characterised by an analytic inequality: the integral over the complementary compositum must be unexpectedly large. This gives a precise meaning to an "exceptional value" of m. These numbers form a set $\mathcal{Z}$, and we consider the exponential sum

$$\sum_{m \in \mathcal{Z}} e(\alpha \nu_m).$$

This is no longer a classical Weyl sum, as was the case with (4.2), but whenever the sum reappears within moment estimates, one can restore the polynomial structure through an enveloping argument. This idea has been explored in I, II, III and V, for diophantine equations, but there is little difficulty to adapt the principal ideas to the wider context. The introductory section of I contains a detailed account of the method to which we have nothing to add. Instead, we introduce the reader to the basic strategy by working through an illustrative example that can be handled from scratch. This will ultimately yield a proof of Theorem 1.2.

Thus, we are now concerned with the weighted counter $\rho_6^*(\tau, \nu)$, we suppose that $1 \leq \nu \leq N$, and that $X \asymp N^{1/3}$ is chosen in accordance with (2.11). We can apply Lemma 2.2, the infrastructure of which also fixes the central interval and the complementary compositum $\mathfrak{c}$. The result is that

$$\rho_6^*(\tau, \nu) = \frac{\Gamma(\tfrac{4}{3})^6 \tau \nu}{(\lambda_1 \cdots \lambda_6)^{1/3}} + \int_{\mathfrak{c}} F(\alpha) e(-\nu \alpha) \mathrm{d}_\tau \alpha + O(1 + \tau N^{8/9}). \quad (5.1)$$

where $F(\alpha) = f(\lambda_1\alpha)\cdots f(\lambda_6\alpha)$. We will attempt to mimic the amplification procedures in §3.3, and also the interference estimate. It is important
to observe that the latter can be performed without averaging, and we
therefore begin with this part.

Let $\mathfrak{m}, \mathfrak{M}$ be defined as in §3.2. Following the pattern of §3.3, we write

$$\mathfrak{D} = \{\alpha \in \mathfrak{c} : \lambda_j\alpha \in \mathfrak{M} \ (1 \leq j \leq 6)\}$$

for the amplifying set, and slice it into the subsets $\mathfrak{D}(Q_1, Q_2)$ introduced
in (3.14). Let

$$\widetilde{T}(N) = \min(T_{\lambda_1/\lambda_2}(\varepsilon_0 N X^{-3/2}), (\log\log N)^4).$$

By (3.13) and Theorem 2.4, the bound

$$\int_{\mathfrak{D}(Q_1, Q_2)} |f(\lambda_1\alpha)f(\lambda_2\alpha)|^4 d_\tau\alpha \ll X^8 N^{-1}(Q_1 Q_2)^{-1/3}\widetilde{T}(N)^{-1}$$

holds throughout the relevant range $1 \leq Q_1, Q_2 \leq X^{3/4}$. We may therefore
sum over dyadic ranges for Q_1 and Q_2 to confirm that

$$\int_{\mathfrak{D}} |f(\lambda_1\alpha)f(\lambda_2\alpha)|^4 d_\tau\alpha \ll N^{5/3}\widetilde{T}(N)^{-1}. \tag{5.2}$$

Also, by (3.9) one has

$$\int_{\mathfrak{D}} |f(\lambda_j\alpha)|^{16/3} d_\tau\alpha \ll X^{7/3} \quad (3 \leq j \leq 6), \tag{5.3}$$

and an application of Hölder's inequality combined with (5.2) and (5.3)
yields the final estimate

$$\int_{\mathfrak{D}} |f(\lambda_1\alpha)\cdots f(\lambda_6\alpha)| d_\tau\alpha \ll N\widetilde{T}(N)^{-1/4}. \tag{5.4}$$

Another subset of $\mathfrak{c}$ needs to be removed before an averaging process
can be launched. Let $I \subset \{1, 2, \ldots, 6\}$ be a set with 4 elements, and let

$$\mathfrak{E}(I) = \{\alpha \in \mathfrak{c}\backslash\mathfrak{D} : \lambda_i\alpha \in \mathfrak{M} \ (i \in I)\}.$$

We write $\mathfrak{E}$ for the union of all $\mathfrak{E}(I)$. By symmetry, there is no loss of
generality in considering the special case where $I = \{3, 4, 5, 6\}$. Then, since
$\alpha \notin \mathfrak{D}$, at least one of $\lambda_1\alpha$ and $\lambda_2\alpha$ lies in $\mathfrak{m}$, and consequently, the bound
$f(\lambda_1\alpha)f(\lambda_2\alpha) \ll X^{7/4+\varepsilon}$ follows from (3.4). By Hölder's inequality and
(3.9), we therefore obtain

$$\int_{\mathfrak{E}(I)} |F(\alpha)| d_\tau\alpha \ll X^{7/4+\varepsilon} \prod_{j=3}^{6} \left(\int_{\lambda_j\alpha \in \mathfrak{M}} |f(\lambda_j\alpha)|^4 d_\tau\alpha\right)^{1/4} \ll X^{11/4+\varepsilon}.$$

We conclude that

$$\int_{\mathfrak{E}} |F(\alpha)| \mathrm{d}_\tau \alpha \ll X^{11/4+\varepsilon}. \tag{5.5}$$

Now let $\mathfrak{e} = \mathfrak{c}\backslash(\mathfrak{D} \cup \mathfrak{E})$ and write

$$H_\tau(\nu) = \int_{\mathfrak{e}} F(\alpha)e(-\nu\alpha)\mathrm{d}_\tau \alpha. \tag{5.6}$$

Inserting (5.4), (5.5) and (5.8) into the initial formula (5.1) for $\rho_6^*(\tau, \nu)$, in the restricted range $(\log N)^{-1} \leq \tau \leq 1$ we deduce that

$$\rho_6^*(\tau, \nu) = \frac{\Gamma(\frac{4}{3})^6 \tau \nu}{(\lambda_1 \cdots \lambda_6)^{1/3}} + H_\tau(\nu) + O(N\widetilde{T}(N)^{-1/4}) \tag{5.7}$$

holds uniformly in $1 \leq \nu \leq N$.

We are ready to define the exceptional set. Let ϕ denote a positive integral quadratic polynomial, as in Theorem 1.2. Also, let M be the positive solution of the equation $\phi(2M) = N$. Then, for large N, the $\phi(m)$ with $M < m \leq 2M$ are positive integers, and we define

$$\mathcal{Z}(M) = \{M < m \leq 2M : |H_\tau(\phi(m))| \geq N/\log\log N\}.$$

We remark that when $M < m \leq 2M$, but $m \notin \mathcal{Z}(M)$, then (5.9) yields the asymptotic formula

$$\rho_6^*(\tau, \phi(m)) = \frac{\Gamma(\frac{4}{3})^6 \tau \phi(m)}{(\lambda_1 \cdots \lambda_6)^{1/3}} + O(N\widetilde{T}(N)^{-1/4}). \tag{5.8}$$

Hence it remains to establish a good bound for the number $Z = \operatorname{card} \mathcal{Z}(M)$.

5.2. *A quadratic average*

When $m \in \mathcal{Z}(M)$, define the complex number η_m by means of the equation $\eta_m H_\tau(\phi(m)) = |H_\tau(\phi(m))|$, and then write

$$K(\alpha) = \sum_{m \in \mathcal{Z}(M)} \eta_m e(-\alpha\phi(m)). \tag{5.9}$$

From the definitions of $H_\tau(\nu)$ and $\mathcal{Z}(M)$, we have

$$\frac{ZN}{\log\log N} \leq \sum_{m \in \mathcal{Z}(M)} |H_\tau(\phi(m))| = \int_{\mathfrak{e}} F(\alpha)K(\alpha)\mathrm{d}_\tau \alpha. \tag{5.10}$$

This is the essential step: an upper bound for the size of the exceptional set is provided by an integral, and $K(\alpha)$ inherits the arithmetical structure of the test sequence.

Lemma 5.1. *Let $\lambda > 0$ be a fixed real number, and let $\mathcal{Z} \subset [M, 2M] \cap \mathbb{N}$ be a set of Z elements. Let ϕ be a positive polynomial of degree at least 2. For $0 < \tau \le 1$, let U_τ denote the number of solutions of the inequality*

$$|\phi(m_1) - \phi(m_2) + \lambda(x_1^3 - x_2^3)| < \tau, \tag{5.11}$$

with $m_j \in \mathcal{Z}$ and $1 \le x_j \le X$. Then

$$U_\tau \ll XZ + X^\varepsilon Z^2. \tag{5.12}$$

If ϕ is an integral polynomial, then one has also

$$U_\tau \ll XZ + X^{2+\varepsilon}. \tag{5.13}$$

Proof. When $m_1 = m_2$, the inequality (5.11) reduces to $|x_1^3 - x_2^3| < \tau/\lambda$, which has $O(X)$ solutions with $1 \le x_j \le X$. The number of solutions of this type to be counted is therefore $O(XZ)$. When $m_1 \ne m_2$, on the other hand, one has $|\phi(m_1) - \phi(m_2)| \gg M$. Hence, for any of the $O(X^2)$ possible choices for x_1, x_2, the inequality (5.11) may have no solution with $m_1 \ne m_2$ and $m_j \in \mathcal{Z}$, or else reduces to at most two equations

$$\phi(m_1) - \phi(m_2) = u, \quad \phi(m_1) - \phi(m_2) = u + 1, \tag{5.14}$$

for some $u \in \mathbb{Z}$ with $M \ll |u| \ll X^3$. When ϕ is an integral polynomial, a divisor function estimate shows that (5.14) leaves at most $O(X^\varepsilon)$ choices for m_1 and m_2. The number of solutions with $m_1 \ne m_2$ is thus at most $O(X^{2+\varepsilon})$ in this case, and this completes the proof of (5.13).

In order to establish (5.12), we count the solutions with $m_1 \ne m_2$ in a different way. There are $O(Z^2)$ possible choices for such m_1, m_2, and for each fixed such pair, the inequality (5.11) reduces to $|x_1^3 - x_2^3 - \kappa| < \tau/\lambda$, for a suitable number κ satisfying $\kappa \gg M$. Again, a divisor function estimate suffices to conclude that the number of solutions of this last inequality, in integers x_1, x_2 satisfying $1 \le x_j \le X$, is at most $O(X^\varepsilon)$. On assembling this bound together with our earlier estimate for the number of diagonal solutions, we confirm (5.12). $\square$

As an immediate consequence of this lemma, we infer from (2.3), (2.6) and (5.11) that when λ is any one of the numbers λ_j, then

$$\int |K(\alpha)f(\lambda\alpha)|^2 \mathrm{d}_\tau\alpha \ll ZX + X^{2+\varepsilon}. \tag{5.15}$$

An estimate for Z can now be obtained through (5.12) and (5.15). Let $\mathfrak{a} \subset \mathbb{R}$ be a measurable set, and write

$$\mathcal{I}(\mathfrak{a}) = \int_\mathfrak{a} |f(\lambda_1\alpha) \cdots f(\lambda_6\alpha)K(\alpha)| \mathrm{d}_\tau\alpha.$$

An estimate for $\mathcal{I}(\mathfrak{e})$ is needed, and this is accomplished by an individual treatment of various subsets of $\mathfrak{e}$, specifically

$$\mathfrak{d} = \{\alpha \in \mathfrak{e} : \lambda_j \alpha \in \mathfrak{m} \ (1 \le j \le 6)\}$$

and, when $i, j, l \in \{1, 2, \ldots, 6\}$ are distinct, also

$$\mathfrak{e}_{ij}(l) = \{\alpha \in \mathfrak{e} : \lambda_i \alpha \in \mathfrak{m}, \ \lambda_j \alpha \in \mathfrak{m}, \ \lambda_l \alpha \in \mathfrak{M}\}.$$

It is important to observe that $\mathfrak{e}$ is the union of $\mathfrak{d}$ and the $\mathfrak{e}_{ij}(l)$. To see this, consider $\alpha \in \mathfrak{e} \backslash \mathfrak{d}$. Then, there exists at least one l with $\lambda_l \alpha \in \mathfrak{M}$. Since α is neither in $\mathfrak{D}$ nor $\mathfrak{E}$, there are two distinct i, j with $\lambda_i \alpha \in \mathfrak{m}$, $\lambda_j \alpha \in \mathfrak{m}$, whence $\alpha \in \mathfrak{e}_{ij}(l)$, as required.

By Hölder's inequality,

$$\mathcal{I}(\mathfrak{d}) \ll \left(\int |Kf_1|^2 d_r \alpha \right)^{1/2} \prod_{j=2}^{5} \left(\int_{\mathfrak{m}} |f_j|^8 d_r \alpha \right)^{1/8} \sup_{\lambda_6 \alpha \in \mathfrak{m}} |f_6(\alpha)|.$$

Here and hereafter, we write $f_j = f(\lambda_j \alpha)$. By (3.4), (3.10) and (5.15), it follows that

$$\mathcal{I}(\mathfrak{d}) \ll X^{13/4} (ZX + X^{2+\varepsilon})^{1/2} (\log X)^{\varepsilon - 5/4}. \tag{5.16}$$

Also, by (3.4), one has $f_2 f_1 \ll X^{3/2+\varepsilon}$ on $\mathfrak{e}_{12}(3)$, so by Hölder's inequality,

$$\mathcal{I}(\mathfrak{e}_{12}(3)) \ll X^{3/2+\varepsilon} \left(\int |Kf_6|^2 d_r \alpha \right)^{1/2} \left(\int |f_5|^8 d_r \alpha \right)^{1/8}$$
$$\times \left(\int |f_4|^8 d_r \alpha \right)^{1/8} \left(\int_{\mathfrak{M}} |f_3|^4 d_r \alpha \right)^{1/4}.$$

Employing (3.10) once again, we find that the eighth moments of f_5 and f_4 are bounded by $O(X^5)$. The other factors can also be estimated via (5.15) and (3.9), and thus

$$\mathcal{I}(\mathfrak{e}_{12}(3)) \ll X^{3+\varepsilon} (ZX + X^{2+\varepsilon})^{1/2}, \tag{5.17}$$

a bound superior to (5.16). Also, by symmetry, this last bound holds for any other $\mathfrak{e}_{ij}(l)$ in place of $\mathfrak{e}_{12}(3)$. Hence, the bound (5.16) remains valid with $\mathfrak{d}$ replaced by $\mathfrak{e}$.

Estimating the right hand side of (5.12) by means of (5.16) and (5.17), we deduce that

$$\frac{ZN}{\log \log N} \ll X^{13/4} (ZX + X^{2+\varepsilon})^{1/2} (\log X)^{\varepsilon - 5/4}.$$

But $N \asymp X^3$, whence it now follows that

$$Z \ll X^{3/2} (\log X)^{\varepsilon - 5/2} \ll M (\log M)^{\varepsilon - 5/2}.$$

In particular, the asymptotic formula (5.8) holds for all but $O(M(\log M)^{\varepsilon-5/2})$ of the integers m with $M < m \leq 2M$. One can now apply the transference principle (2.12), with $s = 6$ and $\delta = (\log \log N)^{-1/2}$, say, to conclude that the expected asymptotic formula

$$\rho_6(\tau, \phi(m)) = \frac{2\Gamma(\frac{4}{3})^6 \tau \phi(m)}{(\lambda_1 \cdots \lambda_6)^{1/3}} + O(N\widetilde{T}(N)^{-1/8})$$

holds for all but $O(M(\log M)^{\varepsilon-5/2})$ of the integers $m \in (M, 2M]$ as well. A dyadic dissection argument completes the proof of Theorem 1.2.

5.3. *Some brief heckling*

Most of the results in I, II, III, V and VI depend on mean value estimates over exceptional sets, often in mixed form. Lemma 5.1 is only a typical example, and one may take our proof of Theorem 1.2 as a model for generalising our results on diophantine equations to the wider class of inequalities. However, it should be stressed that the estimate (5.13) applies to *integral polynomials only*. Its proof crucially depends on a divisor estimate that is otherwise not available. If the test sequence stems from a positive polynomial that is no longer integral, different methods have to be applied, and this is the main reason why in the non-integral case our exceptional set estimates are considerably weaker. We proceed by presenting two mean value estimates relating to general polynomials, and then illustrate their use within the proof of Theorem 1.6.

5.4. *An inequality involving quadratic polynomials*

The sole purpose of this section is to establish the following simple mean value theorem. Although the result is not needed until chapter 6, the method is a model for the work in §5.5 which is more relevant for our immediate needs.

Lemma 5.2. Let $\phi \in \mathbb{R}[t]$ denote a quadratic polynomial. Suppose that $\mathcal{Z} \subset [1, M] \cap \mathbb{Z}$, and write $Z = \operatorname{card} \mathcal{Z}$. Finally, let $\Omega(M, \mathcal{Z})$ denote the number of solutions of the inequality

$$|\phi(m_1) + \phi(m_2) - \phi(m_3) - \phi(m_4)| < 1 \tag{5.18}$$

with $m_j \in \mathcal{Z}$. Then one has $\Omega(M, \mathcal{Z}) \ll M^{1+\varepsilon}Z$.

Proof. We may write $\phi(t) = \lambda_2 t^2 + \lambda_1 t + \lambda_0$ for some real numbers $\lambda_0, \lambda_1, \lambda_2$ with $\lambda_2 \neq 0$. Given a solution of (5.18) counted by $\Omega(M, \mathcal{Z})$, let k_1 and k_2

be defined by means of the equations

$$k_j = m_1^j + m_2^j - m_3^j - m_4^j \quad (j = 1, 2). \tag{5.19}$$

The inequality (5.18) then reduces to

$$|\lambda_2 k_2 + \lambda_1 k_1| < 1. \tag{5.20}$$

Substituting from the linear equation in (5.19) into the quadratic one, we find that

$$2(m_1 - m_3 - k_1)(m_2 - m_3 - k_1) = 2m_3 k_1 + k_1^2 - k_2. \tag{5.21}$$

We note that $|k_1| \leq 2M$, and that for any given k_1, the inequality (5.20) allows only $O(1)$ possible choices for k_2. Thus, when the right hand side of (5.21) is non-zero, there are $O(MZ)$ possible choices for k_1, k_2 and m_3, and for any one of these choices, a divisor function estimate shows that there are $O(M^\varepsilon)$ possible values for m_1 and m_2 satisfying (5.21). The solutions of this type consequently contribute $O(M^{1+\varepsilon}Z)$ to $\Omega(M, \mathcal{Z})$. On the other hand, when the right hand side of (5.21) is zero, one has either $k_1 = 0$ or $m_3 = (k_2 - k_1^2)/(2k_1)$, and this implies that there are at most $O(M)$ possible choices for k_1, k_2 and m_3. For each fixed choice of this type, one finds from (5.21) that $m_j = m_3 + k_1$ for $j = 1$ or 2, whence there are at most $O(Z)$ integers m_1 and m_2 satisfying (5.21). The solutions of this second type therefore contribute at most $O(MZ)$ to $\Omega(M, \mathcal{Z})$. The conclusion of the lemma now follows. $\quad\square$

5.5. *An application of Vinogradov's method*

A version of Lemma 5.2 for polynomials of higher degree can be fabricated through a suitable generalisation of the idea exploited in §5.4. The problem may be addressed through an application of Vinogradov's mean value theorem. Let $J_{k,s}(M)$ denote the number of solutions of the simultaneous equations

$$\sum_{j=1}^{s}(x_j^l - y_j^l) = 0 \quad (1 \leq l \leq k),$$

with $1 \leq x_j, y_j \leq M$.

Lemma 5.3. *Let $\phi \in \mathbb{R}[t]$ denote a polynomial of degree $d \geq 3$. Let $s \geq 2$, and let $U_{\phi,s}(M)$ denote the number of solutions of the inequality*

$$\left| \sum_{j=1}^{s}(\phi(x_j) - \phi(y_j)) \right| < 1, \tag{5.22}$$

in integers x_j, y_j with $1 \leq x_j, y_j \leq M$. Then

$$U_{\phi,s}(M) \ll M^{\frac{1}{2}d(d-1)} J_{d,s}(M).$$

Proof. Motivated by the argument used to prove Lemma 5.2, we begin by writing $\phi(t) = \lambda_d t^d + \cdots + \lambda_1 t + \lambda_0$, with $\lambda_d \neq 0$. Given a solution of (5.22) counted by $U_{\phi,s}(M)$, we define $k_1, \ldots, k_d$ by

$$k_l = \sum_{j=1}^{s} (x_j^l - y_j^l). \tag{5.23}$$

Then (5.22) implies that $|\lambda_d k_d + \cdots + \lambda_1 k_1| < 1$. If $k_1, \ldots, k_{d-1}$ are determined, then this inequality leaves $O(1)$ possibilities for k_d. On noting that $|k_l| < sM^l$, we find that the number of choices for $k_1, \ldots, k_d$ is $O(M^{\frac{1}{2}d(d-1)})$, and hence

$$U_{\phi,s}(M) \ll M^{\frac{1}{2}d(d-1)} \max_{k_1,\ldots,k_d} J_{d,s,\mathbf{k}}(M),$$

where $J_{d,s,\mathbf{k}}(M)$ denotes the number of solutions of the diophantine system (5.23), with $1 \leq x_j, y_j \leq M$. But a well-known argument (see inequality (5.4) of Vaughan [35]) shows that $J_{d,s,\mathbf{k}}(M) \leq J_{d,s}(M)$, and the lemma follows. □

An upper bound for $J_{d,s}(M)$ is now required that is of the expected order of magnitude. According to Theorem 3 of Wooley [42], there exists a constant C with the property that whenever

$$s > d^2(\log d + 2\log\log d + C), \tag{5.24}$$

one has $J_{d,s}(M) \ll M^{2s - \frac{1}{2}d(d+1)}$. Subject to the condition (5.24), one deduces from Lemma 5.3 the bound

$$U_{\phi,s}(M) \ll M^{2s-d}. \tag{5.25}$$

5.6. *Linear forms in primes, yet again*

In this section, we briefly indicate how (5.25) may be applied to establish Theorem 1.6. Thus, we now work under the hypotheses of that theorem. In particular, we suppose that λ_1/λ_2 is an algebraic irrational, and $\phi \in \mathbb{R}[t]$ is a polynomial of degree d. The argument, at the beginning, is largely similar to that proceeding in §4.1, but we need to replace the test sequence by $\nu_m = \phi(m)$, and the parameter M by $\phi(2M) = N$. In all other respects, we use the notation and work from §4.1 that is partly inherited from Lemma 2.3 and §3.4. This defines a complementary compositum, a parameter X

with $X \asymp N$, and $E_\tau(\nu)$ via (3.15). The substitute for Lemma 4.1 is the following estimate.

Lemma 5.4. *Let s denote a natural number, and suppose that* (5.25) *holds. Then, whenever* $0 < \tau \leq 1$ *is fixed, one has*

$$\sum_{M < m \leq 2M} |E_\tau(\phi(m))| \ll MN^{1-1/(6s)+\varepsilon}.$$

This implies Theorem 1.6, as we now demonstrate. By Lemma 5.4, the inequality $|E_\tau(\phi(m))| > N(\log N)^{-A}$ can hold for no more than $O(MN^{-1/(6s)+2\varepsilon})$ of the integers m with $M < m \leq 2M$, and for the remaining values in this range, the relation (4.1) yields the asymptotic formula

$$\sigma^*(\tau, \phi(m)) = \frac{\tau\phi(m)}{\lambda_1\lambda_2} + O(N(\log N)^{-A}).$$

The now familiar transference principle takes this to an asymptotic formula for $\sigma(\tau, \phi(m))$, outside an exceptional set that still has no more than $O(MN^{-1/(6s)+2\varepsilon})$ members. But $M \asymp N^{1/d}$, and (5.25) holds whenever s satisfies (5.24). Consequently, a dyadic dissection argument delivers the conclusion of Theorem 1.6 with a permissible value of δ satisfying $\delta = \frac{1}{6} + O(\log \log d / \log d)$.

It remains to prove Lemma 5.4. Define the numbers η_m by putting $\eta_m = 0$ when $E_\tau(\phi(m)) = 0$, and otherwise via the equation $|E_\tau(\phi(m))| = \eta_m E_\tau(\phi(m))$. Also, write

$$K(\alpha) = \sum_{M < m \leq 2M} \eta_m e(-\alpha\phi(m)).$$

Then, by (3.15) and Hölder's inequality,

$$\sum_{M < m \leq 2M} |E_\tau(\phi(m))| = \int_{\mathfrak{c}} K(\alpha)h(\lambda_1\alpha)h(\lambda_2\alpha)\mathrm{d}_\tau\alpha$$

$$\leq \left(\int |K(\alpha)|^{2s}\mathrm{d}_\tau\alpha\right)^{1/(2s)} J^{1/(2s)}(J_1 J_2)^{\frac{1}{2}(1-1/s)},$$

where

$$J = \int_{\mathfrak{c}} |h(\lambda_1\alpha)h(\lambda_2\alpha)|^2\mathrm{d}_\tau\alpha, \qquad J_l = \int |h(\lambda_l\alpha)|^2\mathrm{d}_\tau\alpha.$$

A consideration of the underlying diophantine inequality reveals that

$$\int |K(\alpha)|^{2s}\mathrm{d}_\tau\alpha \leq U_{\phi,s}(M),$$

and we may now apply (5.25), (3.19) and (3.18) to infer that

$$\sum_{M < m \leq 2M} |E_\tau(\phi(m))| \ll (M^{2s-d})^{1/(2s)} (N^{8/3+\varepsilon})^{1/(2s)} (N \log N)^{1-1/s}.$$

The proof of Lemma 5.4 is completed by recalling that $M \asymp N^{1/d}$.

Note that when ϕ is an integral polynomial, the work of Ford [20] gives much better bounds for $U_{\phi,s}(M)$, and there is then no essential difficulty in improving the exceptional set estimate to a full analogue of Theorem 1 of II. On the other hand, it seems more difficult to relax the hypothesis that λ_1/λ_2 be algebraic. If λ_1/λ_2 is merely supposed to be irrational, one might have to accept weaker exceptional set estimates.

6. Outstanding arts

6.1. *Smooth cubic Weyl sums*

This chapter is devoted to all the remaining results concerned with additive representation in thin sequences. In the next section we consider diagonal cubic forms in six variables, and we establish Theorem 1.3 by applying a complementary compositum estimate, the proof of which is the central focus of §6.3. We then pause to discuss diagonal cubics in five variables, the topic of Theorem 1.4, and a situation in which we are restricted to contemplating only lower bound estimates via the methods of IV. The argument here makes use of an asymptotic lower bound for the number of solutions of the diophantine inequality in question, a matter we defer to §6.5, incorporated into a mean value involving the exceptional set within §6.4. Some preparatory work concerning smooth Weyl sums of higher degree leads from §6.6 to the proof of Theorem 1.8 in §6.7. Finally, in §6.8, we consider prime numbers close to diagonal forms using the methods of VI, completing our journey with the proof of Theorem 1.9. All of these discussions are dependent, in some way or other, on mean value estimates for smooth Weyl sums. We finish this section by recording here those estimates that are needed to establish Theorems 1.3 and 1.4.

Define

$$\mathcal{A}(P, R) = \{x \in \mathbb{N} : x \leq P, \, p|x \Rightarrow p \leq R\}.$$

Also, let $\eta > 0$ denote a small real number, and write

$$g(\alpha) = \sum_{x \in \mathcal{A}(X, X^\eta)} e(\alpha x^3). \tag{6.1}$$

Lemma 6.1. *Let u be one of 6 and $77/10$. Then, there exists a real number $\eta_0 > 0$ such that, whenever $0 < \eta \le \eta_0$, one has the estimate*

$$\int_0^1 |g(\alpha)|^u d\alpha \ll P^{\mu_u},$$

where $\mu_6 = 3.2495$ and $\mu_{77/10} = 4.7$.

The permissible exponent μ_u claimed in Lemma 6.1 is a consequence of Theorem 1.2 of Wooley [43] when $u = 6$, and is a special case of Theorem 2 of Brüdern and Wooley [16] when $u = 77/10$. In fact, in the case $u = 6$, we have rounded up; for a microscopically better bound see [43].

6.2. *Senary cubic forms*

In this and the next section we establish Theorem 1.3. Let $\lambda_1, \ldots, \lambda_6$ be positive real numbers with λ_1/λ_2 irrational. As always, our leading parameter is N, and X is chosen in accordance with (2.11), so that $X^3 \asymp N$. With $f(\alpha)$ and $g(\alpha)$ defined by (2.10) and (6.1), and fixed $0 < \tau \le 1$, we consider the integral

$$\rho^*(\tau, \nu) = \int f(\lambda_1\alpha) f(\lambda_2\alpha) g(\lambda_3\alpha) \cdots g(\lambda_6\alpha) e(-\nu\alpha) d_\tau\alpha. \qquad (6.2)$$

By considering the underlying diophantine inequality, it follows that for all $\nu \in \mathbb{R}$ one has

$$\rho_6(\tau, \nu) \ge \rho^*(\tau, \nu). \qquad (6.3)$$

The central interval is chosen as

$$\mathfrak{C} = \{\alpha \in \mathbb{R} : |\alpha| \le (\log N)^{1/4} N^{-1}\}. \qquad (6.4)$$

Then, as a consequence of Lemma 8.5 of Wooley [39], there exists a number $C = C(\eta) > 0$ such that, whenever $\alpha \in \mathfrak{C}$, one has

$$f(\lambda_1\alpha) f(\lambda_2\alpha) g(\lambda_3\alpha) \cdots g(\lambda_6\alpha) - Cv(\lambda_1\alpha) \cdots v(\lambda_6\alpha) \ll X^6(\log X)^{-1/2},$$

where v is defined in (2.17). Now, much as in the proof of Lemma 2.2, it follows that

$$\int_{\mathfrak{C}} f(\lambda_1\alpha) f(\lambda_2\alpha) g(\lambda_3\alpha) \cdots g(\lambda_6\alpha) e(-\nu\alpha) d_\tau\alpha$$

$$= C \int_{-\infty}^{\infty} v(\lambda_1\alpha) \cdots v(\lambda_6\alpha) e(-\nu\alpha) d_\tau\alpha + O(X^3(\log X)^{-1/4}).$$

The integral on the right hand side here has also been evaluated in the course of the proof of Lemma 2.2, the result being $\Gamma(\tfrac{4}{3})^6 (\lambda_1 \cdots \lambda_6)^{-1/3} \tau\nu$

for $0 \leq \nu \leq N$. In particular, we may conclude from the above that whenever $\frac{1}{10}N \leq \nu \leq N$, then one has

$$\int_{\mathfrak{C}} f(\lambda_1\alpha)f(\lambda_2\alpha)g(\lambda_3\alpha)\cdots g(\lambda_6\alpha)e(-\nu\alpha)\mathrm{d}_\tau\alpha \geq 2cN, \qquad (6.5)$$

where $c > 0$ denotes a certain positive constant independent of ν and N.

For the treatment of the complementary compositum $\mathfrak{c} = \mathbb{R}\backslash\mathfrak{C}$, we engineer an amplification procedure that is quite similar to the one used in §3.3. The amplifier here will be

$$\mathfrak{D} = \{\alpha \in \mathfrak{c} : \lambda_1\alpha \in \mathfrak{M}(X^{3/4}), \ \lambda_2\alpha \in \mathfrak{M}(X^{3/4})\}.$$

We show in the next section that

$$\int_{\mathfrak{D}} |f(\lambda_1\alpha)f(\lambda_2\alpha)g(\lambda_3\alpha)\cdots g(\lambda_6\alpha)|\mathrm{d}_\tau\alpha = o(N), \qquad (6.6)$$

a conclusion that for the remainder of this section we take as granted.

On the complement $\mathfrak{d} = \mathfrak{c}\backslash\mathfrak{D}$, the averaging method described in §5.2 is required. Let ϕ be a positive quadratic polynomial. For large N, let M be the unique positive solution of $\phi(2M) = N$. Then, for any m with $M < m \leq 2M$, we have $\frac{1}{10}N \leq \phi(m) \leq N$. We define $\mathcal{Z}(M)$ to be the set of integers m with $M < m \leq 2M$ for which $\rho_6(\tau, \phi(m)) < cN$, where c is the number introduced in (6.5). In addition, we write $Z(M) = \mathrm{card}\, \mathcal{Z}(M)$. We note that Theorem 1.3 follows from a dyadic dissection argument, once one has established the bound

$$Z(M) \ll M^{23/27}. \qquad (6.7)$$

Write

$$H(\nu) = \int_{\mathfrak{d}} f(\lambda_1\alpha)f(\lambda_2\alpha)g(\lambda_3\alpha)\cdots g(\lambda_6\alpha)e(-\nu\alpha)\mathrm{d}_\tau\alpha.$$

Then, one finds from (6.2), (6.3), (6.5) and (6.6) that for $m \in \mathcal{Z}(M)$, one has $|H(\phi(m))| \gg N$. For $m \in \mathcal{Z}(M)$, define η_m through the equation $|H(\phi(m))| = \eta_m H(\phi(m))$, and then define $K(\alpha)$ by means of (5.11). By the argument that delivered (5.12) we now infer that

$$NZ(M) \ll \int_{\mathfrak{d}} |f(\lambda_1\alpha)f(\lambda_2\alpha)g(\lambda_3\alpha)\cdots g(\lambda_6\alpha)K(\alpha)|\mathrm{d}_\tau\alpha. \qquad (6.8)$$

In the next section we show that

$$\int_{\mathfrak{d}} |f(\lambda_1\alpha)f(\lambda_2\alpha)g(\lambda_3\alpha)\cdots g(\lambda_6\alpha)|^{4/3}\mathrm{d}_\tau\alpha \ll X^{43/9-2\nu}, \qquad (6.9)$$

where $v = 10^{-6}$. Equipped with this estimate, we may apply Hölder's inequality to the right hand side of (6.8) and bound the fourth moment of $K(\alpha)$ by utilising Lemma 5.2. This yields

$$NZ(M) \ll (M^{1+\varepsilon}Z(M))^{1/4}X^{43/12-v},$$

and the desired conclusion (6.7) follows.

6.3. *Two technical estimates*

For notational convenience, we now write $f_j = f(\lambda_j\alpha)$ and $g_j = g(\lambda_j\alpha)$, and we define

$$J(\mathfrak{a}) = \int_{\mathfrak{a}} |f_1 f_2 g_3 g_4 g_5 g_6|^{4/3} d_\tau\alpha.$$

We split $\mathfrak{d}$ into the three subsets

$$\mathfrak{d}_1 = \{\alpha \in \mathfrak{c} : \lambda_1\alpha \in \mathfrak{m}, \lambda_2\alpha \in \mathfrak{M}\},$$
$$\mathfrak{d}_2 = \{\alpha \in \mathfrak{c} : \lambda_2\alpha \in \mathfrak{m}, \lambda_1\alpha \in \mathfrak{M}\},$$
$$\mathfrak{d}_3 = \{\alpha \in \mathfrak{c} : \lambda_1\alpha \in \mathfrak{m}, \lambda_2\alpha \in \mathfrak{m}\},$$

where, as on earlier occasions, we put $\mathfrak{M} = \mathfrak{M}(X^{3/4})$, $\mathfrak{m} = \mathfrak{m}(X^{3/4})$. By making use of (3.4) in order to estimate f_1, an application of Hölder's inequality shows that

$$J(\mathfrak{d}_1) \ll X^{1+\varepsilon}\Big(\int_{\lambda_2\alpha\in\mathfrak{M}} |f_2|^4 d_\tau\alpha\Big)^{1/3} \prod_{j=3}^{6}\Big(\int |g_j|^8 d_\tau\alpha\Big)^{1/6}.$$

The eighth moment of g_j is $O(X^5)$; this can be seen either via (3.10), on considering the underlying diophantine inequality, or by reference to Lemma 6.1. By (3.9), the restricted fourth moment of f_2 is $O(X^{1+\varepsilon})$, and thus we deduce that $J(\mathfrak{d}_1) \ll X^{14/3+\varepsilon}$. By symmetry, the same bound holds also for $J(\mathfrak{d}_2)$. In order to estimate $J(\mathfrak{d}_3)$, meanwhile, we again use (3.4) to bound f_1 and $|f_2|^{8/9}$, and then apply Hölder's inequality, thereby confirming that

$$J(\mathfrak{d}_3) \ll X^{5/3+\varepsilon}\Big(\int |f_2|^4 d_\tau\alpha\Big)^{1/9} \prod_{j=3}^{6}\Big(\int |g_j|^6 d_\tau\alpha\Big)^{2/9}.$$

Here the fourth moment of $|f_2|$ is $O(X^{2+\varepsilon})$, as one finds by considering the underlying diophantine inequality, or by reference to Hua's lemma. The sixth moments of g_j are each $O(X^{\mu_6})$, by Lemma 6.1. This then yields the estimate $J(\mathfrak{d}_3) \ll X^{43/9-2v}$. In combination with our earlier bounds for $J(\mathfrak{d}_1)$ and $J(\mathfrak{d}_2)$, the estimate (6.9) is confirmed on noting that $J(\mathfrak{d}) = J(\mathfrak{d}_1) + J(\mathfrak{d}_2) + J(\mathfrak{d}_3)$.

We now turn to the proof of (6.6). Let $T = T_{\lambda_1/\lambda_2}((\log N)^{1/4})$ be defined via (2.22). Then $T \to \infty$ as $N \to \infty$. Now let

$$\mathfrak{C} = \{\alpha \in \mathfrak{D} : \lambda_j \alpha \in \mathfrak{M}(T^{1/4}) \text{ for } j = 1, 2\}, \quad \mathfrak{c} = \mathfrak{D} \backslash \mathfrak{C}.$$

Then, one finds from Theorem 2.4 that

$$\int_{\mathfrak{C}} |f_1 f_2 g_3 g_4 g_5 g_6| \mathrm{d}_\tau \alpha \ll X^6 \int_{\mathfrak{C}} \mathrm{d}_\tau \alpha \ll X^3 T^{-1/2}, \qquad (6.10)$$

which is acceptable. Moreover, on combining (3.4) with the major arc upper bound for $|f_1 f_2|$ derived from (3.13), we infer that

$$\sup_{\mathfrak{c}} |f_1 f_2| \ll X^2 T^{-1/12}.$$

Now, by Hölder's inequality,

$$\int_{\mathfrak{c}} |f_1 f_2 g_3 g_4 g_5 g_6| \mathrm{d}_\tau \alpha$$

$$\ll (\sup_{\mathfrak{c}} |f_1 f_2|)^{1/44} \left(\int_{\lambda_1 \alpha \in \mathfrak{M}} |f_1|^{301/74} \mathrm{d}_\tau \alpha \right)^{37/154}$$

$$\times \left(\int_{\lambda_2 \alpha \in \mathfrak{M}} |f_2|^{301/74} \mathrm{d}_\tau \alpha \right)^{37/154} \prod_{j=3}^{6} \left(\int |g_j|^{77/10} \mathrm{d}_\tau \alpha \right)^{10/77}.$$

The moments of g_j can be bounded using Lemma 6.1, and the moments of f_1, f_2 by (3.9) (where it is important to note that $\frac{301}{74} > 4$). It follows that the integral in question is $O(X^3 T^{-1/528})$. The desired estimate (6.6) now follows by combining this bound with (6.10).

6.4. *The lower bound variant*

We now embark on the proof of Theorem 1.4. The method derives from IV to which we refer for a discussion of the main idea. Fundamental to its success is a lower bound for the number of solutions of a related equation or inequality in which the test sequence occurs as an additional variable. Thus, we need the following result.

Lemma 6.2. *Let $\phi \in \mathbb{R}[t]$ denote a positive quadratic polynomial, and let $\lambda_1, \ldots, \lambda_5$ denote positive real numbers with $\lambda_1/\lambda_2 \notin \mathbb{Q}$. Let X be sufficiently large, and let M be a positive real number with $M \asymp X^{3/2}$ satisfying the condition that $\phi(2M) < 2\lambda_j X^3$ for all j. Finally, for any fixed $0 < \tau \le 1$, let*

$$V = \sum \widehat{w}_\tau(\lambda_1 x_1^3 + \cdots + \lambda_5 x_5^3 - \phi(m)),$$

with the sum extended over $m, x_1, \ldots, x_5$ *in the ranges*

$$x_1, x_2 \leq X, \quad x_3, x_4, x_5 \in \mathcal{A}(X, X^\eta), \quad M < m \leq 2M.$$

Then, one has $V \gg X^2 M$.

Note that V counts solutions of the diophantine inequality

$$|\lambda_1 x_1^3 + \cdots + \lambda_5 x_5^3 - \phi(m)| < \tau, \tag{6.11}$$

with a certain non-negative weight attached. A related result occurs as Theorem 2 of Brüdern [4], but it does not cover Lemma 6.2. The cited work predates the innovations of Bentkus and Götze, and of Freeman, and therefore, one would find a lower bound for V only for a certain sequence of values of the parameter X. Secondly, in [4] the polynomial ϕ has to be a monomial. It is relatively straightforward to attend to these two problems. More importantly, however, the method in [4] is laid out only for an algebraic irrational coefficient ratio. Therefore, we spell out a proof of Lemma 6.2 in the next section.

It is a delightful exercise to deduce Theorem 1.4. Let N be a large real number, let X be defined in accordance with (2.11), and choose M as in the statement of Lemma 6.2. Next, let $\mathcal{Z}(M)$ denote the set of all m with $M < m \leq 2M$ for which the diophantine inequality (6.11) has at least one solution in positive integers x_j, and put $Z = \operatorname{card} \mathcal{Z}(M)$. We write

$$\Phi(\alpha) = \sum_{M < m \leq 2M} e(\alpha\phi(m)), \quad k(\alpha) = \sum_{m \in \mathcal{Z}(M)} e(\alpha\phi(m)),$$

$$G(\alpha) = f_1(\alpha)f_2(\alpha)g_3(\alpha)g_4(\alpha)g_5(\alpha).$$

Then, on considering the underlying diophantine inequality, one verifies that

$$\int G(\alpha)\Phi(-\alpha)\mathrm{d}_\tau\alpha = \int G(\alpha)k(-\alpha)\mathrm{d}_\tau\alpha. \tag{6.12}$$

By orthogonality, the left hand side here is equal to the quantity V defined in Lemma 6.2. Hence, by that lemma, it follows that the integrals in (6.12) are asymptotically bounded from below by $X^2 M$.

We now estimate the integral on the right hand side from above. With this end in view, we cover the real line by the three sets

$$\begin{aligned} \mathfrak{e}_j &= \{\alpha \in \mathbb{R} : \lambda_j \alpha \in \mathfrak{m}\} \quad (j = 1, 2), \\ \mathfrak{E} &= \{\alpha \in \mathbb{R} : \lambda_1 \alpha \in \mathfrak{M}, \ \lambda_2 \alpha \in \mathfrak{M}\}. \end{aligned}$$

When $\mathfrak{a}$ is any one of these sets, we write

$$I(\mathfrak{a}) = \int_{\mathfrak{a}} |G(\alpha)k(\alpha)| \mathrm{d}_\tau \alpha.$$

By (6.12) and the discussion thereafter, it follows that

$$X^2 M \ll I(\mathfrak{e}_1) + I(\mathfrak{e}_2) + I(\mathfrak{E}). \tag{6.13}$$

It remains to establish upper bounds for $I(\mathfrak{e}_1)$, $I(\mathfrak{e}_2)$ and $I(\mathfrak{E})$. On the set $\mathfrak{E}$, we use the trivial bound $|k(\alpha)| \leq Z$, and apply Hölder's inequality to infer that

$$I(\mathfrak{E}) \leq Z \left(\int_{\lambda_1\alpha\in\mathfrak{M}} |f_1|^4 \mathrm{d}_\tau \alpha \right)^{1/4} \left(\int_{\lambda_2\alpha\in\mathfrak{M}} |f_2|^4 \mathrm{d}_\tau \alpha \right)^{1/4} \\ \times \prod_{j=3}^{5} \left(\int |g_j|^6 \mathrm{d}_\tau \alpha \right)^{1/6}.$$

One may estimate the first two integrals by applying (3.9). An upper bound for the sixth moments of g_j is given in Lemma 6.1, and we deduce that

$$I(\mathfrak{E}) \ll X^{17/8} Z. \tag{6.14}$$

A similar argument may be used to estimate $I(\mathfrak{e}_1)$. By (3.4), we find that $f_1(\alpha) \ll X^{3/4+\varepsilon}$ whenever $\alpha \in \mathfrak{e}_1$. Hölder's inequality now reveals that

$$I(\mathfrak{e}_1) \ll X^{3/4+\varepsilon} \left(\int |kf_2|^2 \mathrm{d}_\tau \alpha \right)^{1/2} \prod_{j=3}^{5} \left(\int |g_j|^6 \mathrm{d}_\tau \alpha \right)^{1/6}.$$

As before, the sixth moments of g_j may be estimated through Lemma 6.1. Moreover, by Lemma 5.1, we have

$$\int |k(\alpha)f_2(\alpha)|^2 \mathrm{d}_\tau \alpha \ll X^\varepsilon Z^2 + XZ.$$

Therefore, it follows that $I(\mathfrak{e}_1) \ll X^{19/8}(Z^2 + XZ)^{1/2}$, and by symmetry, the same bound holds for $I(\mathfrak{e}_2)$. On combining these estimates with (6.13) and (6.14), we conclude that

$$X^2 M \ll X^{17/8} Z + X^{19/8}(Z^2 + XZ)^{1/2}.$$

This implies the lower bound $Z \gg M^{3/4}$, as required to complete the proof of Theorem 1.4.

6.5. An auxiliary inequality

In the notation of the previous section, our object is to evaluate the integral

$$V = \int G(\alpha)\Phi(-\alpha)\mathrm{d}_\tau\alpha. \tag{6.15}$$

Define $\mathfrak{C}$ as in (6.4). Then, just as in the discussion following the latter definition, there is little difficulty in adapting the arguments applied to prove Lemma 2.2 so as to establish here the lower bound

$$\int_{\mathfrak{C}} G(\alpha)\Phi(-\alpha)\mathrm{d}_\tau\alpha \gg X^2M. \tag{6.16}$$

We feel entitled by now to omit the details.

The treatment of the complementary compositum depends on a technique sometimes referred to as pruning, and made available for diophantine inequalities by Brüdern [4].

Lemma 6.3. *Let $\phi(t) = \lambda t^2 + \mu t + \zeta \in \mathbb{R}[t]$, with $\lambda \neq 0$. In addition, let $\mathfrak{K} = \{\alpha \in \mathbb{R} : \lambda\alpha \in \mathfrak{M}(X)\}$, and let $\mathfrak{k} = \mathbb{R}\backslash\mathfrak{K}$. Then given a fixed non-zero real number ω, for any $0 < \tau \leq 1$, one has*

$$\int_{\mathfrak{K}} |\Phi(\alpha)g(\omega\alpha)|^2\mathrm{d}_\tau\alpha \ll X^{2+\varepsilon},$$

$$\int_{\mathfrak{k}} |\Phi(\alpha)|^4|g(\omega\alpha)|^2\mathrm{d}_\tau\alpha \ll M^2X^{1+\varepsilon}.$$

Moreover, these estimates remain valid if f is substituted for g.

Proof. Note that ϕ, ω and τ are fixed in the current context. We may substitute $\beta = \lambda\alpha$ in both integrals. This replaces ω by ω/λ, and τ by τ/λ. Hence, we may assume that $\lambda = 1$ in the proof of this lemma. This is mostly for notational convenience. Note that we now have $\mathfrak{K} = \mathfrak{M}(X)$ and $\mathfrak{k} = \mathfrak{m}(X)$.

Next, we define the function Υ on $\mathbb{R}$ by taking

$$\Upsilon(\alpha) = (q + N|q\alpha - a|)^{-1}, \tag{6.17}$$

when $\alpha \in \mathfrak{M}(\frac{1}{5}M)$, and a and q are the unique coprime integers with $1 \leq q \leq \frac{1}{5}M$ and $|q\alpha - a| \leq M/(5N)$. We put $\Upsilon(\alpha) = 0$ for $\alpha \notin \mathfrak{M}(\frac{1}{5}M)$. Then, by Weyl's inequality and a familiar transference principle (see [35], Lemma 2.4 and Exercise 2 of §2.8), for all $\alpha \in \mathbb{R}$, one has

$$|\Phi(\alpha)|^2 \ll M^{2+\varepsilon}\Upsilon(\alpha) + M^{1+\varepsilon}. \tag{6.18}$$

Next, by Lemma 3 of Brüdern [4], one finds that whenever $1 \leq Q \leq \frac{1}{5}M$, one has

$$\int_{\mathfrak{M}(Q)} \Upsilon(\alpha)|g(\omega\alpha)|^2 \mathrm{d}_\tau\alpha \ll (QX + X^2)N^{\varepsilon-1}. \tag{6.19}$$

We choose $Q = X$ and observe that for $\alpha \in \mathfrak{M}(X)$ one has $\Upsilon(\alpha) \geq \frac{1}{2}X^{-1}$, so that $|\Phi(\alpha)|^2 \ll M^{2+\varepsilon}\Upsilon(\alpha)$. The first bound claimed in the lemma is therefore immediate. In order to establish the second bound, we begin by covering $\mathfrak{m}(X)$ by a dyadic dissection of the form $\mathfrak{M}(2Q)\backslash\mathfrak{M}(Q)$, with $X \leq Q \leq \frac{1}{10}M$, and the residual set $\mathfrak{m}^* = \mathfrak{m}(X) \backslash \mathfrak{M}(\frac{1}{5}M)$. Then one finds from (6.18) that $|\Phi(\alpha)|^2 \ll M^{1+\varepsilon}$ for $\alpha \in \mathfrak{m}^*$, whence

$$\int_{\mathfrak{m}^*} |\Phi(\alpha)|^4|g(\omega\alpha)|^2 \mathrm{d}_\tau\alpha \ll M^{2+\varepsilon}\int|g(\omega\alpha)|^2\mathrm{d}_\tau\alpha \ll XM^{2+\varepsilon}.$$

Also from (6.18), one sees that when $\alpha \in \mathfrak{M}(2Q)\backslash\mathfrak{M}(Q)$ and $Q \leq \frac{1}{10}M$, then one has $|\Phi(\alpha)|^4 \ll M^{4+\varepsilon}Q^{-1}\Upsilon(\alpha)$, and thus (6.19) yields the bound

$$\int_{\mathfrak{M}(2Q)\backslash\mathfrak{M}(Q)} |\Phi(\alpha)|^4|g(\omega\alpha)|^2\mathrm{d}_\tau\alpha \ll (X + X^2Q^{-1})M^{2+\varepsilon}.$$

The second bound of the lemma now follows on adding the contribution arising from the $O(\log N)$ values of Q comprising the aforementioned dyadic dissection, and incorporating our earlier bound for the contribution stemming from $\mathfrak{m}^*$. Finally, when g is replaced by f, then the conclusion of Lemma 6.2 follows *mutatis mutandis*. $\qquad\square$

We now consider the integral

$$\mathcal{I}(\mathfrak{a}) = \int_\mathfrak{a} |G(\alpha)\Phi(\alpha)|\mathrm{d}_\tau\alpha.$$

In view of (6.16), it suffices now to show that $\mathcal{I}(\mathfrak{c}) = o(X^2M)$, for it then follows from (6.15) that

$$V \gg X^2M, \tag{6.20}$$

as required to complete the proof of Lemma 6.2. Recall the notation of the statement of Lemma 6.3. We begin an amplification argument by considering the set

$$\mathfrak{e} = \{\alpha \in \mathfrak{c} : \lambda\alpha \in \mathfrak{m}(X)\}.$$

Then, by Hölder's inequality, one finds that

$$\mathcal{I}(\mathfrak{e}) \leq \left(\int_\mathfrak{e} |\Phi|^4|f_1|^2\mathrm{d}_\tau\alpha\right)^{1/4}\left(\int |f_1|^2|f_2|^4\mathrm{d}_\tau\alpha\right)^{1/4}\prod_{j=3}^5\left(\int |g_j|^6\mathrm{d}_\tau\alpha\right)^{1/6}.$$

By Schwarz's inequality, followed by applications of (3.10) and Hua's Lemma (see Vaughan [35], Lemma 2.5), we deduce that

$$\int |f_1|^2 |f_2|^4 \mathrm{d}_\tau \alpha \leq \left(\int |f_1|^4 \mathrm{d}_\tau \alpha \right)^{1/2} \left(\int |f_2|^8 \mathrm{d}_\tau \alpha \right)^{1/2} \ll X^{7/2+\varepsilon}.$$

Consequently, by Lemmata 6.1 and 6.3, it now follows that

$$\mathcal{I}(\mathfrak{e}) \ll (M^2 X)^{1/4} X^{7/8+\varepsilon} (X^{\mu_6})^{1/2} = o(MX^2),$$

a bound which does not interfere with the desired conclusion (6.20).

On the remaining set, we may suppose that $\lambda \alpha \in \mathfrak{M}(X)$. We next dispose of the subsets

$$\mathfrak{E}_j = \{ \alpha \in \mathfrak{c} : \lambda \alpha \in \mathfrak{M}(X) \text{ and } \lambda_j \alpha \in \mathfrak{m}(X^{3/4}) \}.$$

Then, by (3.4) and Hölder's inequality, we have

$$\mathcal{I}(\mathfrak{E}_1) \ll X^{3/4+\varepsilon} \left(\int_{\mathfrak{K}} |\Phi f_2|^2 \mathrm{d}_\tau \alpha \right)^{1/2} \prod_{j=3}^{5} \left(\int |g_j|^6 \mathrm{d}_\tau \alpha \right)^{1/6}.$$

The first integral may be estimated by applying Lemma 6.3, and for the sixth moments of g_j we again make use of Lemma 6.1. We then find that

$$\mathcal{I}(\mathfrak{E}_1) \ll X^{3/4+\varepsilon} (X^{2+\varepsilon})^{1/2} (X^{\mu_6})^{1/2} \ll MX^{15/8}.$$

By symmetry, the same bound holds for $\mathcal{I}(\mathfrak{E}_2)$.

It remains to consider the amplifying set

$$\mathfrak{D} = \{ \alpha \in \mathfrak{c} : \lambda \alpha \in \mathfrak{M}(X) \text{ and } \lambda_j \alpha \in \mathfrak{M}(X^{3/4}) \ (j = 1, 2) \}.$$

The endgame is almost identical to the one described in §6.3. If T is defined as in the discussion surrounding (6.10), then one may show that the portion of $\mathfrak{D}$, wherein $\lambda_1 \alpha \in \mathfrak{M}(T^{1/4})$ and $\lambda_2 \alpha \in \mathfrak{M}(T^{1/4})$ hold simultaneously, makes a contribution to $\mathcal{I}(\mathfrak{D})$ that is $O(MX^2 T^{-1/2}) = o(MX^2)$. The required argument follows exactly that leading to (6.10). This leaves the two sets

$$\mathfrak{D}_j = \{ \alpha \in \mathfrak{D} : \lambda_j \alpha \in \mathfrak{m}(T^{1/4}) \} \quad (j = 1, 2).$$

Here we apply (3.13) to bound $f(\lambda_j \alpha)$, and thereby conclude that one has $f(\lambda_j \alpha) \ll X T^{-1/12}$ throughout $\mathfrak{D}_j$, whence

$$\sup_{\alpha \in \mathfrak{D}_1 \cup \mathfrak{D}_2} |f_1 f_2| \ll X^2 T^{-1/12}.$$

We now infer from Hölder's inequality that

$$\mathcal{I}(\mathfrak{D}_1 \cup \mathfrak{D}_2) \leq \sup_{\alpha \in \mathfrak{D}_1 \cup \mathfrak{D}_2} |f_1 f_2|^{1/8} \Big(\int_{\mathfrak{D}} |\Phi|^5 \mathrm{d}_\tau \alpha \Big)^{1/5} \prod_{j=3}^{5} \Big(\int |g_j|^8 \mathrm{d}_\tau \alpha \Big)^{1/8}$$

$$\times \Big(\int_{\mathfrak{D}} |f_1|^{70/17} \mathrm{d}_\tau \alpha \Big)^{17/80} \Big(\int_{\mathfrak{D}} |f_2|^{70/17} \mathrm{d}_\tau \alpha \Big)^{17/80}.$$

The last two integrals can be bounded with the aid of (3.9), and for the eighth moments of g_j, one may apply Lemma 6.1 combined with a trivial estimate. In addition, when $\lambda \alpha \in \mathfrak{M}(X)$, one may apply the recent work of Vaughan [36] to show that $\Phi(\alpha) \ll M \Upsilon(\lambda \alpha)^{1/2}$, with Υ defined as in (6.17). Then, a straightforward calculation reveals that

$$\int_{\mathfrak{D}} |\Phi(\alpha)|^5 \mathrm{d}_\tau \alpha \ll M^3.$$

Note here the important feature that there is no inflation of the estimate by an unacceptable logarithmic factor. Collecting together the estimates of this paragraph, we find that $\mathcal{I}(\mathfrak{D}_1 \cup \mathfrak{D}_2) \ll M X^2 T^{-1/96}$. In summary, we have shown that $\mathcal{I}(\mathfrak{D}) \ll M X^2 T^{-1/96}$, and hence also $\mathcal{I}(\mathfrak{c}) = o(M X^2)$. The proof of Lemma 6.2 is therefore complete.

6.6. *Additive forms of large degree*

In the last three sections of this chapter, we discuss the distribution of the values of the additive form (1.1) for larger degree. Theorem 1.8 will be established in §6.7, and Theorem 1.9 in §6.8. Here, we summarize results on smooth Weyl sums over k-th powers. The definition of $g(\alpha)$ in (6.1) is now to be replaced by the more general

$$g(\alpha) = \sum_{x \in \mathcal{A}(X, X^\eta)} e(\alpha x^k).$$

It will also be convenient to define

$$t_0(k) = \lceil \tfrac{1}{2} k (\log k + \log \log k + 1) \rceil + \lceil \tfrac{1}{2} k (1 + 1/\sqrt{\log k}) \rceil.$$

Lemma 6.4. *For any* $\varepsilon > 0$ *there exists* $\eta_0(\varepsilon) > 0$ *such that whenever* $0 < \eta \leq \eta_0$ *and* $t \in \mathbb{N}$, *one has*

$$\int_0^1 |g(\alpha)|^{2t} \mathrm{d}\alpha \ll X^{2t - k + \Delta_t + \varepsilon},$$

where the real number Δ_t satisfies $\Delta_t e^{\Delta_t/k} \leq k e^{1-2t/k}$. Moreover, when $t \geq t_0(k)$, then

$$\int_0^1 |g(\alpha)|^{2t} d\alpha \ll X^{2t-k}.$$

Proof. The first estimate is the corollary to Theorem 2.1 of Wooley [41], the second is (5.2) of V. □

Lemma 6.5. *Let $Y = X^{2/3}$. Then, uniformly for $\alpha \in \mathfrak{m}(Y)$, one has $g(\alpha) \ll X^{1-\mu}$ for some $\mu = \mu_k > 0$. Moreover, for $1 \leq T \leq (\log X)^2$ and $\alpha \in \mathfrak{m}(T)$, one has $g(\alpha) \ll XT^{\varepsilon-1/(2k)}$. Finally, for any real number $t > 4k$, one has*

$$\int_{\mathfrak{N}(Y)} |g(\alpha)|^t d\alpha \ll X^{t-k}.$$

Proof. The first bound, of Weyl's type, follows from Theorem 1.4 of Wooley [40], for example. By combining Lemmata 7.2 and 8.5 of Vaughan and Wooley [37], one may confirm that

$$|g(\alpha)| \ll X(q + X^k|q\alpha - a|)^{\varepsilon-1/(2k)}$$

whenever $q \leq Y$, $|q\alpha - a| \leq YX^{-k}$ and $(a,q) = 1$. The second upper bound for $g(\alpha)$ is now transparent, and the major arc estimate follows via a straightforward calculation. □

We also present a rather general treatment for the expected main terms. Suppose that $\lambda_1, \ldots, \lambda_t$ are positive real numbers. Also, define

$$X = 2(\lambda_1^{-1/k} + \cdots + \lambda_t^{-1/k} + 1)N^{1/k}. \tag{6.21}$$

Then, whenever $1 \leq \nu \leq N$ and $0 < \tau \leq 1$, and $x_j \in \mathbb{N}$ satisfy

$$|\lambda_1 x_1^k + \cdots + \lambda_t x_t^k - \nu| < \tau,$$

one finds that $x_j \leq X$. The central interval $\mathfrak{C}$ remains defined by (6.4). Then for any fixed $\tau \in (0,1]$, uniformly in $1 \leq \nu \leq N$, it follows that whenever $t > k$ one has

$$\int_{\mathfrak{C}} g(\lambda_1\alpha) \cdots g(\lambda_t\alpha)e(-\nu\alpha)d_\tau\alpha = c\nu^{t/k-1} + O(X^{t-k}(\log X)^{-1/4}). \tag{6.22}$$

Here, the constant $c > 0$ is independent of ν and X. This is readily established by following the method of proof of Lemma 2.2, but the approximation of $g(\lambda_j\alpha)$ is accomplished via Lemma 8.5 of Wooley [39], for example. The reader is entitled to be spared further details.

6.7. *Proof of Theorem 1.8*

Consider the situation described in Theorem 1.8. Suppose that $1 \leq \nu \leq N$, and that $X \asymp N^{1/k}$ is chosen in accordance with (6.21). In particular, let $s \in \mathbb{N}$ with $s \geq \max(\frac{3}{2}t_0(k) + 1, 4k + 3)$, and consider the integral

$$\mathfrak{J}(\nu) = \int g(\lambda_1\alpha) \cdots g(\lambda_s\alpha) e(-\nu\alpha) \mathrm{d}_\tau\alpha, \qquad (6.23)$$

where $0 < \tau \leq 1$ is fixed from now on. Then by (6.22), uniformly for $\frac{1}{10}N < \nu \leq N$, the central interval $\mathfrak{C}$ contributes $\gg X^s N^{-1}$ to the integral in (6.23). On the complementary compositum $\mathfrak{c}$, we first define $T = T_{\lambda_1/\lambda_2}(\log N)$ via (2.22), and then the amplifying set by

$$\mathfrak{D} = \{\alpha \in \mathfrak{c} : \lambda_1\alpha \in \mathfrak{M}(T^{1/4}), \ \lambda_2\alpha \in \mathfrak{M}(T^{1/4})\}.$$

Then, as in (6.10), one may employ Theorem 2.4 to establish that the contribution of $\mathfrak{D}$ to the integral (6.23) does not exceed $O(X^s N^{-1} T^{-1/2})$, uniformly in ν.

We are now reduced to the set $\mathfrak{d} = \mathfrak{c} \backslash \mathfrak{D}$, and here we average over the quadratic polynomial ϕ. Let M be the positive solution of $\phi(2M) = N$. Suppose that m is an integer with $M < m \leq 2M$ for which

$$|\lambda_1 x_1^k + \cdots + \lambda_s x_s^k - \phi(m)| < \tau$$

has no solution in natural numbers x_j. Then $\mathfrak{J}(\phi(m)) = 0$. Let $\mathcal{Z}(M)$ be the set of all such m, and write $Z(M)$ for its cardinality. Recalling what has just been said concerning the contributions of $\mathfrak{C}$ and $\mathfrak{D}$ to (6.23), it follows that for all $m \in \mathcal{Z}(M)$, one has the lower bound

$$\left| \int_{\mathfrak{d}} g(\lambda_1\alpha) \cdots g(\lambda_s\alpha) e(-\phi(m)\alpha) \mathrm{d}_\tau\alpha \right| \gg X^s N^{-1}.$$

We sum over these exceptional m. Then, for suitable $\eta_m \in \mathbb{C}$ with $|\eta_m| = 1$, and with $K(\alpha)$ defined by (5.11), we infer that

$$X^s N^{-1} Z(M) \ll \int_{\mathfrak{d}} |g(\lambda_1\alpha) \cdots g(\lambda_s\alpha) K(\alpha)| \mathrm{d}_\tau\alpha.$$

Let $Y = X^{2/3}$, and let $\mathfrak{d}_j$ be the set of all $\alpha \in \mathfrak{d}$ with $\lambda_j\alpha \in \mathfrak{m}(Y)$. Then, by Hölder's inequality and Lemma 6.5,

$$\int_{\mathfrak{d}_1} |g_1 \cdots g_s K| \mathrm{d}_\tau\alpha$$

$$\ll X^{1-\mu} \left(\int |K|^4 \mathrm{d}_\tau\alpha \right)^{1/4} \prod_{j=2}^{s} \left(\int |g_j|^{(4s-4)/3} \mathrm{d}_\tau\alpha \right)^{3/(4s-4)}.$$

We estimate the first integral using Lemma 5.2. Also, since $\frac{4}{3}(s-1) \geq 2t_0(k)$, one may apply Lemma 6.4 to the moments of g_j. Thus we obtain the upper bound

$$\int_{\mathfrak{d}_1} |g_1 \cdots g_s K| d_\tau \alpha \ll (M^{1+\varepsilon} Z(M))^{1/4} X^{s-3k/4-\mu}.$$

By symmetry, the same bound holds for any other $\mathfrak{d}_j$. It therefore remains only to discuss the contribution from the set

$$\mathfrak{e} = \{\alpha \in \mathfrak{d} : \lambda_j \alpha \in \mathfrak{M}(Y) \, (1 \leq j \leq s)\}.$$

The definition of $\mathfrak{D}$ combined with Lemma 6.5 shows that for $\alpha \in \mathfrak{d}$ one has $g(\lambda_1 \alpha) g(\lambda_2 \alpha) \ll X^2 T^{\varepsilon-1/(8k)}$. Moreover, by applying Hölder's inequality in combination with the major arc estimate from Lemma 6.5, it is clear that

$$\int_{\mathfrak{e}} |g_3 g_4 \cdots g_s| d_\tau \alpha \ll X^{s-2-k},$$

whence

$$\int_{\mathfrak{e}} |g_1 \cdots g_s K| d_\tau \alpha \ll X^{s-k} T^{\varepsilon-1/(8k)} Z(M).$$

Assembling together the estimates of this section, we find that

$$X^s N^{-1} Z(M) \ll X^{s-k} T^{\varepsilon-1/(8k)} Z(M) + (M^{1+\varepsilon} Z(M))^{1/4} X^{s-3k/4-\mu},$$

and we may therefore conclude that $Z(M) \ll M^{1+\varepsilon} X^{-4\mu/3}$. This establishes Theorem 1.8.

6.8. *Proof of Theorem 1.9*

In this section, we discuss the diophantine inequality

$$|\lambda_1 x_1^k + \cdots + \lambda_s x_s^k - p - \tfrac{1}{2}| < \tfrac{1}{2} \qquad (6.24)$$

by the methods of §6.7. If p is a prime and $x_1, \ldots, x_s$ are natural numbers satisfying (6.24), then

$$[\lambda_1 x_1^k + \cdots + \lambda_s x_s^k] = p.$$

We choose N and X in accordance with (6.21), and set $\tau = \frac{1}{2}$. In the present context, it is appropriate to modify the definition of $h(\alpha)$ so that

$$h(\alpha) = \sum_{p \leq N} (\log p) e(p\alpha).$$

We then put

$$\mathcal{J} = \int g(\lambda_1\alpha)\cdots g(\lambda_s\alpha)h(-\alpha)e(-\tfrac{1}{2}\alpha)\mathrm{d}_\tau\alpha. \qquad (6.25)$$

This integral provides a weighted count of the solutions of (6.24) with $x_j \le X$ and $p \le N$, in which the weight is non-negative. We assume the Riemann hypothesis for Dirichlet L-functions, and proceed to show that for $s \ge \tfrac{8}{3}k + 2$, one has $\mathcal{J} \gg X^s$. This suffices to establish Theorem 1.9.

It will be useful to denote the contribution to (6.25) from a measurable set $\mathfrak{a} \subset \mathbb{R}$ by $\mathcal{J}(\mathfrak{a})$. In order to estimate $\mathcal{J}(\mathfrak{C})$, we choose $\nu = p + \tfrac{1}{2}$ and $\tau = \tfrac{1}{2}$ in (6.22). Multiplying by $\log p$ and summing over primes $p \le N$, we confirm the lower bound $\mathcal{J}(\mathfrak{C}) \gg X^s$. Hence, if $\mathfrak{c}$ denotes the complementary compositum, it now suffices to show that $\mathcal{J}(\mathfrak{c}) = o(X^s)$. It is only at this point that the Riemann hypothesis is required, and it is invoked through Lemma 2 of Brüdern and Perelli [14]. As a consequence of the latter, we have

$$\sup_{\alpha\in\mathfrak{m}(N^{1/6})} |h(\alpha)| \ll N^{5/6+\varepsilon}.$$

Also, when $\alpha \in \mathfrak{M}(N^{1/6})$, one has $h(\alpha) \ll \Upsilon^*(\alpha)$, where for $\alpha = a/q + \beta$ with $(a,q) = 1$, $1 \le q \le N^{1/6}$ and $|\beta| \le q^{-1}N^{-5/6}$, the function Υ^* is defined by

$$\Upsilon^*(\alpha) = N\varphi(q)^{-1}(1 + N|\beta|)^{-1}.$$

Here $\varphi(q)$ denotes Euler's totient.

We split $\mathfrak{c}$ into various subsets of which we first treat

$$\mathfrak{e} = \{\alpha \in \mathfrak{c} : \alpha \in \mathfrak{m}(N^{1/6})\}.$$

On this set, Hölder's inequality yields

$$\mathcal{J}(\mathfrak{e}) \ll N^{5/6+\varepsilon} \prod_{j=1}^{s} \left(\int |g_j|^s \mathrm{d}_\tau\alpha\right)^{1/s},$$

and since $s \ge \tfrac{8}{3}k + 2$, we may apply Lemma 6.4 to establish the bound

$$\int |g_j|^s \mathrm{d}_\tau\alpha \ll X^{s-\frac{5}{6}k-\mu},$$

for some $\mu > 0$. Consequently, one deduces that $\mathcal{J}(\mathfrak{e}) \ll X^{s-\mu+\varepsilon}$, which is acceptable.

It remains to consider the set $\mathfrak{c} \cap \mathfrak{M}(N^{1/6})$. The auxiliary estimate

$$\int_{\mathfrak{M}(N^{1/6})} \Upsilon^*(\alpha)|g(\lambda_j\alpha)|^{2t}\mathrm{d}_\tau\alpha \ll N^\varepsilon\left(N^{1/6}\int_0^1 |g(\alpha)|^{2t}\mathrm{d}\alpha + X^{2t}\right) \qquad (6.26)$$

is now required, and this may be verified by a route paralleling that which arrives at (6.19). Note here that the irritating factor $\varphi(q)^{-1}$ can be replaced by q^{-1} at the cost of an inflationary factor $O(\log \log N)$ that can be absorbed into the term N^ε. The amplification is now similar to the work in §6.7. Let

$$\mathfrak{d}_j = \{\alpha \in \mathfrak{c} \cap \mathfrak{M}(N^{1/6}) : \lambda_j \alpha \in \mathfrak{m}(Y)\}.$$

Then, by Lemma 6.5 and Hölder's inequality, one obtains

$$\mathcal{J}(\mathfrak{d}_1) \ll X^{\mathsf{T}-\mu} \prod_{j=2}^{2k+1} \left(\int_{\mathfrak{d}_1} \Upsilon^*(\alpha) |g(\lambda_j \alpha)|^{2k} \mathrm{d}_\tau \alpha \right)^{1/(2k)} X^{s-2k-1}.$$

On combining (6.26) with the first estimate of Lemma 6.4, it readily follows that $\mathcal{J}(\mathfrak{d}_1) \ll X^{s-\mu+\varepsilon}$. By symmetry, the same is true for $\mathcal{J}(\mathfrak{d}_j)$ when $2 \le j \le s$.

It now remains to discuss the set

$$\mathfrak{E} = \{\alpha \in \mathfrak{c} \cap \mathfrak{M}(N^{1/6}) : \lambda_j \alpha \in \mathfrak{M}(Y) \text{ for } 1 \le j \le s\}.$$

Define $T = T_{\lambda_1/\lambda_2}(\log N)$ as in (2.22) once again, and put

$$\mathfrak{D} = \{\alpha \in \mathfrak{E} : \lambda_1 \alpha \in \mathfrak{M}(T^{1/4}) \text{ and } \lambda_2 \alpha \in \mathfrak{M}(T^{1/4})\}.$$

Then, again as in §6.7, one finds that $\mathcal{J}(\mathfrak{D}) \ll X^s T^{-1/2}$, which is again acceptable. For $\alpha \in \mathfrak{E} \backslash \mathfrak{D}$, it follows from Lemma 6.5 that

$$g(\lambda_1 \alpha) g(\lambda_2 \alpha) \ll X^2 T^{\varepsilon - 1/(8k)}.$$

The estimate

$$\int_{\mathfrak{E}} \Upsilon^*(\alpha)^3 \mathrm{d}_\tau \alpha \le \int_{\mathfrak{M}(N^{1/6})} \Upsilon^*(\alpha)^3 \mathrm{d}_\tau \alpha \ll N^2$$

follows by a simple calculation. In addition, on noting that $\frac{3}{2}(s-1) > 4k$ and $\lambda_j \alpha \in \mathfrak{M}(Y)$ for all j, an application of Hölder's inequality in alliance with Lemma 6.5 confirms the upper bound

$$\int_{\mathfrak{E}} (|g_1 g_2|^{1/2} |g_3 \cdots g_s|)^{3/2} \mathrm{d}_\tau \alpha \ll X^{3(s-1)/2-k}.$$

Here, and in what follows, we have written g_i for $g(\lambda_i \alpha)$. It now follows by Hölder's inequality that

$$\int_{\mathfrak{E} \backslash \mathfrak{D}} |\Upsilon^*(\alpha) g_1 \cdots g_s| \mathrm{d}_\tau \alpha \le \sup_{\alpha \in \mathfrak{E} \backslash \mathfrak{D}} |g_1 g_2|^{1/2} \left(\int_{\mathfrak{E}} \Upsilon^*(\alpha)^3 \mathrm{d}_\tau \alpha \right)^{1/3}$$

$$\times \left(\int_{\mathfrak{E}} (|g_1 g_2|^{1/2} |g_3 \cdots g_s|)^{3/2} \mathrm{d}_\tau \alpha \right)^{2/3}.$$

We therefore deduce that $\mathcal{J}(\mathfrak{E}\backslash\mathfrak{D}) \ll X^s T^{\varepsilon-1/(16k)}$, a bound that in concert with our earlier estimates yields $\mathcal{J}(\mathfrak{c}) = o(X^s)$. This suffices to complete the proof of Theorem 1.9.

7. An appendix on inhomogenous polynomials

7.1. *The counting integral*

In this final chapter we sketch a proof of Theorem 1.10. Freeman's work [24] will be invoked when appropriate, but the argument relies heavily on Lemma 5.3, and is otherwise largely standard.

We keep as much notation from earlier chapters as is possible, and in particular apply the conventions of §1.5. We can then rewrite the polynomials ϕ_j as $\phi_j(t) = \lambda_{jd_j}t^{d_j} + \cdots + \lambda_{j1}t$, and rearrange the indices of $\phi_1, \ldots, \phi_s$ so as to assure that

$$\lambda_{1l_1} \text{ and } \lambda_{2l_2} \text{ are not in rational ratio} \tag{7.1}$$

for some $1 \le l_1 \le d_1$, $1 \le l_2 \le d_2$. To see this, suppose first that all the ϕ_j are multiples of rational polynomials. Then, there exist non-zero real numbers μ_j such that $\mu_j\phi_j \in \mathbb{Q}[t]$. Under the current hypotheses, there must be an index j with $\mu_1/\mu_j \notin \mathbb{Q}$. Exchanging j with 2, we find that (7.1) holds. In the contrary case, at least one of the polynomials ϕ_j is irrational, and we may assume that this is so for ϕ_1. Then $d_1 \ge 2$, and $\lambda_{1d_1}/\lambda_{1i}$ is irrational for some i with $1 \le i \le d_1 - 1$. Hence, one of the numbers $\lambda_{1d_1}/\lambda_{2d_2}$, $\lambda_{1i}/\lambda_{2d_2}$ is also irrational, as required.

From now on, suppose that (7.1) holds, and that ν is large. Define X_j to be the unique positive solution of $\phi_j(X_j) = \nu$. Then, for any positive solution of (1.9) with $0 < \tau \le 1$, one has $x_j \le 2X_j$. Define the Weyl sums

$$f_j(\alpha) = \sum_{x \le 2X_j} e(\alpha\phi_j(x)), \tag{7.2}$$

and the integral

$$\rho_\phi^*(\tau, \nu) = \int f_1(\alpha) \cdots f_s(\alpha) e(-\nu\alpha)\mathrm{d}_\tau\alpha.$$

In view of (7.1) and (2.6), this is a counting integral with weight, of the type considered in (2.6), and can therefore be compared with the number $\rho_\phi(\tau, \nu)$ through the now familiar mechanism based on Lemma 2.1. In particular, the proof of Theorem 1.10 is made complete with the verification of the asymptotic formula

$$\rho_\phi^*(\tau, \nu) = \tfrac{1}{2}c(\phi)\tau\nu^{D-1} + o(\nu^{D-1}). \tag{7.3}$$

7.2. *The central interval*

In the interests of brevity, we put

$$\mathcal{I}(\mathfrak{a}) = \int_{\mathfrak{a}} f_1(\alpha) \cdots f_s(\alpha) e(-\nu\alpha) \mathrm{d}_\tau \alpha.$$

Set $d = \max d_j$, and write $Y = \nu^{1/(2d)-1}$ and $\mathfrak{C} = [-Y, Y]$. One may closely follow the proof of Lemma 2.2, or the arguments of Freeman [24], pp. 239–243, in order to evaluate $\mathcal{I}(\mathfrak{C})$. One replaces $f_j(\alpha)$ with the function

$$v_j(\alpha) = \int_0^{2X_j} e(\alpha \phi_j(t)) \mathrm{d}t,$$

and then completes the singular integral

$$\mathcal{I}_\infty(\nu) = \int v_1(\alpha) \cdots v_s(\alpha) e(-\nu\alpha) \mathrm{d}_\tau \alpha.$$

The error terms in these processes can be controlled by appealing to Lemma 4.4 of Baker [1] (see also Lemma 4 of [24]) and Theorem 7.3 of Vaughan [35]. In this way, one may confirm that

$$\mathcal{I}(\mathfrak{C}) = \mathcal{I}_\infty(\nu) + O(\nu^{D-1-1/(2d)}),$$

provided only that $s > 2d$. This asymptotic relation is more than we need later. Under the same condition on s, the concluding part of the proof of Lemma 2.2 is readily modified to yield

$$\mathcal{I}_\infty(\nu) = \tfrac{1}{2}c(\phi)\tau\nu^{D-1}\big(1 + O(\nu^{-1/(2d)})\big),$$

where $c(\phi)$ is the positive real number defined in the statement of Theorem 1.10. If $\mathfrak{c} = \mathbb{R} \setminus \mathfrak{C}$, then, in order to confirm (7.5), it now suffices to show that $\mathcal{I}(\mathfrak{c}) = o(\nu^{D-1})$.

7.3. *The complementary compositum*

The prearrangement in (7.1) is required only within the following lemma, which combines Lemmata 8 and 9 of Freeman [24]. We choose $\delta = 1/(2d)$ in Lemma 9 of [24], and then conclude as follows.

Lemma 7.1. *Suppose that (7.1) holds. Then there exists a monotone function $T(\nu)$, with $T(\nu) \to \infty$ as $\nu \to \infty$, such that*

$$\sup_{Y \le |\alpha| \le T(\nu)} |f_1(\alpha)f_2(\alpha)| = o(X_1 X_2).$$

We remark that this is an estimate of Bentkus-Götze-Freeman type. Instead, it would be possible to work with Theorem 2.4, but at this stage the tidy reference for Lemma 7.1 saves some effort.

Now let $C > 1$ be a suitably large positive number with the property that $t(d) = d^2(\log d + 2\log\log d + C)$ is an integer. From (5.25) we have the bound

$$\int |f_j(\alpha)|^{2t(d)}\mathrm{d}_\tau\alpha \ll X_j^{2t(d)-d_j} \ll X_j^{2t(d)}\nu^{-1}. \tag{7.4}$$

By Lemma 11 of Freeman [24] (which is essentially already in Davenport and Roth [18], Lemma 2), it then also follows that

$$\int_{|\alpha|>T(\nu)} |f_j(\alpha)|^{2t(d)}\mathrm{d}_\tau\alpha \ll X_j^{2t(d)}\nu^{-1}T(\nu)^{-1}. \tag{7.5}$$

Write

$$\mathfrak{L} = \{\alpha : Y < |\alpha| < T(\nu)\} \quad \text{and} \quad \mathfrak{l} = \{\alpha : |\alpha| \geq T(\nu)\}.$$

Then for $s \geq 2t(d)+2$ and $j \geq 3$, we may apply (7.7), together with Hölder's inequality and Lemma 7.1, to confirm the estimate

$$\mathcal{I}(\mathfrak{L}) = o(X_1 X_2 \cdots X_s \nu^{-1}).$$

Likewise, by Hölder's inequality and (7.8), we infer that

$$\mathcal{I}(\mathfrak{l}) = o(X_1 X_2 \cdots X_s \nu^{-1}).$$

These two bounds combine to deliver the conlusion $\mathcal{I}(\mathfrak{c}) = o(\nu^{D-1})$, which was all that was required to complete the proof of Theorem 1.10.

Acknowledgements. The third author was supported in part by a Royal Society Wolfson Research Merit Award. Research on this project was conducted at Shandong University in Weihai, during the 5th Japan-China Conference on Number Theory, Kinki University, Osaka, and at the Hausdorff Institute for Mathematics, Bonn. The authors are grateful for the hospitality and excellent working conditions provided by their hosts.

References

1. R. C. Baker, *Diophantine inequalities*. London Mathematical Society Monographs. New Series, vol. 1, Oxford University Press, Oxford, 1986.
2. V. Bentkus and F. Götze, Lattice point problems and distribution of values of quadratic forms. *Ann. of Math.* (2) **150** (1999), 977–1027.
3. K. D. Boklan, A reduction technique in Waring's problem, I. *Acta Arith.* **65** (1993), 147–161.

4. J. Brüdern, The Davenport-Heilbronn Fourier transform method, and some Diophantine inequalities. *Number theory and its applications* (Kyoto, 1997), 59–87, Dev. Math., vol. 2, Kluwer Acad. Publ., Dordrecht, 1999.

5. J. Brüdern, Counting diophantine approximations. *Funct. Approx. Comment. Math.* **39** (2008), 237–260.

6. J. Brüdern, R. J. Cook, and A. Perelli, The values of binary linear forms at prime arguments. *Sieve methods, exponential sums, and their applications in number theory* (Cardiff, 1995), 87–100, London Math. Soc. Lecture Note Ser., vol. 237, Cambridge Univ. Press, Cambridge, 1997.

7. J. Brüdern, K. Kawada and T. D. Wooley, Additive representation in thin sequences, I: Waring's problem for cubes. *Ann. Sci. École Norm. Sup.* (4) **34** (2001), 471–501.

8. J. Brüdern, K. Kawada and T. D. Wooley, Additive representation in thin sequences, II: The binary Goldbach problem. *Mathematika* **47** (2000), 117–125.

9. J. Brüdern, K. Kawada and T. D. Wooley, Additive representation in thin sequences, III: Asymptotic formulae. *Acta Arith.* **100** (2001), 267–289.

10. J. Brüdern, K. Kawada and T. D. Wooley, Additive representation in thin sequences, IV: Lower bound methods. *Quart. J. Math. Oxford* **52** (2001), 423–436.

11. J. Brüdern, K. Kawada and T. D. Wooley, Additive representation in thin sequences, V: Mixed problems of Waring's type. *Math. Scand.* **92** (2003), 181–209.

12. J. Brüdern, K. Kawada and T. D. Wooley, Additive representation in thin sequences, VI: Representing primes, and related problems. *Glasg. Math. J.* **44** (2002), 419–434.

13. J. Brüdern, K. Kawada and T. D. Wooley, Additive representation in thin sequences, VII: Restricted moments of the number of representations. *Tsukuba J. Math.* **32** (2008), 383–406.

14. J. Brüdern and A. Perelli, The addition of primes and powers. *Canad. J. Math.* **48** (1996), 512–526.

15. J. Brüdern and A. Perelli, Goldbach numbers in sparse sequences. *Ann. Inst. Fourier (Grenoble)* **48** (1998), 353–378.

16. J. Brüdern and T. D. Wooley, On Waring's problem for cubes and smooth Weyl sums. *Proc. London Math. Soc.* (3) **82** (2001), 89–109.

17. H. Davenport and H. Heilbronn, On indefinite quadratic forms in five variables. *J. London Math. Soc.* **21** (1946), 185–193.

18. H. Davenport and K. F. Roth, The solubility of certain Diophantine inequalities. *Mathematika* **2** (1955), 81–96.

19. A. Eskin, G. A. Margulis and S. Mozes, Upper bounds and asymptotics in a quantitative version of the Oppenheim conjecture. *Ann. of Math.* (2) **147** (1998), 93–141.

20. K. B. Ford, New estimates for mean values of Weyl sums. *Internat. Math. Res. Notices* 1995, 155–171.

21. D. E. Freeman, Asymptotic lower bounds for Diophantine inequalities. *Mathematika* **47** (2000), 127–159.

22. D. E. Freeman, Asymptotic lower bounds and formulas for Diophantine inequalities. *Number theory for the millennium, II* (Urbana, IL, 2000), 57–74, Edited by M. A. Bennett, B. C. Berndt, N. Boston, H. G. Diamond, A. J. Hildebrand and W. Philipp. A K Peters, Natick, MA, 2002.

23. D. E. Freeman, Systems of diagonal Diophantine inequalities. *Trans. Amer. Math. Soc.* **355** (2003), 2675–2713.

24. D. E. Freeman, Additive inhomogeneous Diophantine inequalities. *Acta Arith.* **107** (2003), 209–244.

25. F. Götze, Lattice point problems and values of quadratic forms. *Invent. Math.* **157** (2004), 195–226.

26. R. R. Hall and G. Tenenbaum, *Divisors.* Cambridge Tracts in Mathematics, vol. 90. Cambridge University Press, Cambridge, 1988.

27. M. C. Liu and K. M. Tsang, On pairs of linear equations in three prime variables and an application to Goldbach's problem. *J. Reine Angew. Math.* **399** (1989), 109–136.

28. G. A. Margulis, Indefinite quadratic forms and unipotent flows on homogeneous spaces. *Dynamical systems and ergodic theory* (Warsaw, 1986), 399–409, Banach Center Publ., vol. 23, PWN, Warsaw, 1989.

29. H. L. Montgomery and R. C. Vaughan, The exceptional set in Goldbach's problem. *Acta Arith.* **27** (1975), 353–370.

30. S. T. Parsell, Irrational linear forms in prime variables. *J. Number Theory* **97** (2002), 144–156.

31. S. T. Parsell and T. D. Wooley, Exceptional sets for Diophantine inequalities, to appear.

32. A. Perelli, Goldbach numbers represented by polynomials. *Rev. Mat. Iberoamericana* **12** (1996), 477–490.

33. R. C. Vaughan, On Waring's problem for cubes. *J. Reine Angew. Math.* **365** (1986), 122–170.

34. R. C. Vaughan, On Waring's problem for cubes II. *J. London Math. Soc.* (2) **39** (1989), 205–218.

35. R. C. Vaughan, *The Hardy-Littlewood method.* Second edition. Cambridge Tracts in Mathematics, vol. 125. Cambridge University Press, Cambridge, 1997.

36. R. C. Vaughan, *On generating functions in additive number theory, I.* Analytic Number Theory, Essays in Honour of Klaus Roth, edited by W. W. L. Chen, W. T. Gowers, H. Halberstam, W. M. Schmidt and R. C. Vaughan, 436–448, Cambridge University Press, Cambridge, 2009.

37. R. C. Vaughan and T. D. Wooley, On Waring's problem: some refinements. *Proc. London Math. Soc.* (3) **63** (1991), 35–68.

38. G. L. Watson, On indefinite quadratic forms in five variables. *Proc. London Math. Soc.* (3) **3** (1953), 170–181.

39. T. D. Wooley, On simultaneous additive equations, II. *J. Reine Angew. Math.* **419** (1991), 141–198.

40. T. D. Wooley, Large improvements in Waring's problem. *Ann. of Math.* (2) **135** (1992), 131–164.

41. T. D. Wooley, The application of a new mean value theorem to the fractional

parts of polynomials. *Acta Arith.* **65** (1993), 163–179.

42. T. D. Wooley, Some remarks on Vinogradov's mean value theorem and Tarry's problem. *Monatsh. Math.* **122** (1996), 265–273.
43. T. D. Wooley, Sums of three cubes. *Mathematika* **47** (2000), 53–61.
44. T. D. Wooley, *On Diophantine inequalities: Freeman's asymptotic formulae.* Proceedings of the Session in Analytic Number Theory and Diophantine Equations, 32 pp., Bonner Math. Schriften, 360, Univ. Bonn, Bonn, 2003.

RECENT PROGRESS ON DYNAMICS OF A
SPECIAL ARITHMETIC FUNCTION

CHAOHUA JIA

Institute of Mathematics, Academia Sinica,
Beijing 100190, China
E-mail: jiach@math.ac.cn

In this paper we introduce some basic concepts of dynamics and give a survey on the recent progress on the dynamics of the w function.

1. Introduction

In calculus, when studying a function $f(x)$, we let x run through some set S. In functional analysis, we often fix $x \in S$ and let f vary in some family of functions. This is a symmetric relationship in which we can exchange the roles of x and f. In the theory of dynamics, we should manage the family of functions

$$f(x), \qquad f(f(x)), \qquad f(f(f(x))), \qquad \cdots\cdots, \qquad (1.1)$$

where x runs in some set S and the condition $f(x) \in S$ is needed. For simplicity, we write

$$f^0(x) = x, \quad f^1(x) = f(x), \quad f^2(x) = f(f(x)), \quad \cdots\cdots. \qquad (1.2)$$

If $f^i(x) = x$ for some $i \geq 1$, x is called periodic. If some $f^j(x)(j \geq 1)$ is periodic, x is called preperiodic. The principal goal of dynamics is to classify the elements $x \in S$ according to the behavior of the sequence (1.1).

If S is a set of positive integers, f is an arithmetic function. The dynamics of arithmetic function is an important subject. In 2006 Wushi Goldring [4] introduced a special arithmetic function named w function. The study on the dynamics of the w function requires a lot of knowledge on the distribution of prime numbers.

Now we introduce some notations. Let $\mathcal{P}$ be the set of prime numbers and $P(n)$ denote the largest prime factor of the integer $n > 1$. Write

$$C_3 = \{p_1 p_2 p_3 : \ p_i \in \mathcal{P} \ (i = 1, 2, 3), \ p_i \neq p_j \ (i \neq j)\},$$
$$B_3 = \{p_1 p_2 p_3 : \ p_i \in \mathcal{P} \ (i = 1, 2, 3), \ p_1 = p_2 \text{ or } p_1 = p_3$$
$$\text{or } p_2 = p_3, \text{ but not } p_1 = p_2 = p_3\},$$
$$D_3 = \{p^3 : \ p \in \mathcal{P}\}.$$

Then

$$\{p_1 p_2 p_3 : \ p_i \in \mathcal{P} \ (i = 1, 2, 3)\} = C_3 \cup B_3 \cup D_3,$$

where no two of C_3, B_3 and D_3 intersect. Let

$$A_3 = C_3 \cup B_3.$$

For $n = p_1 p_2 p_3 \in A_3$, we define the w function by

$$w(n) = P(p_1 + p_2)P(p_1 + p_3)P(p_2 + p_3). \tag{1.3}$$

The value of $w(n)$ does not depend on the order of p_i so that it is well defined. Wushi Goldring [4] showed that if $n \in A_3$, then $w(n) \in A_3$ (Lemma 2.1 in [4]). Therefore we can consider the dynamics of the w function in A_3.

2. Dynamics of the w function

The calculation shows that

$$w^1(20) = 98, \quad w^2(20) = 63, \quad w^3(20) = 75, \quad w^4(20) = 20,$$

which means that every element in the set $\{20, 98, 63, 75\}$ is periodic. For the other elements in A_3, Wushi Goldring [4] proved that they are preperiodic in the following theorem.

Theorem 2.1. *If $n \in A_3$, then there exists a positive integer i such that*

$$w^i(n) = 20.$$

It is natural to ask how many steps there are to go from n to 20. We define the periodicity index as the least positive integer i such that $w^i(n) = 20$ and denote it by $\mathrm{ind}(n)$. Wushi Goldring [4] first gave the following estimate for the periodicity index.

Theorem 2.2. *If $n \in A_3$, then*

$$\mathrm{ind}(n) = O(\pi(P(n))).$$

By the prime number theorem, for sufficiently large $P(n)$, we know that

$$\pi(P(n)) \sim \frac{P(n)}{\log P(n)},$$

where $f(m) \sim g(m)$ means $\frac{f(m)}{g(m)} \to 1$ as $m \to \infty$. Wushi Goldring [4] put forward the following conjecture.

Conjecture 2.1. *If* $n \in A_3$, *then*

$$\text{ind}(n) = O(\log P(n)).$$

In 2008 Yong-Gao Chen and Ying Shi [2] proved the following result which is a big improvement on Theorem 2.2 and is quite approximate to Conjecture 2.1.

Theorem 2.3. *If* $n \in A_3$, *then*

$$\text{ind}(n) = O(\log^2 P(n)).$$

In the opposite direction, Wushi Goldring [4] made the following conjecture on the periodicity index.

Conjecture 2.2. *There is a subset in* A_3 *in which the elements have arbitrarily large periodicity index.*

Yong-Gao Chen and Ying Shi [2] gave a positive answer to Conjecture 2.2. One can further ask the following problem: Are there infinitely many $n \in A_3$ such that

$$\text{ind}(n) \gg \log P(n)? \tag{2.1}$$

3. Inverse problem

Inverse problems are important in the dynamics of the w function. Wushi Goldring [4] asked three inverse problems:

1) For $n \in A_3$, can we find $m \in A_3$ such that $w(m) = n$?
2) If so, how many such elements are there?
3) What form do they have?

For $n \in A_3$, if there is $m \in A_3$ such that $w(m) = n$, we call m a parent of n. If this $m \in S \subset A_3$, we call it S-parent of n. It is obvious that the element of C_3 has no B_3-parent. Wushi Goldring [4] put forward the following conjecture.

Conjecture 3.1.

1) *Every element of C_3 has infinitely many C_3-parents;*
2) *every element of B_3 has infinitely many C_3-parents;*
3) *every element of B_3 has infinitely many B_3-parents.*

Wushi Goldring [4] obtained the first result on the inverse problem in the following theorem.

Theorem 3.1. *There are infinitely many elements of B_3 which have at least seven parents.*

In 2008 Yong-Gao Chen and Ying Shi [3] got the following improvement on Theorem 3.1.

Theorem 3.2. *For any given positive integer k, there are infinitely many elements of B_3 which have at least k B_3-parents.*

On the other hand, Yong-Gao Chen and Ying Shi [3] proved the following conclusion which denies the assertion in part 3) of Conjecture 3.1.

Theorem 3.3. *There are infinitely many elements of B_3 which have no B_3-parent.*

The dynamics of the w function is concerned with the distribution of prime numbers. In the process of their study, Yong-Gao Chen and Ying Shi [3] put forward the following conjecture.

Conjecture 3.2. *For any integer $a \neq 0$, there are infinitely many $p \in \mathcal{P}$ such that $p + a$ have the same largest prime factor.*

In [5] Chaohua Jia studied inverse problems, especially on the C_3-parents of elements in C_3 which had not been attacked by previous authors. The novelty in [5] is an application of the following theorem.

Theorem 3.4. *Let $r \in \mathcal{P}$, $n_j \, (1 \leq j \leq Z)$ be distinct positive integers not exceeding N and $Z(N; r, a)$ denote the number of those n_j which are congruent to $a \,(\mathrm{mod}\, r)$. If $X \geq 2$, then we have*

$$\sum_{r \leq X} r \sum_{a=1}^{r} \left(Z(N; r, a) - \frac{Z}{r} \right)^2 \ll (N + X^2)Z.$$

Theorem 3.4 is Theorem 1 in [1], which is obtained by the large sieve method. The large sieve method has been playing an important role in analytic number theory for forty years and more.

Chaohua Jia [5] proved the following theorem which asserts that there are infinitely many elements of C_3 which have enough C_3-parents. In the following, p, p_1, p_2, p_3, q, r, r_1, $r_2 \in \mathcal{P}$ and c_1, c_2, c_3, c_4 denote positive constants and x is sufficiently large.

Theorem 3.5. *There exists an element $r_1 r_2 q$ of C_3 which satisfies $x^{\frac{1}{2}} \log x < r_i \le 2x^{\frac{1}{2}} \log x \, (i = 1, 2)$, $q \le 4x$ and has at least $c_1 \frac{x}{\log^4 x}$ different C_3-parents $p_1 p_2 p_3$ with $x < p_i \le 2x \, (i = 1, 2, 3)$.*

By similar means, Chaohua Jia [5] proved the following theorem which asserts that there are infinitely many elements of B_3 which have enough C_3-parents.

Theorem 3.6. *There exists an element $q r^2$ of B_3 which satisfies $q \le 4x$, $x^{\frac{1}{2}} \log x < r \le 2x^{\frac{1}{2}} \log x$ and has at least $c_2 \frac{x}{\log^4 x}$ different C_3-parents $p_1 p_2 p_3$ with $x < p_i \le 2x \, (i = 1, 2, 3)$.*

Chaohua Jia [5] also proved the following theorem which asserts that there are infinitely many elements of B_3 which have enough B_3-parents. This is a quantitative improvement on Theorem 3.2.

Theorem 3.7. *There exists an element $q r^2$ of B_3 which satisfies $x < q \le 2x$, $x^{\frac{1}{2}} \log x < r \le 2x^{\frac{1}{2}} \log x$ and has at least $c_3 \frac{x^{\frac{1}{2}}}{\log^2 x}$ different B_3-parents $p q^2$ with $x < p \le 2x$.*

4. The sketch of the proof of Theorem 3.5

Firstly it is deduced from Theorem 3.4 that

$$\sum_{x^{\frac{1}{2}} \log x < r \le 2x^{\frac{1}{2}} \log x} \sum_{x < p_1 \le 2x} \left(\sum_{\substack{x < p \le 2x \\ p \equiv -p_1 \,(\mathrm{mod}\, r)}} 1 - \frac{1}{r} \sum_{x < p \le 2x} 1 \right)^2 \qquad (4.1)$$

$$\ll \frac{x^2}{\log x}.$$

We estimate

$$N(x) = |\{p_1 p_2 p_3 : \ x < p_i \le 2x \, (i = 1, 2, 3), \ x^{\frac{1}{2}} \log x < P(p_1 + p_2)$$
$$\le 2x^{\frac{1}{2}} \log x, \ x^{\frac{1}{2}} \log x < P(p_1 + p_3) \le 2x^{\frac{1}{2}} \log x\}|.$$

Note the fact that if $p > n^{\frac{1}{2}}$, then $p \mid n \iff P(n) = p$. Then we write

$$N(x) = \sum_{\substack{x<p_1\leq 2x}} \sum_{\substack{x<p_2\leq 2x \\ x^{\frac{1}{2}}\log x<P(p_1+p_2)\leq 2x^{\frac{1}{2}}\log x}} \sum_{\substack{x<p_3\leq 2x \\ x^{\frac{1}{2}}\log x<P(p_1+p_3)\leq 2x^{\frac{1}{2}}\log x}} 1$$

$$= \sum_{\substack{x<p_1\leq 2x}} \left(\sum_{\substack{x<p\leq 2x \\ x^{\frac{1}{2}}\log x<P(p+p_1)\leq 2x^{\frac{1}{2}}\log x}} 1 \right)^2. \qquad (4.2)$$

The inner sum can be transformed into

$$\sum_{x^{\frac{1}{2}}\log x<r\leq 2x^{\frac{1}{2}}\log x} \sum_{\substack{x<p\leq 2x \\ P(p+p_1)=r}} 1 = \sum_{x^{\frac{1}{2}}\log x<r\leq 2x^{\frac{1}{2}}\log x} \sum_{\substack{x<p\leq 2x \\ p\equiv -p_1 \,(\mathrm{mod}\, r)}} 1.$$

By the estimate (4.1), $N(x)$ is equal to the sum

$$\sum_{\substack{x<p_1\leq 2x}} \left(\sum_{x^{\frac{1}{2}}\log x<r\leq 2x^{\frac{1}{2}}\log x} \frac{1}{r} \sum_{x<p\leq 2x} 1 \right)^2 \qquad (4.3)$$

with a small error term. The lower bound of sum (4.3) is $\gg \frac{x^3}{\log^5 x}$ by the prime number theorem. Therefore we have

$$N(x) \gg \frac{x^3}{\log^5 x}.$$

By an elementary discussion, $w(p_1 p_2 p_3)$ can be confined to C_3 so that we have

$$|\{p_1 p_2 p_3 \in C_3 : \; x < p_i \leq 2x \,(i=1,2,3), \; x^{\frac{1}{2}}\log x < P(p_1+p_2)$$
$$\leq 2x^{\frac{1}{2}}\log x, \; x^{\frac{1}{2}}\log x < P(p_1+p_3) \leq 2x^{\frac{1}{2}}\log x, \; w(p_1 p_2 p_3) \in C_3\}| \quad (4.4)$$
$$\gg \frac{x^3}{\log^5 x}.$$

On the other hand, the number of triples (r_1, r_2, q) is $O(\frac{x^2}{\log x})$, where $x^{\frac{1}{2}}\log x < r_i \leq 2x^{\frac{1}{2}}\log x \,(i=1,2)$, $q \leq 4x$. Therefore in the set in (4.4), there are at least $c_4 \frac{x}{\log^4 x}$ different triples (p_1, p_2, p_3) satisfying $P(p_1+p_2) = r_1$, $P(p_1+p_3) = r_2$, $P(p_2+p_3) = q$ for some (r_1, r_2, q). In other words, there are at least $\frac{1}{3!} c_4 \frac{x}{\log^4 x}$ different numbers $n = p_1 p_2 p_3 \in C_3$ such that $w(n) = r_1 r_2 q$. Thus Theorem 3.5 holds true.

5. Acknowledgements

This paper is basically my lecture notes at the fifth Japan-China seminar on number theory in Osaka in August 2008. I would like to thank Professors Shigeru Kanemitsu and Jianya Liu for their kind invitation and excellent organization.

The results in Theorems 3.5, 3.6 and 3.7 were reported at the "combinatorial and analytic number theory" seminar in Nanjing Normal University of China in September 2007.

References

1. E. Bombieri, On the large sieve, *Mathematika* **12** (1965), 201-225.
2. Yong-Gao Chen and Ying Shi, Dynamics of the *w* function and the Green-Tao Theorem on arithmetic progressions in the primes, *Proc. Amer. Math. Soc.* **136** (2008), 2351-2357.
3. Yong-Gao Chen and Ying Shi, Distribution of primes and dynamics of the *w* function, *J. Number Theory* **128** (2008), 2085-2090.
4. Wushi Goldring, Dynamics of the *w* function and primes, *J. Number Theory* **119** (2006), 86-98.
5. Chaohua Jia, On the inverse problem relative to dynamics of the *w* function, *arXiv*: 0901.3834v1.

SOME DIOPHANTINE PROBLEMS
ARISING FROM THE ISOMORPHISM PROBLEM
OF GENERIC POLYNOMIALS

AKINARI HOSHI*

*Department of Mathematics, Rikkyo University,
3-34-1 Nishi-Ikebukuro Toshima-ku, Tokyo, 171-8501, Japan
E-mail: hoshi@rikkyo.ac.jp*

KATSUYA MIYAKE[†]

*Department of Mathematics,
School of Fundamental Science and Engineering, Waseda University
3-4-1 Ohkubo Shinjuku-ku, Tokyo, 169-8555, Japan
E-mail: miyakek@aoni.waseda.jp*

Let k be a field and G a finite group. A k-generic polynomial for G covers all G-Galois extensions L/K over k by specializing parameters to elements of the bottom field K of the G-Galois extension. We study the field isomorphism problem of k-generic polynomials for various G's to determine the 'moduli' of G-Galois extensions given by them over K, and announce some results of cubic, quartic and quintic cases. In cubic and quartic cases over infinite field, we see that the generic polynomials give the same splitting fields for infinitely many specialized values of the parameters. For cubic generic polynomials with the least possible number of parameters, furthermore, we show by Siegel's Theorem for curves of genus 0 that there are only finitely many integral values of the parameters which may give the same number fields. We will also show a correspondence between integral solutions of some parametric cubic Thue equations and isomorphism classes of Shanks' simplest cubic fields. Some of the results of this survey article will be published elsewhere.

Keywords: Generic polynomial, parametric Thue equations, field isomorphism problem, simplest cubic fields.

---

*Supported partially by Rikkyo University Special Fund for Research.
[†]Supported partially by Grant-in-Aid for Scientific Research (C) 19540057 of Japan Society for the Promotion of Science.

1. Introduction

Let G be a finite group, k a field of an arbitrary characteristic and $k(\mathbf{t})$ the rational function field over k with n indeterminates $\mathbf{t} = (t_1, \ldots, t_n)$. We call a Galois extension L/K a G-Galois extension if $\mathrm{Gal}(L/K)$ is isomorphic to G.

A k-generic polynomial for G supplies us all G-Galois extensions of K for every those fields K which contain k (cf. [46], [22]).

Definition 1.1 (Generic polynomial). *A monic separable polynomial* $f_{\mathbf{t}}^G(X) \in k(t_1, \ldots, t_n)[X]$ *is called k-generic for G if it has the following two properties:*

(G_i) *the Galois group of $f_{\mathbf{t}}^G(X)$ over $k(\mathbf{t})$ is isomorphic to G;*
(G_{ii}) *every G-Galois extension L/K, $K \supset k$, may be obtained as* $L = \mathrm{Spl}_K f_{\mathbf{a}}^G(X)$, *the splitting field of $f_{\mathbf{a}}^G(X)$ over K, for some* $\mathbf{a} = (a_1, \ldots, a_n) \in K^n$.

For example, Kummer theory shows us that if char k, the characteristic of k, dose not divide n and if k contains a primitive n-th root of unity, then $X^n - t$ is k-generic for the cyclic group C_n of order n. Similarly if char $k = p$ then $X^p - X - t$ is k-generic for C_p by Artin-Schreier theory.

A standard way to construct a k-generic polynomial for G is known as Noether's strategy (cf. [22]). Noether [41] asked whether the invariant field $k(t_1, \ldots, t_n)^G$ under the action of permutation group $G \leq S_n$ is rational (= purely transcendental) over k or not. Indeed if this problem has an affirmative answer then we may obtain a k-generic polynomial for G.

More generally Kemper-Mattig [24] showed that if the invariant field $k(V)^G$ of a faithful linear representation V of G is rational over k then we may get a k-generic polynomial for G in $m = \dim(V)$ parameters as a generating polynomial of the field extension $k(V)/k(V)^G$. Examples of k-generic polynomials for G are known for various pairs of k and G (for example, see [23], [24], [22], [44]).

Let $f_{\mathbf{t}}^G(X) \in k(\mathbf{t})[X]$ be a k-generic polynomial for G. Since a k-generic polynomial $f_{\mathbf{t}}^G(X)$ for G covers all G-Galois extensions over $K \supset k$ by specializing parameters to elements of K, it is natural to ask the following problem:

Problem 1.1 (Field isomorphism problem of a generic polynomial). *For a field $K \supset k$ and $\mathbf{a}, \mathbf{b} \in K^n$, determine whether $\mathrm{Spl}_K f_{\mathbf{a}}^G(X)$ and $\mathrm{Spl}_K f_{\mathbf{b}}^G(X)$ are isomorphic over K or not.*

In particular, it would be desirable to give an answer to the problem within the base field K by using the data $\mathbf{a}, \mathbf{b} \in K^n$.

The minimal number of parameters in k-generic polynomials for G is called the generic dimension for G over k, which is denoted by $\mathrm{gd}_k(G)$ (cf. [22]). In the case of $k = \mathbb{Q}$, it is known that $\mathrm{gd}_{\mathbb{Q}}(G) = 1$ if and only if $G = C_2$, C_3, S_3 via the essential dimension of finite groups given by Buhler-Reichstein [1] (see also [22], [32], [3]). For a transitive subgroup G of S_n ($n = 4, 5$) we also have $\mathrm{gd}_{\mathbb{Q}}(G) = 2$ (cf. [22, Chapter 8]).

To investigate the field isomorphism problem of $f_t^G(X)$, i.e. to determine all G-Galois extensions over $K \supset k$, we take $f_t^G(X)$ with exactly $\mathrm{gd}_k(G)$ parameters, because a large number of parameters in $f_t^G(X)$ give rise to many overlaps of splitting fields of $f_{\mathbf{a}}^G(X)$ over K by specializing parameters $\mathbf{t} \mapsto \mathbf{a} \in K^n$.

For example, we take $\mathbb{Q}$-generic polynomials $f_t^1(X) = X^3 + t_1 X + t_1$, $f_t^2(X) = X^3 + t_1 X + t_2$, $f_t^3(X) = X^3 + t_1 X^2 + t_2 X + t_3$ for the symmetric group S_3 of degree 3 with 1, 2, 3 parameters respectively and consider the overlap of the splitting fields $L_{\mathbf{a}}^i := \mathrm{Spl}_{\mathbb{Q}} f_{\mathbf{a}}^i(X)$ ($i = 1, 2, 3$) by specializing parameters $\mathbf{t} \mapsto \mathbf{a} \in \mathbb{Z}^i$. For $i = 2, 3$, we see that for a given $\mathbf{a} \in \mathbb{Z}^i$ there exist infinitely many $\mathbf{b} \in \mathbb{Z}^i$ such that $L_{\mathbf{a}}^i = L_{\mathbf{b}}^i$. However, for $f_t^1(X) = X^3 + t_1 X + t_1$, we obtained a result which claims that for a given $a \in \mathbb{Z}$ there exist only finitely many overlaps $L_a^1 = L_b^1$ with $b \in \mathbb{Z}$ (see Corollary 2.2).

Let S_n (resp. D_n, C_n) be the symmetric (resp. the dihedral, the cyclic) group of degree n. In this article, for a field k with char $k \neq 2$, we take k-generic polynomials

$$f_t^{S_3}(X) := X^3 + tX + t \in k(t)[X],$$

$$f_t^{C_3}(X) := X^3 - tX^2 - (t+3)X - 1 \in k(t)[X],$$

$$f_{s,t}^{D_4}(X) := X^4 + sX^2 + t \in k(s,t)[X],$$

$$f_{s,t}^{C_4}(X) := X^4 + sX^2 + s^2/(u^2+4) \in k(s,u)[X],$$

$$f_{s,t}^{D_5}(X) := X^5 + (t-3)X^4 + (s-t+3)X^3$$
$$+ (t^2 - t - 2s - 1)X^2 + sX + t \in k(s,t)[X],$$

for $G = S_3$, C_3, D_4, C_4, D_5, respectively. We note that $f_t^{S_3}(X)$, $f_t^{C_3}(X)$ and $f_{s,t}^{D_5}(X)$ are also k-generic even in the case of char $k = 2$.

Kemper showed in [25] that if k is infinite, the condition $(\mathrm{G_{ii}})$ of generic polynomials is equivalent to the following much stronger condition:

$(\mathrm{G_{iii}})$ for each subgroup $H \leq G$, every H-Galois extension L/K,

$K \supset k$, may be obtained as $L = \mathrm{Spl}_K f_{\mathbf{a}}^G(X)$, the splitting field of $f_{\mathbf{a}}^G(X)$ over K, for some $\mathbf{a} = (a_1, \ldots, a_n) \in K^n$.

DeMeyer [5] gave a definition of a generic polynomial in terms of the conditions $(\mathrm{G_i})$ and $(\mathrm{G_{iii}})$ instead of $(\mathrm{G_i})$ and $(\mathrm{G_{ii}})$ (see also [31]). Note that, when k is a finite field, the generic polynomial $f_{\mathbf{t}}^G(X)$ might not satisfy the condition $(\mathrm{G_{iii}})$ in general (see [6, Example 4]). For instance, $\mathbb{F}_2$-generic polynomial $f_t^{C_3}(X)$ does not satisfy $(\mathrm{G_{iii}})$ when $K = \mathbb{F}_2$ and $H = \{1\}$ because both of $f_0^{C_3}(X)$ and of $f_1^{C_3}(X)$ are irreducible over $\mathbb{F}_2$.

With regard to the condition $(\mathrm{G_{iii}})$, the following two problems naturally arise:

Problem 1.2 (Subfield problem of a generic polynomial). *For a field $K \supset k$ and $\mathbf{a}, \mathbf{b} \in K^n$, determine whether $\mathrm{Spl}_K f_{\mathbf{b}}^G(X)$ is a subfield of $\mathrm{Spl}_K f_{\mathbf{a}}^G(X)$ or not.*

Problem 1.3 (Field intersection problem of a generic polynomial). *For a field $K \supset k$ and $\mathbf{a}, \mathbf{b} \in K^n$, determine the intersection of $\mathrm{Spl}_K f_{\mathbf{a}}^G(X)$ and $\mathrm{Spl}_K f_{\mathbf{b}}^G(X)$.*

In this article we only treat the field isomorphism problem of generic polynomials; for the subfield problem and the field intersection problem see the papers [17], [18], [19], [20].

2. Some results: the cubic case

In [16], we gave an answer to the field isomorphism problem of $f_t^{S_3}(X) = X^3 + tX + t$ via formal Tschirnhausen transformation (see also [17], [18], [19], [20]). The discriminant of $f_t^{S_3}(X)$ being $-t^2(4t + 27)$, we assume that for $a \in K$, $f_a^{S_3}(X)$ is separable over K, i.e., $a \neq 0, -27/4$.

Theorem 2.1. *Assume that char $K \neq 3$. For $a, b \in K \setminus \{0, -27/4\}$ with $a \neq b$, the splitting field over K of $f_a^{S_3}(X) = X^3 + aX + a$ and that of $f_b^{S_3}(X) = X^3 + bX + b$ coincide if and only if there exists $u \in K$ such that*

$$b = \frac{a(u^2 + 9u - 3a)^3}{(u^3 - 2au^2 - 9au - 2a^2 - 27a)^2}.$$

We note that $b = 0$ (resp. $b = -27/4$) corresponds to $u^2 + 9u - 3a = 0$ (resp. $u^3 + 9au + 27a = 0$). Hence for a given $a \in K \setminus \{0, -27/4\}$ we have $\mathrm{Spl}_K f_b^{S_3}(X) = \mathrm{Spl}_K f_a^{S_3}(X)$ where b is given as in Theorem 2.1 except for finitely many $u \in K$ with $(u^3 - 2au^2 - 9au - 2a^2 - 2a)(u^2 + 9u - 3a)(u^3 + 9au + 27a) = 0$. Hence we see the following:

Corollary 2.1. *Let K be an infinite field of char $K \neq 3$. For a fixed $a \in K \setminus \{0, -27/4\}$, there exist infinitely many $b \in K$ such that $\mathrm{Spl}_K f_a^{S_3}(X) = \mathrm{Spl}_K f_b^{S_3}(X)$.*

The discriminant of $f_s^{C_3}(X) = X^3 - sX^2 - (s+3)X - 1$ is given by $(s^2 + 3s + 9)^2$; we assume that for $m \in K$, $f_m^{C_3}(X)$ is separable over K, i.e. $m^2 + 3m + 9 \neq 0$. For $f_s^{C_3}(X)$, we obtained in [17] the following theorem which is an analogue to the results of Morton [40] and Chapman [2].

Theorem 2.2. *Assume that char $K \neq 2$. For $m, n \in K$ with $(m^2 + 3m + 9)(n^2 + 3n + 9) \neq 0$, the splitting field over K of $f_m^{C_3}(X) = X^3 - mX^2 - (m+3)X - 1$ and that of $f_n^{C_3}(X) = X^3 - nX^2 - (n+3)X - 1$ coincide if and only if there exists $z \in K$ such that either*

$$n = \frac{m(z^3 - 3z - 1) - 9z(z+1)}{mz(z+1) + z^3 + 3z^2 - 1} \quad or$$

$$n = -\frac{m(z^3 + 3z^2 - 1) + 3(z^3 - 3z - 1)}{mz(z+1) + z^3 + 3z^2 - 1}.$$

By a similar argument to the one for $f_t^{S_3}(X)$ given above, we obtain the following corollary:

Corollary 2.2. *Let K be an infinite field of char $K \neq 2$. For a fixed $m \in K$ with $m^2 + 3m + 9 \neq 0$, there exist infinitely many $n \in K$ such that $\mathrm{Spl}_K f_m^{C_3}(X) = \mathrm{Spl}_K f_n^{C_3}(X)$.*

The following theorem is well-known as Siegel's theorem for curves of genus 0 (cf. [28, Theorem 6.1], [29, Chapter 8, Section 5], [14, Theorem D.8.4]).

Theorem 2.3 (Siegel). *Let K be a number field and $\mathcal{O}_K$ the ring of integers in K. If a rational function $\varphi(t) \in K(t)$ has at least three distinct poles, then there are only finitely many $u \in K$ such that $\varphi(u) \in \mathcal{O}_K$.*

In contrast with Corollaries 2.1 and 2.2, by applying Siegel's theorem to Theorems 2.1 and 2.2 we get the followings:

Corollary 2.3. *Let K be a number field and $\mathcal{O}_K$ the ring of integers in K. For $f_a^{S_3}(X) = X^3 + aX + a$ with a given non-zero integer $a \in \mathcal{O}_K$, there exist only finitely many integers $b \in \mathcal{O}_K$ such that $\mathrm{Spl}_K f_a^{S_3}(X) = \mathrm{Spl}_K f_b^{S_3}(X)$.*

Corollary 2.4. *Let K be a number field and $\mathcal{O}_K$ the ring of integers in K. For $f_m^{C_3}(X) = X^3 - mX^2 - (m+3)X - 1$ with a given integer $m \in \mathcal{O}_K$ $(m^2 + 3m + 9 \neq 0)$, there exist only finitely many integers $n \in \mathcal{O}_K$ such that $\mathrm{Spl}_K f_m^{C_3}(X) = \mathrm{Spl}_K f_n^{C_3}(X)$.*

In the next section, we investigate this phenomenon in detail for $f_t^{C_3}(X)$ and $K = \mathbb{Q}$.

3. A parametric family of Thue equations

We consider the following parametric family of Thue equations

$$F_m(X,Y) := X^3 - mX^2Y - (m+3)XY^2 - Y^3 = j \qquad (3.1)$$

for $m \in \mathbb{Z}$ and $j \in \mathbb{Z} \setminus \{0\}$. We first note that

$$F_{-m-3}(X,Y) = F_m(-Y,-X), \qquad (3.2)$$

$$-F_m(X,Y) = F_m(-X,-Y). \qquad (3.3)$$

From (3.2) and (3.3), we may suppose that $-1 \le m$ and $0 < j$ without loss of generality. In the case of $j = \pm 1$, the Thue equations (3.1) is solved completely. Thomas [47] showed that if $m + 1 \ge 1.365 \times 10^7$ then the equation $F_m(X,Y) = 1$ has only the trivial solutions $(0,-1),(-1,1),(-1,0)$. Moreover he also showed that for $0 \le m + 1 \le 10^3$ non-trivial solutions exist only for $m = -1, 0, 2$. Mignotte [38] solved completely the equations $F_m(X,Y) = 1$. He proved that for $-1 \le m$ non-trivial solutions occur only for $m = -1, 0, 2$.

For $-1 \le m \in \mathbb{Z}$ and $j = 1$, all integer solutions of the equations (3.1) are given by trivial solutions $(x,y) = (1,0), (0,-1), (-1,1)$ for an arbitrary m and additionally

$$\begin{array}{ll} (x,y) = (-1,-1),(-1,2),(2,-1) & \text{for} \quad m = -1, \\ (x,y) = (5,4),(4,-9),(-9,5) & \text{for} \quad m = -1, \\ (x,y) = (2,1),(1,-3),(-3,2) & \text{for} \quad m = 0, \\ (x,y) = (-7,-2),(-2,9),(9,-7) & \text{for} \quad m = 2 \end{array}$$

(cf. also the textbook [9, p. 54]). Note that if (x,y) is an integer solution of (3.1) then $(y, -x-y)$ and $(-x-y, x)$ are also integer solutions of (3.1) because $F_m(X,Y)$ is invariant under the action of $C_3 = \langle \sigma \rangle$ where $\sigma :$ $X \longmapsto Y \longmapsto -X - Y$.

Mignotte-Pethö-Lemmermeyer [39] studied Thue equation $F_m(X,Y) = j$ for general $j \in \mathbb{Z}$ and gave a complete solution to Thue inequality $|F_m(X,Y)| \le 2m+3$ with a result of [35]. For $-1 \le m$ and $1 < j \le 2m+3$, all solutions to $F_m(X,Y) = j$ are given by trivial solutions for $j = c^3$ and

$$(x,y) \in \{(-1,-1),(-1,2),(2,-1),$$
$$(-m-1,-1),(-1,m+2),(m+2,-m-1)\}$$

for $j = 2m+3$, except for $m = 1$ in which case (3.1) has the extra solutions:

$$(x,y) \in \{(3,1),(1,-4),(-4,3),(8,3),(3,-11),(-11,8)\}$$

for $j = 5$ ‡. The Thue inequality $|F_m(X,Y)| \le j(m)$ for $j : \mathbb{Z} \to \mathbb{N}$ is also investigated by Lettl-Pethö-Voutier [36] and Xia-Chen-Zhang [49] under some conditions.

Take the k-generic polynomial $f_t^{C_3}(X) = X^3 - tX^2 - (t+3)X - 1 \in k(t)[X]$ for C_3. We consider the case of $k = K = \mathbb{Q}$ and $m \in \mathbb{Z}$. Then we have $f_m^{C_3}(X) = F_m(X,1)$. For $m \in \mathbb{Z}$, we call the splitting fields

$$L_m := \mathrm{Spl}_{\mathbb{Q}} f_m^{C_3}(X)$$

the simplest cubic fields (cf. [45]). Note that $F_m(X,1) = F_{-m-3}(-1,-X)$. Thus if z is a root of $f_m^{C_3}(X)$ then $1/z$ becomes a root of $f_{-m-3}^{C_3}(X)$, and hence $L_m = L_{-m-3}$ for $m \in \mathbb{Z}$.

By using Theorem 2.2 we obtained in [15] the following theorem which asserts a correspondence between some integral solutions of Thue equations (3.1) and isomorphism classes of the simplest cubic fields.

Theorem 3.1. *Let m be an integer. There exists an integer $n \in \mathbb{Z} \setminus \{m, -m-3\}$ such that $L_m = L_n$ if and only if there exists a solution $(x,y) \in \mathbb{Z}^2$ with $xy(x+y) \ne 0$ to the cubic Thue equation*

$$F_m(x,y) = j \quad \text{for some} \quad j > 0 \quad \text{with} \quad j \mid m^2 + 3m + 9.$$

Moreover integers m, n and solutions $(x,y) \in \mathbb{Z}^2$ above satisfy either

$$\text{(I)} \qquad n = m + \frac{(m^2 + 3m + 9)xy(x+y)}{F_m(x,y)} \qquad \text{or}$$

$$\text{(II)} \qquad -n - 3 = m + \frac{(m^2 + 3m + 9)xy(x+y)}{F_m(x,y)}.$$

Note that the discriminant of $F_m(X,Y)$ equals $(m^2 + 3m + 9)^2$.

Ennola [8] verified that for integers $-1 \le m < n \le 10^4$ the overlap $L_m = L_n$ occurs only in the case of $L_{-1} = L_5 = L_{12} = L_{1259}$, $L_0 = L_3 = L_{54}$, $L_1 = L_{66}$ and $L_2 = L_{2389}$. In [17], we checked that this claim is also valid for $-1 \le m < n \le 10^5$ by using Theorem 2.2.

Okazaki [42] studied Thue equation $f(X,Y) = 1$ for an irreducible cubic form $f(X,Y)$ with positive discriminant and established a very strong result on gaps between solutions. For example, applying Theorem 1.3 in [42] to $F_m(X,Y)$, we see that if $m + 1 \ge 4 \times 10^4$ then

---

‡In [39, Theorem 3], there is a misprint: it should be added $(-11,8)$.

$F_m(X, Y) = 1$ has only three trivial solutions (instead of Thomas' estimate $m + 1 \geq 1.365 \times 10^7$ as we mentioned above). Using the gap principle given in [42], Okazaki showed the following theorem. Although the result seems to be unpublished yet (up to now), a brief sketch of the proof is available at http://www1.doshisha.ac.jp/~rokazaki/papers.html as a presentation sheet (cf. also Wakabayashi's paper [48]).

Theorem 3.2 (Okazaki). *For integers $m, n \geq -1$, if $L_m = L_n$ $(m \neq n)$ then $m, n \in \{-1, 0, 1, 2, 3, 5, 12, 54, 66, 1259, 2389\}$. In particular,*

$$L_{-1} = L_5 = L_{12} = L_{1259}, \quad L_0 = L_3 = L_{54}, \quad L_1 = L_{66}, \quad L_2 = L_{2389}.$$

As a consequence of Theorem 3.1 and Theorem 3.2 we obtain the followings:

Theorem 3.3. *For $m \geq -1$, all integer solutions $(x, y) \in \mathbb{Z}^2$ with $xy(x + y) \neq 0$ of Thue equations $F_m(x, y) = j$ with $j \in \mathbb{N}$ and $j \mid m^2 + 3m + 9$ are given on Table 1.*

Table 1

m	n	Type	j	$m^2 + 3m + 9$	(x, y)
-1	12	(II)	1	7	$(-1, -1), (-1, 2), (2, -1)$
-1	1259	(I)	1	7	$(5, 4), (4, -9), (-9, 5)$
-1	5	(I)	7	7	$(2, 1), (1, -3), (-3, 2)$
0	54	(I)	1	3^2	$(2, 1), (1, -3), (-3, 2)$
0	3	(II)	3	3^2	$(-1, -1), (-1, 2), (2, -1)$
1	66	(II)	13	13	$(-5, -2), (-2, 7), (7, -5)$
2	2389	(II)	1	19	$(-7, -2), (-2, 9), (9, -7)$
3	0	(II)	3^2	3^3	$(-1, -1), (-1, 2), (2, -1)$
3	54	(II)	3^2	3^3	$(-4, -1), (-1, 5), (5, -4)$
5	-1	(I)	7^2	7^2	$(3, -1), (-1, -2), (-2, 3)$
5	12	(II)	7^2	7^2	$(-4, -1), (-1, 5), (5, -4)$
5	1259	(I)	7^2	7^2	$(19, 3), (3, -22), (-22, 19)$
12	-1	(II)	3^3	$3^3 \cdot 7$	$(-1, -1), (-1, 2), (2, -1)$
12	1259	(II)	3^3	$3^3 \cdot 7$	$(-13, -1), (-1, 14), (14, -13)$
12	5	(II)	$3^3 \cdot 7$	$3^3 \cdot 7$	$(-4, -1), (-1, 5), (5, -4)$
54	0	(I)	7^3	$3^2 \cdot 7^3$	$(3, -1), (-1, -2), (-2, 3)$
54	3	(II)	$3 \cdot 7^3$	$3^2 \cdot 7^3$	$(-4, -1), (-1, 5), (5, -4)$
66	1	(II)	$3^3 \cdot 13^2$	$3^3 \cdot 13^2$	$(-5, -2), (-2, 7), (7, -5)$
1259	-1	(I)	61^3	$7 \cdot 61^3$	$(9, -4), (-4, -5), (-5, 9)$
1259	12	(II)	61^3	$7 \cdot 61^3$	$(-13, -1), (-1, 14), (14, -13)$
1259	5	(I)	$7 \cdot 61^3$	$7 \cdot 61^3$	$(22, -3), (-3, -19), (-19, 22)$
2389	2	(II)	67^3	$19 \cdot 67^3$	$(-7, -2), (-2, 9), (9, -7)$

Corollary 3.1. *For $m \geq -1$, integer solutions $(x, y) \in \mathbb{Z}^2$ with $xy(x + y) \neq 0$ of Thue equations $F_m(x, y) = m^2 + 3m + 9$ exist only for $m = -1, 1, 5, 12, 66, 1259$ as on Table 1.*

We provide more detailed data of Table 1 in the last section (Appendix) of this paper for the reader's convenient.

4. The case of D_4

For a field k with char $k \neq 2$, we take a k-generic polynomial $f_{s,t}^{D_4}(X) = X^4 + sX^2 + t \in k(s,t)[X]$ for D_4 with discriminant $16t(s^2 - 4t)^2$. We assume that for $\mathbf{a} = (a,b) \in K^2$, $f_{a,b}^{D_4}(X)$ is separable over K, i.e., $b(a^2 - 4b) \neq 0$.

In this case, for an infinite field $K \supset k$ and a fixed $\mathbf{a} = (a,b) \in K^2$, there exist infinitely many $\mathbf{a}' = (a',b') \in K^2$ such that $\mathrm{Spl}_K f_{\mathbf{a}}^{D_4}(X) = \mathrm{Spl}_K f_{\mathbf{a}'}^{D_4}(X)$ because $f_{a,b}^{D_4}(X)$ and $f_{a',b'}^{D_4}(X) = f_{ac^2,bc^4}^{D_4}(X) = f_{a,b}^{D_4}(X/c) \cdot c^4$ have the same splitting field over K for arbitrary $c \in K$. Hence we consider the non-trivial case of $a^2 b' - a'^2 b \neq 0$ or $b'/b \neq c^4$ for any $c \in K$.

In [18], we gave the following theorem which is a generalization of the result of van der Ploeg [43]:

Theorem 4.1. *For $\mathbf{a} = (a,b)$, $\mathbf{a}' = (a',b') \in K^2$ with $bb'(a^2 - 4b)(a'^2 - 4b') \neq 0$ we assume that $C_4 \leq \mathrm{Gal}(f_{\mathbf{a}}^{D_4}/K)$ and $C_4 \leq \mathrm{Gal}(f_{\mathbf{a}'}^{D_4}/K)$. We also assume that $a^2 b' - a'^2 b \neq 0$ or b'/b is not a fourth power in K. Then the quotient fields $K[X]/(f_{\mathbf{a}}^{D_4}(X))$ and $K[X]/(f_{\mathbf{a}'}^{D_4}(X))$ are K-isomorphic if and only if there exist $u, w \in K$ such that*

$$a' = (a^3 - 3ab - 2a^2 u + 4bu + au^2)w^2, \quad b' = b(b - au + u^2)^2 w^4.$$

Suppose that K is a Hilbertian field; that is to say, suppose that Hilbert's irreducibility theorem holds for K. Then, as a consequence of Theorem 4.1, we get

Corollary 4.1. *Let $K \supset k$ be a Hilbertian field. For $\mathbf{a} = (a,b) \in K^2$ with $b(a^2 - 4b) \neq 0$ and $C_4 \leq \mathrm{Gal}(f_{\mathbf{a}}^{D_4}/K)$, there exist infinitely many $\mathbf{a}' = (a',b') \in K^2$ which satisfy the condition that b'/b is not a fourth power in K and that $\mathrm{Spl}_K f_{\mathbf{a}}^{D_4}(X) = \mathrm{Spl}_K f_{\mathbf{a}'}^{D_4}(X)$.*

Proof. Applying Theorem 4.1 to $w = 1$, it follows that for arbitrary $u \in K$ $\mathrm{Spl}_K f_{a,b}^{D_4}(X) = \mathrm{Spl}_K f_{a',b'}^{D_4}(X)$ where $a' = (a^3 - 3ab - 2a^2 u + 4bu + au^2)$, $b' = b(b - au + u^2)^2$. By Hilbert's irreducibility theorem, there exist infinitely many $u \in K$ such that $X^2 - (b - au + u^2) = X^2 - (u^2 - a/2)^2 + (a^2 - 4b)/4$ is irreducible over K, since $a^2 - 4b \neq 0$. For such infinitely many $u \in K$, we have $b'/b = (b - au + u^2)^2$ which is not a fourth power in K. $\qquad\square$

It follows from the assumption $C_4 \leq \mathrm{Gal}(f_{\mathbf{a}}^{D_4}/K)$ that b is not a square in K, i.e., $\mathrm{Gal}(f_{\mathbf{a}}^{D_4}/K) \not\leq A_4$. Applying Theorem 4.1 to $(u,w) = (0,1)$, we have $\mathrm{Spl}_K f_{a,b}^{D_4}(X) = \mathrm{Spl}_K f_{a',b'}^{D_4}(X)$ for $(a',b') = (a(a^2 - 3b), b^3)$ with $b'/b = b^2$ which is not a fourth power in K. By repeating this process, we see that for a given $(a,b) \in \mathbb{Z}^2$ with $b(a^2 - 4b) \neq 0$ there exist infinitely many $(a',b') \in \mathbb{Z}^2$ such that b'/b is not a fourth power in $\mathbb{Z}$ and that $\mathrm{Spl}_{\mathbb{Q}} f_{a,b}^{D_4}(X) = \mathrm{Spl}_{\mathbb{Q}} f_{a',b'}^{D_4}(X)$; this shows a fact in contrast with Corollaries 2.3 and 2.4 although $f_{s,t}^{D_4}(X)$ have the least possible number of parameters over $\mathbb{Q}$.

5. Numerical examples: the case of C_4

We take the k-generic polynomial $f_{s,u}^{C_4}(X)$ for C_4:

$$f_{s,u}^{C_4}(X) = X^4 + sX^2 + \frac{s^2}{u^2 + 4} \in k(s,u)[X]$$

with discriminant $16s^6u^4/(u^2 + 4)^3$. We assume that for $\mathbf{a} = (a,c) \in K^4$, $f_{a,c}^{C_4}(X)$ is well-defined and separable over K, i.e., $ac(c^2 + 4) \neq 0$.

In [18], we gave the following theorem:

Theorem 5.1. For $\mathbf{a} = (a,c)$, $\mathbf{a}' = (a',c') \in K^2$ with $aa'cc'(c^2 + 4)(c'^2 + 4) \neq 0$, we assume that $c \neq \pm c'$ and $c \neq \pm 4/c'$. Then the splitting field of $f_{a,c}^{C_4}(X)$ over K and that of $f_{a',c'}^{C_4}(X)$ coincide if and only if either $f_{A,C^+}^{C_4}(X)$ or $f_{A,C^-}^{C_4}(X)$ has a linear factor over K where $A = -aa'$ and $C^{\pm} = (cc' \mp 4)/(c \pm c')$.

Example 5.1. For $f_{a,c}^{C_4}(X) = X^4 + aX^2 + a^2/(c^2 + 4)$, we first note that

$$\mathrm{Spl}_K f_{a,c}^{C_4}(X) = \mathrm{Spl}_K f_{a,-c}^{C_4}(X) \quad \text{and} \quad \mathrm{Spl}_K f_{a,c}^{C_4}(X) = \mathrm{Spl}_K f_{ae^2,c}^{C_4}(X)$$

for $a,c,e \in K$. By Theorem 5.1, we have

$$\mathrm{Spl}_K f_{a,c}^{C_4}(X) = \mathrm{Spl}_K f_{(c^2+4)/a,c}^{C_4}(X) \tag{5.4}$$

since $f_{A,C^+}^{C_4}(X) = (X-2)(X+2)(X-c)(X+c)$ for $(a',c') = ((c^2+4)/a, c)$. Although Theorem 5.1 is not applicable to the case of $c' = 4/c$, it follows from $\mathrm{Spl}_K(X^4 + aX^2 + b) = \mathrm{Spl}_K(X^4 + 2aX^2 + a^2 - 4b)$ that

$$\mathrm{Spl}_K f_{a,c}^{C_4}(X) = \mathrm{Spl}_K f_{2a,4/c}^{C_4}(X) \tag{5.5}$$

(cf. [18]). For example, by (5.4) and (5.5), the polynomials $f_{a,c}^{C_4}(X)$ and $f_{2(c^2+4)/a,4/c}^{C_4}(X)$ have the same splitting field over K.

Example 5.2. We take $K = \mathbb{Q}$ and the simplest quartic polynomial

$$h_n(X) = X^4 - nX^3 - 6X^2 + nX + 1 \in \mathbb{Q}[X], \quad (n \in \mathbb{Z})$$

with discriminant $4(n^2+16)^3$ whose Galois group over $\mathbb{Q}$ is isomorphic to C_4 except for $n = 0, \pm3$ (cf. for example, [11], [12], [30], [35], [27], [13], [7], [37], and the references therein). We see that $h_n(X)$ and

$$H_n(X) := f^{C_4}_{-(n^2+16),n/2}(X) = X^4 - (n^2 + 16)X^2 + 4(n^2 + 16)$$

have the same splitting field over $\mathbb{Q}$ (cf. [18]). For $n \in \mathbb{Z}$, we may assume $1 \leq n$ because $\mathrm{Spl}_{\mathbb{Q}} H_n(X) = \mathrm{Spl}_{\mathbb{Q}} H_{-n}(X)$ and $H_n(X)$ splits over $\mathbb{Q}$ only for $n = 0, \pm3$. For $1 \leq m < n$, we apply Theorem 5.1 to $H_m(X)$ and $H_n(X)$ as $(a, c, a', c') = (-(m^2 + 16), m/2, -(n^2 + 16), n/2)$, then we see for $(m, n) = (2, 22), (1, 103), (4, 956)$,

$$f^{C_4}_{A,C^+}(X) = (X - 60)(X + 60)(X - 80)(X + 80),$$
$$f^{C_4}_{A,C^-}(X) = (X - 255)(X + 255)(X - 340)(X + 340),$$
$$f^{C_4}_{A,C^+}(X) = (X - 2080)(X + 2080)(X - 4992)(X + 4992)$$

respectively. Hence we get

$$\mathrm{Spl}_K h_m(X) = \mathrm{Spl}_K h_n(X) \quad \text{for} \quad (m, n) \in \{(1, 103), (2, 22), (4, 956)\}.$$

We may not apply Theorem 5.1 to two cases $(m, n) = (1, 16), (2, 8)$. In these cases, we may use the transformation (5.5) in the previous example.

We checked by Theorem 5.1 that for integers m, n in the range $1 \leq m < n \leq 10^5$, $f^{C_4}_{A,C^\pm}(X)$ has a linear factor over $\mathbb{Q}$, i.e. $\mathrm{Spl}_K h_m(X) = \mathrm{Spl}_K h_n(X)$, only for the values of $(m, n) = (1, 103), (2, 22), (4, 956)$.

6. Numerical examples: the case of D_5

We take the Brumer's quintic generic polynomial for D_5:

$$f^{D_5}_{\mathbf{s}}(X) = X^5 + (t - 3)X^4 + (s - t + 3)X^3$$
$$+ (t^2 - t - 2s - 1)X^2 + sX + t \in k(\mathbf{s})[X]$$

with discriminant $t^2 \cdot \delta^2_{\mathbf{s}}$ where $\mathbf{s} = (s, t)$ and

$$\delta_{\mathbf{s}} := s^2 - 4s^3 + 4t - 14st - 30s^2t - 91t^2 - 34st^2 + s^2t^2 + 40t^3 + 24st^3 + 4t^4 - 4t^5.$$

We assume that for $\mathbf{a} = (a, b) \in K^2$, $f^{D_5}_{\mathbf{a}}(X)$ is separable over K, i.e., $b \cdot \delta_{\mathbf{a}} \neq 0$. We first see the following lemma (cf. [19]).

Lemma 6.1. *For* $\mathbf{a} = (a, b) \in K^2$ *with* $b \cdot \delta_{\mathbf{a}} \neq 0$, *we have* $\mathrm{Spl}_K f^{D_5}_{\mathbf{a}}(X) = \mathrm{Spl}_K f^{D_5}_{(a+5b)/b^2, -1/b}(X)$.

In the case of char $k \neq 2$, we take

$$F^1_{\mathbf{s},\mathbf{s}'}(X) := \left(X^5 - (t-3)(t'-3)X^4 + c_3 X^3 + \frac{c_2}{2}X^2 + \frac{c_1}{2}X + \frac{c_0}{2}\right)^2$$
$$- \frac{\delta_{\mathbf{s}}\,\delta_{\mathbf{s}'}}{4}\left(X^2 + (t+t'-1)X + (s-t+s'-t'+tt'+2)\right)^2,$$

$$F^2_{\mathbf{s},\mathbf{s}'}(X) := F^1_{(s+5t)/t^2,\,-1/t,\,s',\,t'}(X)$$

where $\mathbf{s} = (s,t)$, $\mathbf{s}' = (s',t')$ and

$$c_3 = \left[2s - 21t + 3t^2 - t(2-t)s' - t^2 t'\right] + 31 - 3ss' + 5tt',$$

$$c_2 = \left[2(-10s + 56t + 4st - 16t^2 + t^3) + t(5 - 13s - 12t + 4t^2)s' - t(15s\right.$$
$$\left. - 14t - 2t^2)t' + 8t^2 s't' - 2t^3 t'^2\right] - 102 + 27ss' - 119tt' - sts't' + 6t^2 t'^2,$$

$$c_1 = \left[2(16s + s^2 - 64t - 13st + 30t^2 + 2st^2 - 4t^3) + (-6s^2 - 7t + 38st\right.$$
$$+ 9t^2 - 5st^2 - 12t^3 + 2t^4)s' + t(2s - 77t + 3st + 8t^2)t' + 2t(-10 - 4s$$
$$\left. + 3t + t^2)s'^2 + t^2(-29 + s + 18t)s't' - 2(s - 5t)t^2 t'^2\right] + 80 - 37ss'$$
$$+ 145tt' - 45sts't' + 24t^2 t'^2 - 8t^3 t'^3,$$

$$c_0 = \left[2(-8s - s^2 + 28t + 12st + s^2 t - 19t^2 - 4st^2 + 4t^3) + (5s^2 - 2t\right.$$
$$- 38st - 7s^2 t + 5t^2 + 13st^2 + 8t^3 + 2st^3 - 4t^4)s' - t(104s + 33s^2$$
$$- 105t - 35st - 4t^2 - 16st^2 + 6t^3 + 2t^4)t' + t(-21 - 11s - 2t + 2st$$
$$+ 4t^2)s'^2 - t(s^2 - 36t + 14st + 6t^2 - 6t^3)s't' - t^2(37s - 22t + 2st$$
$$\left. - 8t^2)t'^2 + 8t^3 s't'^2 + 8t^2 s'^2 t' - 2t^4 t'^3\right] - 24 + 14ss' - 8s^2 s'^2 - 224tt'$$
$$+ sts't' - 101t^2 t'^2 - st^2 s't'^2 - 8t^3 t'^3$$

with $[u] := u + \iota(u)$, $(u \in k(s,t,s',t'))$, for $\iota : s \leftrightarrow s'$, $t \leftrightarrow t'$.

In [19], we gave the following result (for the case of char $K = 2$, see [19]):

Theorem 6.1. *Let $K \supset k$ be a field of char $K \neq 2$. For $\mathbf{a} = (a,b)$, $\mathbf{a}' = (a',b') \in K^2$ with $aa'\delta_{\mathbf{a}}\delta_{\mathbf{a}'} \neq 0$, we assume that $F^1_{\mathbf{a},\mathbf{a}'}(X)F^2_{\mathbf{a},\mathbf{a}'}(X)$ has no repeated factors. Then the splitting field of $f^{D_5}_{\mathbf{a}}(X)$ over K and that of $f^{D_5}_{\mathbf{a}'}(X)$ coincide if and only if either $F^1_{\mathbf{a},\mathbf{a}'}(X)$ or $F^2_{\mathbf{a},\mathbf{a}'}(X)$ has a root in K.*

Remark 6.1. Note that we may assume that $F^1_{\mathbf{a},\mathbf{a}'}(X)F^2_{\mathbf{a},\mathbf{a}'}(X)$ has no repeated factors after suitable Tschirnhausen transformation of $f^{D_5}_{\mathbf{a}'}(X)$ (cf. [10], [4], [19], [21]). For example, we may apply Lemma 6.1 for $\mathbf{a} = (0,1)$, $\mathbf{a}' = (4,-1)$ as follows: the polynomial $F^1_{0,1,4,-1}(X)$ splits into irreducible factors over $\mathbb{Q}$ as

$$F_{0,1,4,-1}^1(X) = (X-1)^2(X-3)(X^2 - 3X + 14)$$
$$\cdot (X^5 - 8X^4 + 36X^3 - 47X^2 + 60X + 193).$$

It follows from Lemma 6.1 that $\mathrm{Spl}_{\mathbb{Q}} f_{4,-1}^{D_5}(X) = \mathrm{Spl}_{\mathbb{Q}} f_{-1,1}^{D_5}(X)$, and we have the $\mathbb{Q}$-irreducible decomposition of $F_{0,1,-1,1}^2(X)$ as

$$F_{0,1,-1,1}^2(X) = X(X^2 - 3X + 14)(X^2 - 5X + 18)$$
$$\cdot (X^5 - 8X^4 + 47X^3 - 171X^2 + 299X - 235).$$

Thus we see $\mathrm{Spl}_{\mathbb{Q}} f_{0,1}^{D_5}(X) = \mathrm{Spl}_{\mathbb{Q}} f_{4,-1}^{D_5}(X)$ $(= \mathrm{Spl}_{\mathbb{Q}} f_{-1,1}^{D_5}(X))$. This conclusion may directly be obtained also from the $\mathbb{Q}$-decomposition type of $F_{0,1,4,-1}^1(X)$ (see [19]).

By Theorem 6.1, we give some numerical examples of the field isomorphism problem of $f_{s,t}^{D_5}(X)$ over $K = \mathbb{Q}$ and for integral points $\mathbf{a}, \mathbf{a}' \in \mathbb{Z}^2$. We do not know, however, for a given $\mathbf{a} \in \mathbb{Z}^2$ whether there exist only finitely many $\mathbf{a}' \in \mathbb{Z}^2$ such that $\mathrm{Spl}_{\mathbb{Q}} f_{\mathbf{a}}^{D_5}(X) = \mathrm{Spl}_{\mathbb{Q}} f_{\mathbf{a}'}^{D_5}(X)$ or not.

As in Remark 6.1, if $F_{\mathbf{a}, \mathbf{a}'}^i(X)$ has a repeated factor then we retake another (modified) $F_{\mathbf{a}, \mathbf{a}''}^i(X)$ which is squarefree and arises from some Tschirnhausen transformation of $f_{\mathbf{a}'}^{D_5}(X)$.

Example 6.1. Take $K = \mathbb{Q}$ and $t := 1$. Then we have $f_{s,1}^{D_5}(X) = X^5 - 2X^4 + (s+2)X^3 - (2s+1)X^2 + sX + 1$. For $a, a' \in \mathbb{Z}$ in the range $-10^4 \leq a < a' \leq 10^4$, $\mathrm{Spl}_{\mathbb{Q}} f_{a,1}^{D_5}(X) = \mathrm{Spl}_{\mathbb{Q}} f_{a',1}^{D_5}(X)$ if and only if $(a, a') \in X_1 \cup X_2$ where

$$X_1 = \{(-6,0), (-1,41), (-94,-10)\},$$
$$X_2 = \{(-1,0), (-6,-1), (-18,-7), (1,34), (0,41), (-6,41), (-167,-8)\}.$$

It was directly checked by Theorem 6.1 that, for $a, a' \in \mathbb{Z}$ in the range $-10^4 \leq a < a' \leq 10^4$, $(a, a') \in X_i$ if and only if (modified) $F_{a,1,a',1}^i(X)$ has a root in $\mathbb{Q}$ for each of $i = 1, 2$.

Example 6.2. Kida-Renault-Yokoyama [26] showed that there exist infinitely many $b \in \mathbb{Q}$ such that $\mathrm{Spl}_{\mathbb{Q}} f_{0,1}^{D_5}(X) = \mathrm{Spl}_{\mathbb{Q}} f_{b,1}^{D_5}(X)$. Their method enables us to construct such b's explicitly via rational points of an associated elliptic curve (cf. [26]).

They also pointed out that in the range $-400 \leq a, b \leq 400$ there are 25 pairs $(a, b) \in \mathbb{Z}^2$ such that $\mathrm{Spl}_{\mathbb{Q}} f_{0,1}^{D_5}(X) = \mathrm{Spl}_{\mathbb{Q}} f_{a,b}^{D_5}(X)$. We may classify the 25 pairs by the polynomials $F_{0,1,a,b}^1(X)$ and $F_{0,1,a,b}^2(X)$. In the range above, for $i = 1, 2$, (modified) $F_{0,1,a,b}^i(X)$ has a root in $\mathbb{Q}$ if and only if $(a, b) \in X_i$ where

$$X_1 = \{(0, 1), (4, -1), (4, 5), (-6, 1), (-24, 19), (34, 11), (36, -5),$$
$$(46, -1), (-188, 23), (264, 31), (372, -5), (378, 43)\},$$
$$X_2 = \{(-1, -1), (-1, 1), (5, -1), (41, 1), (-43, 5), (47, 13), (59, -5),$$
$$(59, 19), (101, 19), (125, -23), (149, 11), (155, 25), (-169, 55)\}.$$

By Theorem 6.1, we see that if $F_{0,1,a,b}^1(X)$ (resp. $F_{0,1,a,b}^2(X)$), $(a, b) \in \mathbb{Z}^2$, has a root in $\mathbb{Q}$ then $(a, b) = (\text{even}, \text{odd})$ (resp. $(a, b) = (\text{odd}, \text{odd})$) because $F_{0,1,a,b}^1(X) \in \mathbb{Z}[X]$ splits into irreducible factors over the field $\mathbb{F}_2$ of two elements as

$$F_{0,1,0,0}^1(X) = (X^5 + X^3 + 1)^2,$$
$$F_{0,1,0,1}^1(X) = X(X + 1)^4(X^5 + X^2 + 1),$$
$$F_{0,1,1,0}^1(X) = X^{10} + X^7 + X^4 + X^3 + 1,$$
$$F_{0,1,1,1}^1(X) = (X^5 + X^3 + 1)(X^5 + X^3 + X^2 + X + 1)$$

and $F_{0,1,a,b}^2(X) \in \mathbb{Z}[X]$ also splits into irreducible factors over $\mathbb{F}_2$ as

$$F_{0,1,0,0}^2(X) = (X^5 + X^3 + 1)^2,$$
$$F_{0,1,0,1}^2(X) = (X^5 + X^3 + 1)(X^5 + X^3 + X^2 + X + 1),$$
$$F_{0,1,1,0}^2(X) = X^{10} + X^7 + X^6 + X^4 + X^2 + X + 1,$$
$$F_{0,1,1,1}^2(X) = X^3(X + 1)^2(X^5 + X^3 + X^2 + X + 1).$$

By Theorem 6.1, we have checked pairs $(a, b) \in \mathbb{Z}^2$ in the range $-2 \times 10^4 \le a, b \le 2 \times 10^4$ and added just $\{(526, 41), (952, 113), (2302, 95), (6466, 311), (7180, 143), (7480, -169)\}$ to X_1, and $\{(785, -25), (3881, 29), (-11215, 299), (19739, -281)\}$ to X_2.

Example 6.3. Let $h_n(X)$ be Lehmer's simplest quintic polynomial Tschirnhausen transformation

$$h_n(X) = X^5 + n^2 X^4 - (2n^3 + 6n^2 + 10n + 10)X^3$$
$$+ (n^4 + 5n^3 + 11n^2 + 15n + 5)X^2 + (n^3 + 4n^2 + 10n + 10)X + 1$$

(cf. [33]), and take $K = \mathbb{Q}$. We regard n as an independent parameter over $\mathbb{Q}$. For Brumer's quintic $f_{s,t}^{D_5}(X)$, we see by the result in [21] that $\mathrm{Spl}_{\mathbb{Q}(n)} h_n(X) = \mathrm{Spl}_{\mathbb{Q}(n)} f_{s,t}^{D_5}(X)$ where $s = -20 - 5n + 10n^2 + 12n^3 + 5n^4 + n^5$, $t = -7 - 10n - 5n^2 - n^3$. By Theorem 6.1, we checked pairs $(m, n) \in \mathbb{Z}^2$ in the range $-10^4 \le m < n \le 10^4$ to confirm that $\mathrm{Spl}_{\mathbb{Q}} h_m(X) = \mathrm{Spl}_{\mathbb{Q}} h_n(X)$ if and only if $(m, n) = (-2, -1)$.

7. Appendix

For the reader's convenience, we give more detailed data of Table 1 of Theorem 3.3 as the following table (Table 2) which is meant to be compared with the known results of Thomas [47], Mignotte [38] and Mignotte-Pethö-Lemmermeyer [39].

Table 2

m	n	$-n-3$	$2m+3$	j	m^2+3m+9	$xy(x+y)$	(x,y)
-1	-15	12	1	1	7	-2	$(-1,-1),\ (-1,2),\ (2,-1)$
-1	1259	-1262	1	1	7	180	$(5,4),\ (4,-9),\ (-9,5)$
-1	5	-8	1	7	7	6	$(2,1),\ (1,-3),\ (-3,2)$
0	54	-57	3	1	9	6	$(2,1),\ (1,-3),\ (-3,2)$
0	-6	3	3	3	9	-2	$(-1,-1),\ (-1,2),\ (2,-1)$
1	-69	66	5	13	13	-70	$(-5,-2),\ (-2,7),\ (7,-5)$
2	-2392	2389	7	1	19	-126	$(-7,-2),\ (-2,9),\ (9,-7)$
3	-3	0	9	9	27	-2	$(-1,-1),\ (-1,2),\ (2,-1)$
3	-57	54	9	9	27	-20	$(-4,-1),\ (-1,5),\ (5,-4)$
5	-1	-2	13	49	49	-2	$(3,-1),\ (-1,-2),\ (-2,3)$
5	-15	12	13	49	49	-20	$(-4,-1),\ (-1,5),\ (5,-4)$
5	1259	-1262	13	49	49	1254	$(19,3),\ (3,-22),\ (-22,19)$
12	-2	-1	27	27	189	-2	$(-1,-1),\ (-1,2),\ (2,-1),$
12	-1262	1259	27	27	189	-182	$(-13,-1),\ (-1,14),\ (14,-13)$
12	-8	5	27	$189=3^3\cdot7$	$189=3^3\cdot7$	-20	$(-4,-1),\ (-1,5),\ (5,-4)$
54	0	-3	111	$343=7^3$	$3087=3^2\cdot7^3$	-6	$(3,-1),\ (-1,-2),\ (-2,3)$
54	-6	3	111	$1029=3\cdot7^3$	$3087=3^2\cdot7^3$	-20	$(-4,-1),\ (-1,5),\ (5,-4)$
66	-4	1	135	$4563=3^3\cdot13^2$	$4563=3^3\cdot13^2$	-70	$(-5,-2),\ (-2,7),\ (7,-5)$
1259	-1	-2	2521	$226981=61^3$	$1588867=7\cdot61^3$	-180	$(9,-4),\ (-4,-5),\ (-5,9)$
1259	-15	12	2521	$226981=61^3$	$1588867=7\cdot61^3$	-182	$(-13,-1),\ (-1,14),\ (14,-13)$
1259	5	-8	2521	$1588867=7\cdot61^3$	$1588867=7\cdot61^3$	-1254	$(22,-3),\ (-3,-19),\ (-19,22)$
2389	-5	2	4781	$300763=67^3$	$5714497=19\cdot67^3$	-126	$(-7,-2),\ (-2,9),\ (9,-7)$

References

1. J. Buhler and Z. Reichstein, On the essential dimension of a finite group, *Compositio Math.* **106** (1997), 159–179.
2. R. J. Chapman, Automorphism polynomials in cyclic cubic extensions, *J. Number Theory* **61** (1996), 283–291.
3. H. Chu, S. Hu, M. Kang, J. Zhang, Groups with essential dimension one, *Asian J. Math.* **12** (2008), 177–191.
4. H. Cohen, *A course in computational algebraic number theory*, Graduate Texts in Mathematics, 138, Springer, Heidelberg, 1993.
5. F. DeMeyer, Generic Polynomials, *J. Algebra* **84** (1983), 441–448.
6. F. DeMeyer, T. McKenzie, On generic polynomials, *J. Algebra* **261** (2003), 327–333.
7. S. Duquesne, Elliptic curves associated with simplest quartic fields, *J. Théor. Nombres Bordeaux* **19** (2007), 81–100.
8. V. Ennola, Cubic number fields with exceptional units, *Computational number theory* (Debrecen, 1989), 103–128, de Gruyter, Berlin, 1991.
9. I. Gaál, *Diophantine equations and power integral bases. New computational methods*, Birkhäuser Boston, Inc., Boston, MA, 2002.
10. K. Girstmair, On the computation of resolvents and Galois groups, *Manuscripta Math.* **43** (1983), 289–307.
11. M. N. Gras, *Table numérique du nombre de classes et des unités des extensions cycliques réelles de degré 4 de* **Q**, Publ. Math. Fac. Sci. Besancon, fasc 2 (1977/1978).
12. M. N. Gras, Special units in real cyclic sextic fields, *Math. Comp.* **48** (1987), 179–182.
13. K. Hashimoto, A. Hoshi, Families of cyclic polynomials obtained from geometric generalization of Gaussian period relations, *Math. Comp.* **74** (2005), 1519–1530.
14. M. Hindry, J. H. Silverman, *Diophantine geometry. An introduction*, Graduate Texts in Mathematics, 201, Springer-Verlag, New York, 2000.
15. A. Hoshi, *On correspondence between solutions of a parametric family of cubic Thue equations and isomorphic simplest cubic fields*, preprint arXiv:0810.3374v2.
16. A. Hoshi, K. Miyake, Tschirnhausen transformation of a cubic generic polynomial and a 2-dimensional involutive Cremona transformation, *Proc. Japan Acad. Ser. A* **83** (2007), 21–26.
17. A. Hoshi, K. Miyake, *A geometric framework for the subfield problem of generic polynomials via Tschirnhausen transformation*, Number Theory and Applications: Proceedings of the International Conferences on Number Theory and Cryptography, 65–104, Hindustan Book Agency, 2009.
18. A. Hoshi, K. Miyake, On the field intersection problem of quartic generic polynomials via formal Tschirnhausen transformation, *Comment. Math. Univ. St. Pauli* **58** (2009), 51–86.
19. A. Hoshi, K. Miyake, On the field intersection problem of solvable quintic generic polynomials, to appear in *Int. J. Number Theory*. Also available as preprint arXiv:0804.4875v2.

20. A. Hoshi, K. Miyake, On the field intersection problem of generic polynomials: a survey, to appear in *RIMS Kôkyûroku Bessatsu*. Also available as preprint arXiv:0810.0382v2.

21. A. Hoshi, Y. Rikuna, On a transformation from dihedral quintic polynomials into Brumer's form, in preparation.

22. C. Jensen, A. Ledet, N. Yui, *Generic polynomials, constructive aspects of the inverse Galois problem*, Mathematical Sciences Research Institute Publications, Cambridge, 2002.

23. G. Kemper, *Das Noethersche Problem und generische Polynome*, Dissertation, University of Heidelberg, 1994, also available as: IWR Preprint 94-49, Heidelberg.

24. G. Kemper, E. Mattig, Generic polynomials with few parameters, Algorithmic methods in Galois theory. *J. Symbolic Comput.* **30** (2000), 843–857.

25. G. Kemper, Generic polynomials are descent-generic, *Manuscripta Math.* **105** (2001), 139–141.

26. M. Kida, G. Renault, K. Yokoyama, Quintic polynomials of Hashimoto-Tsunogai, Brumer, and Kummer, *Int. J. Number Theory* **5** (2009), 555–571.

27. H. K. Kim, *Evaluation of zeta functions at $s = -1$ of the simplest quartic fields*, Proceedings of the 2003 Nagoya Conference "Yokoi-Chowla Conjecture and Related Problems", 63–73, Saga Univ., Saga, 2004.

28. S. Lang, *Elliptic curves: Diophantine analysis*, Grundlehren der Mathematischen Wissenschaften 231, Springer, 1978.

29. S. Lang, *Fundamentals of Diophantine geometry*, Springer-Verlag, New York, 1983.

30. A. J. Lazarus, On the class number and unit index of simplest quartic fields, *Nagoya Math. J.* **121** (1991), 1–13.

31. A. Ledet, Generic extensions and Generic polynomials, *J. Symbolic Comput.* **30** (2000), 867–872.

32. A. Ledet, On groups with essential dimension one, *J. Algebra*, **311** (2007), 31–37.

33. E. Lehmer, Connection between Gaussian periods and cyclic units, *Math. Comp.* **50** (1988), 535–541.

34. F. Lemmermeyer, A. Pethö, Simplest cubic fields, *Manuscripta Math.* **88** (1995), 53–58.

35. G. Lettl, A. Pethö, Complete solution of a family of quartic Thue equations, *Abh. Math. Sem. Univ. Hamburg* **65** (1995), 365–383.

36. G. Lettl, A. Pethö, P. Voutier, Simple families of Thue inequalities, *Trans. Amer. Math. Soc.* **351** (1999), 1871–1894.

37. S. R. Louboutin, Efficient computation of root numbers and class numbers of parametrized families of real abelian number fields, *Math. Comp.* **76** (2007), 455–473.

38. M. Mignotte, Verification of a conjecture of E. Thomas, *J. Number Theory* **44** (1993), 172–177.

39. M. Mignotte, A. Pethö, F. Lemmermeyer, On the family of Thue equations $x^3 - (n-1)x^2y - (n+2)xy^2 - y^3 = k$, *Acta Arith.* **76** (1996), 245–269.

40. P. Morton, Characterizing cyclic cubic extensions by automorphism polyno-

mials, *J. Number Theory* **49** (1994), 183–208.

41. E. Noether, Gleichungen mit vorgeschriebener Gruppe, *Math. Ann* **78** (1916), 221–229.

42. R. Okazaki, Geometry of a cubic Thue equation, *Publ. Math. Debrecen* **61** (2002), 267–314.

43. C. E. van der Ploeg, Duality in nonnormal quartic fields, *Amer. Math. Monthly* **94** (1987), 279–284. Errata: Amer. Math. Monthly **94** (1987), 994.

44. Y. Rikuna, *Explicit constructions of generic polynomials for some elementary groups*, Galois theory and modular forms, 173–194, Dev. Math., 11, Kluwer Acad. Publ., Boston, MA, 2004.

45. D. Shanks, The simplest cubic fields, *Math. Comp.* **28** (1974), 1137–1152.

46. G. W. Smith, Generic cyclic polynomials of odd degree, *Comm. Algebra* **19** (1991), 3367–3391.

47. E. Thomas, Complete solutions to a family of cubic Diophantine equations, *J. Number Theory* **34** (1990), 235–250.

48. I. Wakabayashi, Number of solutions for cubic Thue equations with autor-mophisms, *Ramanujan J.* **14** (2007), 131–154.

49. J. Xia, J. Chen, S. Zhang, On the family of Thue equation $|x^3 + mx^2y - (m+3)xy^2 + y^3| = k$, *Wuhan Univ. J. Nat. Sci.* **11** (2006), 481–485.

A STATISTICAL RELATION OF ROOTS OF A POLYNOMIAL IN DIFFERENT LOCAL FIELDS II

YOSHIYUKI KITAOKA *

Department of Mathematics, Meijo University,
Nagoya, 468-8502, Japan
E-mail: kitaoka@ccmfs.meijo-u.ac.jp

Let $f(x)$ be a monic polynomial in $\mathbb{Z}[x]$. We make an observation on the statistical relation of roots of $f(x)$ modulo p for different primes p, for which $f(x)$ decomposes completely. Based on it, we propose several conjectures.

1. Introduction and Conjectures

Let

$$f(x) = x^n + a_{n-1}x^{n-1} + ... + a_1 x + a_0 \in \mathbb{Z}[x]$$

be a monic polynomial with integer coefficients. We put

$$Spl(f) = \{p \mid f(x) \bmod p \text{ is completely decomposable}\},$$

where p denotes prime numbers. Let $r_1, ..., r_n$ ($r_i \in \mathbb{Z}$, $0 \leq r_i \leq p-1$) be roots of $f(x) \equiv 0 \bmod p$ for $p \in Spl(f)$; then clearly $a_{n-1} + \sum r_i \equiv 0 \bmod p$. Whence there exists an integer $C_p(f)$ such that

$$a_{n-1} + \sum_{i=1}^{n} r_i = C_p(f)p. \tag{1.1}$$

We stress that the local roots are supposed to satisfy

$$0 \leq r_i \leq p-1 \quad (r_i \in \mathbb{Z}).$$

Hence the integer $C_p(f)$ satisfies $0 \leq C_p(f) \leq n$ for sufficiently large primes p.

---

*Partially supported by Grant-in-Aid for Scientific Research (C), The Ministry of Education, Science, Sports and Culture.

Example 1.1. Let $f(x) = x + a$ ($a \in \mathbb{Z}$); then we have, for primes p with finitely many exceptions

$$C_p(f) = \begin{cases} 1 & \text{if } a > 0, \\ 0 & \text{if } a \leq 0. \end{cases}$$

Proof. We confine ourselves to primes p larger than $|a|$. If $a > 0$, then the local root mod p is $p - a \in [0, p-1]$, and so $C_p(f) = 1$. If $a \leq 0$, then the local root is $-a \in [0, p-1]$, and $C_p(f) = 0$. $\qquad\square$

The following is the second exceptional case where we can evaluate $C_p(f)$ explicitly ([5]).

Theorem 1.1. *Let m be a natural number and let*

$$f(x) = \sum_{i=0}^{2m} a_i x^i \in \mathbb{Z}[x]$$

be a monic polynomial without integer roots which decomposes as $f(x) = f_1(f_2(x))$ (cf. (1.2) below) with $f_1(x), f_2(x) \in \mathbb{Q}[x]$ and $\deg f_2(x) = 2$. Then we have

$$C_p(f) = m \ \left(= \frac{1}{2} \deg f(x)\right)$$

for primes $p \in \mathrm{Spl}(f)$ with finitely many exceptions.

Although we do not know the real meaning of $C_p(f)$, we know the following characterization in a particular case.

Theorem 1.2. *Let p ($\neq 2, 5$) be a prime number, and let the purely periodic decimal expansion of $1/p$ be*

$$0.\dot{c_1} \cdots \dot{c_e} = 0.c_1 \cdots c_e c_1 \cdots c_e \cdots, \quad (0 \leq c_i \leq 9)$$

where e is the minimal length of periods, i.e. $e = $ the order of 10 mod p. Suppose $e = nk$ for natural numbers n (> 1), k. We divide the period into n parts of equal length k, and add them. Then we have

$$c_1 \cdots c_k + c_{k+1} \cdots c_{2k} + \cdots + c_{(n-1)k+1} \cdots c_{nk}$$
$$= (10^k - 1)C_p(f)$$
$$= 9 \cdots 9 \times C_p(f),$$

for $f(x) = x^n - 1$, and $C_p(f) = n/2$ if n is even, and $1 \leq C_p(f) \leq n - 2$ if n is odd.

The study of the distribution of $C_p(f)$ in this theorem was the starting point ([4], [3]) for our research.

Except two cases above, the values $C_p(f)$ seem random. Hereafter, we study the distribution of $C_p(f)$ for a given monic polynomial $f(x)$ in $\mathbb{Z}[x]$.

First, the range of $C_p(f)$ is more precisely described by the following ([5]).

Proposition 1.1. *Suppose that* $f(x) = x^n + a_{n-1}x^{n-1} + ... + a_1x + a_0 \in \mathbb{Z}[x]$ *is a monic polynomial without integer roots. Then we have, for* $p \in Spl(f)$

$$1 \le C_p(f) \le n - 1$$

except finitely many primes.

Proposition 1.2. *Let* $f(x)$ *be a monic polynomial of degree* n *without integer roots. Then we have*

$$C_p((-1)^n f(-x)) = n - C_p(f(x)) \ and \ C_p(f(x+a)) = C_p(f(x)) \ (a \in \mathbb{Z})$$

except finitely many primes.

Proof. Let $r_1, \cdots, r_n$ be roots of $f(x)$ modulo p with $0 \le r_i \le p-1$. We note that there are only finitely many primes p such that $r_i \ne 0$, hence for such primes $0 \le p - r_i \le p - 1$ and $p - r_i$'s are the roots of $(-1)^n f(-x)$ modulo p. Now (1.1) reads $-a_{n-1} + \sum_{i=1}^{n}(p - r_i) = (n - C_p(f))p$, and so the first equality follows.

Next, noting that $f(x + a) = x^n + (a_{n-1} + na)x^{n-1} + \cdots$ and $r_i - a$'s are the roots of $f(x+a)$, we have $C_p(f(x+a)) = (a_{n-1} + na + \sum(r_i - a))/p = C_p(f(x))$ under the assumption $0 \le r_i - a < p$. If there are infinitely many primes such that $r_i - a \ge p$, i.e. $1 \le p - r_i \le -a$, then there is an integer b such that $1 \le b \le -a$ for which $p - r_i = b$ holds for infinitely many primes. Hence $f(-b) \equiv f(r_i) \equiv 0 \bmod p$ for infinitely many primes. This implies a contradiction $f(-b) = 0$. Similarly, we have $0 \le r_i - a$ except finitely many primes. $\square$

To state the main conjecture based on our observation in the next section, we prepare notations and terminologies.

(1) For a monic polynomial $f(x) \in \mathbb{Z}[x]$ and a positive number X, we define the natural density by

$$Pr(k, f, X) = \frac{\#\{p \mid p \in Spl(f), p \le X, C_p(f) = k\}}{\#\{p \mid p \in Spl(f), p \le X\}}.$$

(2) Let $f(x)$ be a monic polynomial in $\mathbb{Z}[x]$. If there are polynomials $g(x), h(x) \in \mathbb{Q}[x]$ such that

$$f(x) = g(h(x)), \tag{1.2}$$

then we call this expression a decomposition of f. We may suppose that $h(x)$ is in $\mathbb{Z}[x]$ and primitive; then $g(x), h(x)$ are in $\mathbb{Z}[x]$ and monic.

Proof. Suppose that there is a prime number p such that $g(x) \notin \mathbb{Z}_p[x]$ where $\mathbb{Z}_p$ signifies the ring of p-adic integers; then there is a natural number a such that $p^a g(x)$ is primitive in $\mathbb{Z}_p[x]$. Therefore we have

$$p^a g(x) \equiv a_n x^n + \cdots \bmod p, \ a_n \not\equiv 0 \bmod p,$$
$$h(x) \equiv b_m x^m + \cdots \bmod p, \ b_m \not\equiv 0 \bmod p,$$

which yields

$$p^a f(x) \equiv p^a g(h(x)) \equiv a_n (b_m x^m)^n + \cdots \bmod p.$$

Comparing both sides, we have a contradiction $0 \equiv a_n b_m^n \bmod p$. Hence $g(x) \in \mathbb{Z}_p[x]$ for all p and so $g(x) \in \mathbb{Z}[x]$. Now $g(x)$ and $h(x)$ are clearly monic since $f(x)$ is monic. □

Therefore, for a decomposition $f(x) = g(h(x))$, we may assume that $g(x), h(x)$ are in $\mathbb{Z}[x]$ and monic. Moreover, putting $h_1(x) = h(x) - h(0)$, we have $f(x) = g(h_1(x) + h(0)) = g_1(h_1(x))$ for another monic polynomial $g_1(x) \in \mathbb{Z}[x]$. Consequently, we may assume in (1.2) that

$$g(x), h(x) \in \mathbb{Z}[x] \text{ are monic and } h(0) = 0, \qquad (1.3)$$

which we call a **reduced decomposition** of $f(x)$. Then $h(x)$ divides $f(x) - f(0)$. A reduced decomposition is not unique. Indeed,

$$\begin{cases} g(x) = f(x), \\ h(x) = x, \end{cases} \quad \begin{cases} g(x) = x + f(0), \\ h(x) = f(x) - f(0) \end{cases}$$

are reduced decompositions. We call these **trivial**. If a polynomial $f(x)$ has a non-trivial reduced decomposition, we say that $f(x)$ is **reduced**, otherwise **non-reduced**. When $f(x)$ is reduced, we call the degree of $h(x)$ a **reduced degree** of $f(x)$. Thus every reduced degree is larger than 1 and smaller than $\deg f$. For example, the reduced degrees of $f(x) = x^n$ are all proper divisors m of n because of $f(x) = (x^m)^{n/m}$. Note that $\deg f(x) = \deg g(x) \cdot \deg h(x)$, hence each reduced degree divides the degree of $f(x)$. Therefore, if $\deg f(x)$ is a prime number, $f(x)$ is non-reduced.

Every reduced polynomial treated in Section 5 of [5] has only one non-trivial reduced decomposition. Therefore, the definition of the reduced degree adopted there to be the minimum of reduced degrees coincides with the above definition, and causes no trouble to study examples there. But it turned out to be inappropriate when $f(x)$ has many reduced decompositions, as examples in case of $n = 12$ below exhibit and we are to modify the definition of the reduced degree as above.

(3) A mapping p from the set of integers $\mathbb{Z}$ to the set of non-negative real numbers is called a **table of frequency distribution** if it satisfies (i) $p(n) \neq 0$ for only finitely many integers n, (ii) $p(n) = 0$ if $n < 0$, and (iii) $\sum_{n \in \mathbb{Z}} p(n) = 1$.

The mean μ and the variance σ^2 of p are defined as usual by

$$\mu(p) = \sum_{n \in \mathbb{Z}} np(n), \quad \sigma^2(p) = \sum_{n \in \mathbb{Z}} n^2 p(n) - \mu(p)^2.$$

To a table of frequency distribution p, we associate the **generating polynomial** d_p by

$$d_p(x) = \sum_{n=0}^{\infty} p(n)x^n.$$

Then we have

$$d_p'(x) = \sum p(n)nx^{n-1},$$

$$d_p'(x) + xd_p''(x) = (xd_p'(x))' = \sum p(n)n^2 x^{n-1}.$$

These yield

$$d_p(1) = 1, \; d_p'(1) = \mu(p), \; d_p''(1) = \sigma^2(p) + \mu(p)^2 - \mu(p). \tag{1.4}$$

Let p, q be tables of frequency distribution. The convolution $p * q$ is defined by

$$p * q(n) = \sum_{i+j=n} p(i)q(j),$$

which is indeed a finite sum. It is easy to see that $p*q$ is a table of frequency distribution, and that

$$d_{p*q}(x) = d_p(x)d_q(x).$$

(1.4) yields

$$\mu(p * q) = \mu(p) + \mu(q), \; \sigma^2(p * q) = \sigma^2(p) + \sigma^2(q). \tag{1.5}$$

We define the table of frequency distribution p^m inductively by $p^1 = p$, $p^m = p * p^{m-1}$ for a natural number m.

(4) Let us introduce **Eulerian numbers** $A(n, k)$. Let $A(1, 1) = 1$ and we define inductively $A(n, k)$ $(1 \leq k \leq n)$ by

$$A(n, k) = (n - k + 1)A(n - 1, k - 1) + kA(n - 1, k).$$

We suppose that $A(n,k) = 0$ unless $1 \le k \le n$. Their values are

$n \backslash k$	1	2	3	4	5	6	7	8	9
2	1	1							
3	1	4	1						
4	1	11	11	1					
5	1	26	66	26	1				
6	1	57	302	302	57	1			
7	1	120	1191	2416	1191	120	1		
8	1	247	4293	15619	15619	4293	247	1	
9	1	502	14608	88234	156190	88234	14608	502	1

The values for $n = 7, 9$ in [5] are incorrect and 12, 15619 there should be 120, 156190, respectively.

For natural numbers $n\ (\ge 2)$, we define the normalization E_n by

$$E_n(k) = \frac{A(n-1, k)}{(n-1)!}. \tag{1.6}$$

Since $A(n,k) \ge 0$ and $\sum_{k=1}^{n} A(n,k) = n!$, E_n is a table of frequency distribution, which we call an **Eulerian table**, and

$$\begin{cases} \mu(E_2) = 1, \\ \sigma^2(E_2) = 0, \end{cases} \quad \begin{cases} \mu(E_n) = n/2, \\ \sigma^2(E_n) = n/12 \end{cases} \text{ for } n \ge 3, \tag{1.7}$$

which is easily proved by induction on n. Therefore, for a natural number $n = mr$ we have, by (1.5)

$$\begin{cases} \mu(E_2^m) = m, \\ \sigma^2(E_2^m) = 0, \end{cases} \quad \begin{cases} \mu(E_r^m) = n/2, \\ \sigma^2(E_r^m) = n/12 \end{cases} \text{ for } r \ge 3.$$

Obviously, $d_{E_2}(x) = x$ and $d_{E_3}(x) = \frac{1}{2}x + \frac{1}{2}x^2$. Hence we get, by $d_{E_2^m}(x) = d_{E_2}(x)^m = x^m$ and $d_{E_3^m}(x) = d_{E_3}(x)^m = (\frac{1}{2}x(1+x))^m$, i.e.

$$E_2^m(k) = \begin{cases} 1 & \text{if } k = m, \\ 0 & \text{otherwise,} \end{cases}$$

$$E_3^m(k) = \begin{cases} 2^{-m}\binom{m}{k-m} & \text{if } m \le k \le 2m, \\ 0 & \text{otherwise.} \end{cases}$$

Eulerian tables E_r^m, for which $E_r^m(k)$ is by definition above equal to

$$\sum_{k_1 + \cdots + k_m = k} \prod_{i=1}^{m} E_r(k_i) \tag{1.8}$$

have the properties that

(1) they are symmetric and unimodal, i.e. $E_r^m(k) = E_r^m(rm - k)$ for $^\forall k \in \mathbb{Z}$ and $E_r^m(1) \leq E_r^m(2) \leq \cdots \geq E_r^m(mr - 2) \geq E_r^m(mr - 1)$.

(2) $E_r(1) = 1/(r - 1)! \neq 0$ and $E_r^m(1) = 0$ if $m > 1$.

1.1. Irreducible case

Now we can state our main conjecture.

Conjecture 1.1. *Let $f(x) \in \mathbb{Z}[x]$ be a monic irreducible polynomial of $\deg f(x) = n \geq 2$ and suppose that $f(x)$ is non-reduced. Then for any integer k, the limit of natural density*

$$\lim_{x \to \infty} Pr(k, f, x) \, (= Pr(k, f) \ say)$$

exists and is equal to

$$E_n(k);$$

in other words the table of frequency distribution $Pr(f) : k \mapsto Pr(k, f)$ is equal to E_n :

$$Pr(k, f) = \lim_{x \to \infty} Pr(k, f, x) = E_n(k).$$

Remark 1.1. *The conjecture 1.7 in [5] should be modified as follows, because the concept of the reduced degree therein has been modified as before.*

> *Let $f(x)$ be a monic irreducible polynomial of degree $n \ (\geq 3)$ in $\mathbb{Z}[x]$. We assume that 2 is not a reduced degree of $f(x)$. Then the table of frequency distribution defined by $Pr(f) : c \mapsto Pr(c, f)$ satisfies*
>
> $$\mu(Pr(f)) = n/2, \ \sigma^2(Pr(f)) = n/12,$$
> $$Pr(k, f) = Pr(n - k, f) \ for \ ^\forall k,$$
> $$Pr(1, f) \leq Pr(2, f) \leq \cdots \geq Pr(n - 2, f) \geq Pr(n - 1, f).$$

A classical example of a random variable with mean $1/2$ and variance $1/12$ is a uniform distribution on the interval $[0, 1)$: If $x_1, \cdots, x_n$ are independent uniform distributions on $[0, 1)$, the mean and the variance of the sum $x_1 + \cdots + x_n$ is $n/2$ and $n/12$, respectively. Hence recalling (1.7), we see that the above conjecture suggests the existence of uniformity. The following conjecture stated in Remark 2 in [3] may fall in this category.

Conjecture 1.2. *Let $F = \mathbb{Q}(\alpha) \, (\neq \mathbb{Q})$ be an algebraic number field generated by an algebraic integer α, and let k be a non-negative integer. For a* *prime number p which decomposes fully in F and for a prime ideal $\mathfrak{p}$ lying above p, we write in $F_\mathfrak{p} = \mathbb{Q}_p$*

$$\alpha = c_\mathfrak{p}(0) + c_\mathfrak{p}(1)p + \cdots \quad (c_\mathfrak{p}(i) \in \mathbb{Z}, 0 \le c_\mathfrak{p}(i) < p).$$

Then the points $(c_\mathfrak{p}(0)/p, c_\mathfrak{p}(1)/p, \cdots, c_\mathfrak{p}(k)/p) \, (\in [0,1)^{k+1})$ are uniformly distributed when $p, \mathfrak{p}$ run over prime numbers and prime ideals described above.

Remark 1.2. *Let $f(x)$ be the minimal polynomial of α over $\mathbb{Q}$; then $c_\mathfrak{p}(0)$ is a root of $f(x) \equiv 0 \bmod p$. Conjecture 1.2 is true if α is quadratic irrational and $k = 0$ ([1], [6]).*

Let us give remarks related to Conjecture 1.1 from the view point of the uniform distribution. From now on, we assume that $f(x) \in \mathbb{Z}[x]$ is a monic and irreducible polynomial of $\deg f \ge 2$.

(1) The definition (1.1), i.e. $\sum r_i/p = C_p(f) - a_{n-1}/p$ shows that the set of points $(r_1/p, \cdots, r_n/p)$ are in a proper subset of $[0,1)^n$ defined by $k - \epsilon \le x_1 + x_2 + \cdots + x_n \le k + \epsilon \, (1 \le k \le n, \, \epsilon > 0)$ by $|a_{n-1}/p| < \epsilon$ for large primes p, hence they are not uniformly distributed in $[0,1)^n$.

(2) We denote by $\lceil x \rceil$ the least integer which is larger than or equal to x, i.e. $\lceil x \rceil - 1 < x \le \lceil x \rceil$, and put $\lceil \boldsymbol{x} \rceil = \lceil x_1 + \cdots + x_a \rceil$ for a vector $\boldsymbol{x} = (x_1, \cdots, x_a) \in \mathbb{R}^a$.

Then $E_n(k)$ in (1.6) is the volume of the set

$$S_k = \left\{ \boldsymbol{x} \in [0,1)^{n-1} \mid \lceil \boldsymbol{x} \rceil = k \right\},$$

which was suggested by Dr. Satoshi Fukutani (cf. [2]).

Hence, if $\boldsymbol{x}_m \in [0,1)^{n-1}$ are uniformly distributed, then we have

$$\lim_{m \to \infty} \frac{\#\{m \mid \boldsymbol{x}_m \in S_k\}}{m} = E_n(k).$$

(3) For n local roots $r_1, \cdots, r_n$ of $f(x) \equiv 0 \bmod p$, the condition

$$(r_1/p, \cdots, r_{n-1}/p) \in S_k \tag{1.9}$$

is equivalent to $C_p(f) = k$ except finitely many primes p. For, the condition (1.9) implies that

$$k - 1 + (a_{n-1} + r_n)/p < C_p(f) \le k + (a_{n-1} + r_n)/p.$$

If $C_p(f) \ge k + 1$, then we obtain $0 < p - r_n \le a_{n-1}$. This can occur for finitely many primes, otherwise there exists some integer a with $0 < a \le a_{n-1}$ such that $p - r_n = a$, i.e. $f(-a) \equiv 0 \bmod p$ for infinitely many primes p, hence $f(-a) = 0$, which is a contradiction. Similarly $C_p(f) \leq k - 1$ can occur for only a finite number of primes. Hence $C_p(f) = k$ holds except finitely many primes.

Our conjecture 1.1 for a non-reduced polynomial f suggests that the points $(r_{\sigma(1)}/p, \cdots, r_{\sigma(n-1)}/p) \in [0,1)^{n-1}$ for all permutations σ of the set $\{1, 2, \cdots, n\}$, to make local roots distribute impartially are uniformly distributed when $p \to \infty$. This may explain the background why Eulerian numbers appear in this case.

(4) Suppose that $f(x) = g(h(x))$ is a reduced decomposition (1.3). Then putting

$$g(x) = x^m + a_{m-1}x^{m-1} + \cdots \quad (m > 1),$$
$$h(x) = x^r + b_{r-1}x^{r-1} + \cdots \quad (r > 1),$$

we have

$$f(x) = x^{rm} + mb_{r-1}x^{mr-1} + \cdots.$$

For a prime $p \in Spl(f)$, we group local roots as follows:

$$\{r_i \mid f(r_i) \equiv 0 \bmod p\} = \cup_{i=1}^m \{r_{i,1}, \cdots, r_{i,r}\},$$

where $r_{i,j}$ satisfies

$$h(r_{i,j}) \equiv {}^\exists s_i \bmod p \quad (1 \leq {}^\forall j \leq r), \ g(s_i) \equiv 0 \bmod p.$$

Hence $r_{i,j}$ $(1 \leq j \leq r)$ are roots of a polynomial $h(x) - s_i$ over $\mathbb{Z}/(p)$. Then, notifying that

$$C_p(f)p = mb_{r-1} + \sum_{i,j} r_{i,j} = \sum_{i=1}^m (b_{r-1} + \sum_{j=1}^r r_{i,j}),$$

we see that the summand is $C_p(h(x) - s_i)p$, whence

$$C_p(f) = \sum_{i=1}^m C_p(h(x) - s_i).$$

Proposition 1.3. *For any choice of $r - 1$ elements in $\{r_{i,1}, \cdots, r_{i,r}\}$, say $\{r_{i,1}, \cdots, r_{i,r-1}\}$, we have*

$$C_p(h(x) - s_i) = \lceil (r_{i,1} + \cdots + r_{i,r-1})/p \rceil$$

except finitely many primes p.

Proof. Put $k = \lceil (r_{i,1} + \cdots + r_{i,r-1})/p \rceil$. Then noting the identity

$$(r_{i,1} + \cdots + r_{i,r-1})/p = C_p(h(x) - s_i) - b_{r-1}/p - r_{i,r}/p,$$

we obtain

$$k - 1 < C_p(h(x) - s_i) - b_{r-1}/p - r_{i,r}/p \leq k.$$

If $C_p(h(x) - s_i) \geq k+1$ holds for infinitely many primes p, then we have $k + 1 - b_{r-1}/p - r_{i,r}/p \leq k$, whence $1 \leq p - r_{i,r} \leq b_{r-1}$ for infinitely many p. Consequently, there is an integer a ($1 \leq a \leq b_{r-1}$) such that $a = p - r_{i,r}$ for infinitely many p, which implies $f(-a) \equiv f(r_{i,r}) \equiv 0 \bmod p$ for infinitely many p. Thus $-a$ is a root of $f(x)$, which is a contradiction. Similarly $C_p(h(x) - s_i) \leq k-1$ for infinitely many primes p leads to a contradiction. Thus we have $C_p(h(x) - s_i) = k$. $\qquad\square$

Putting

$$S_{m,r}(k) = \{(\boldsymbol{x}_1, \cdots, \boldsymbol{x}_m) \mid \boldsymbol{x}_i \in [0,1)^{r-1}, \sum_{i=1}^{m} \lceil \boldsymbol{x}_i \rceil = k\},$$

we see that

$$vol(S_{m,r}(k))$$
$$= \sum_{k_1 + \cdots + k_m = k} vol\left\{(\boldsymbol{x}_1, \cdots, \boldsymbol{x}_m) \mid \boldsymbol{x}_i \in [0,1)^{r-1}, \lceil \boldsymbol{x}_i \rceil = k_i\right\}.$$

Now the summand can be written as

$$\prod_{i=1}^{m} vol\{\boldsymbol{x} \in [0,1)^{r-1} \mid \lceil \boldsymbol{x} \rceil = k_i\},$$

which is

$$\prod_{i=1}^{m} E_r(k_i).$$

Hence

$$vol(S_{m,r}(k)) = E_r^m(k)$$

by (1.8). Therefore, if the points $(\boldsymbol{r}_1(\mu, \sigma_1), \cdots, \boldsymbol{r}_m(\mu, \sigma_m)) \in [0,1)^{m(r-1)}$ with $\boldsymbol{r}_k(\mu, \sigma_k) = (r_{\mu(k), \sigma_k(1)}/p, \cdots, r_{\mu(k), \sigma_k(r-1)}/p)$ for all permutations μ of $\{1, 2, \cdots, m\}$ and all permutations σ_k of $\{1, 2, \cdots, r\}$ are uniformly distributed, then by (2) we have $Pr(f) = E_r^m$. If a reduced decomposition is unique and 2 is not a reduced degree, this seems true. But, as the case of $\deg f = 12$ below shows, the situation is not simple if there are plural reduced decompositions. By Theorem 1.1, $Pr(f) = E_2^{\frac{1}{2}\deg(f)}$ is true in case that 2 is a reduced degree of $f(x)$, even if there exist other reduced decompositions. How does one neglect reduced degrees other than 2?

Examples of $\deg f = 15, 21, 35, 105$ suggest $Pr(f) = E_r^m$ for the least reduced degree.

How are reduced degrees or reduced decompositions involved and what is $Pr(f)$ if f has plural reduced decompositions ? Is it likely that points $(r_1/p, \cdots, r_n/p)$ are uniformly distributed up to relations induced by $r_{i,1}/p + \cdots + r_{i,r}/p \in \mathbb{Z}$ $(1 \le i \le m)$ for all reduced decompositions with $h(x) \ne x$?

(5) Conjecture 1.2 immediately implies $\mu(Pr(f)) = \frac{1}{2}\deg f(x)$ (cf. [3]).

1.2. *Reducible case*

For reducible polynomials, let us give some observations.

(1) Let g, h be monic polynomials in $\mathbb{Z}[x]$; then $C_p(gh) = C_p(g) + C_p(h)$ for $p \in Spl(gh)$ is easy to see. Considering $Pr(f)$ as a random variable defined by f, the probability theory suggests that $Pr(gh) = Pr(g) * Pr(h)$ (the convolution) if $Pr(g)$ and $Pr(h)$ are independent. Number-theoretically, it may mean that the intersection of the fields defined by g, h is the rational number field. Hence, we expect that if the intersection of minimal splitting fields of g, h is the rational number field, then $Pr(gh) = Pr(g) * Pr(h)$ holds.

(2) In view of $C_p(g^m) = mC_p(g)$, the identity $Pr(k, g^m) = Pr(k/m, g)$ clearly holds, whence we have $\mu(Pr(g^m)) = m\mu(Pr(g))$ and $\sigma^2(Pr(g^m)) = m^2\sigma^2(Pr(g))$, so that the identity $\sigma^2 = \deg /12$ does not hold in this case.

(3) If 2 is a reduced degree of g, then $Pr(gh) = Pr(g) * Pr(h)$.

(4) If g is a power of an irreducible polynomial of degree > 1 and h is irreducible over the field $\mathbb{Q}(\alpha)$, α being a root of g then $Pr(gh) = Pr(g) * Pr(h)$.

(5) Let g, h be irreducible polynomials of degree 3. If either $\mathbb{Q}[x]/(g)$ is not isomorphic to $\mathbb{Q}[x]/(h)$, or $\phi(x)$ is not linear in $\mathbb{Q}[x]/(h)$ for any isomorphisms $\phi : \mathbb{Q}[x]/(g) \cong \mathbb{Q}[x]/(h)$, then $Pr(gh) = Pr(g) * Pr(h) = E_3^2$.

(6) If F is an irreducible polynomials of degree 3, we put $g = a^3 F(x/a), h = b^3 F(x/b)$ where a, b are any non-zero integers.
Put $d = \gcd(a, b)$, $A = a/d$ and $B = b/d$; then we have $[Pr(1, gh), \cdots,$

$Pr(5, gh)] = [0, 1/4 + 1/(4AB), 1/2 - 1/(2AB), 1/4 + 1/(4AB), 0]$. The mean is $3 (= \deg(gh)/2)$, but the variance is $1 (\neq \deg(gh)/12)$.

(7) Even if $f(x)$ is reducible, the properties $\mu(Pr(f)) = \frac{1}{2} \deg f$ and the symmetry $Pr(k, f) = Pr(\deg f - k, f)$ are possible if $f(x)$ has no linear factors. But the unimodality fails in general.

1.3. *Generalization*

Although we considered prime numbers p for which $f(x) \bmod p$ is completely decomposable, the following generalization is possible. Let n be a natural number and let natural numbers $d_1 \leq d_2 \leq \cdots \leq d_g$ satisfy $\sum d_i = n$. Consider prime numbers p such that $f(x) \equiv f_1(x) \cdots f_g(x) \bmod p$ where each $f_i(x) \bmod p$ is irreducible and $\deg f_i(x) = d_i$. Put $f(x) = x^n + a_{n-1}x^{n-1} + \cdots$ and $f_i(x) = x^{d_i} + a_{i,d_i-1}x^{d_i-1} + \cdots$; then clearly $a_{n-1} - \sum_i a_{i,d_i-1} \equiv 0 \bmod p$. Hence in this situation, we may define the probability Pr in a similar way and it seems to exist.

2. Numerical data

2.1. $n = 3$

In the following table, $Pr(c)$ is the abbreviation of $Pr(k, f, 10^9)$ and $\#Spl = \#Spl(f, 10^9)$. The expected values E_3 of $Pr(k, f)$ are listed in the last line. We use these abbreviations hereafter, and the values are rounded off to four decimal places.

f	$Pr(1)$	$Pr(2)$	$\#Spl$
$x^3 - x - 1$	0.4998	0.5002	8474030
$x^3 + x^2 + x - 1$	0.5002	0.4998	8472910
$x^3 - 3x + 1$	0.4999	0.5001	16949354
$x^3 + x^2 - 4x + 1$	0.4999	0.5001	16948980
E_3	1/2	1/2	

2.2. $n = 4$

We put

$$f_1 = x^4 - x^3 - x^2 - x - 1,$$
$$f_2 = x^4 - x^3 - x^2 + x + 1,$$
$$f_3 = x^4 + x^3 + x^2 + x + 1.$$

These polynomials are non-reduced.

f	$Pr(1)$	$Pr(2)$	$Pr(3)$	$\#Spl$
f_1	0.1664	0.6667	0.1669	2118177
f_2	0.1667	0.6667	0.1666	6354490
f_3	0.1666	0.6667	0.1667	12711386
E_4	$1/6 = 0.1667$	$4/6 = 0.6667$	$1/6 = 0.1667$	

2.3. $n = 5$

We put

$$f_1 = x^5 - x^3 - x^2 - x + 1,$$
$$f_2 = x^5 - x^4 - x^2 - x + 1,$$
$$f_3 = x^5 + x^4 - x^2 - x + 1,$$
$$f_4 = x^5 + x^4 - 4x^3 - 3x^2 + 3x + 1.$$

f	$Pr(1)$	$Pr(2)$	$Pr(3)$	$Pr(4)$	$\#Spl$
f_1	0.04160	0.4578	0.4587	0.04187	423981
f_2	0.04228	0.4600	0.4561	0.04157	423719
f_3	0.04110	0.4590	0.4579	0.04193	422711
f_4	0.04180	0.4582	0.4584	0.04167	10169695
E_5	$1/24 = 0.04167$	$11/24 = 0.4583$	$11/24$	$1/24$	

2.4. $n = 6$

We put

$$f_1 = x^6 + x + 1,$$
$$f_2 = x^6 + x^5 + x^4 + x^3 + x^2 + x + 1,$$
$$f_3 = y^2 + y + 1 \quad (y = x^3 + x^2),$$
$$f_4 = y^2 + y + 2 \quad (y = x^3 + x).$$

f_1, f_2 are non-reduced.

f	$Pr(1)$	$Pr(2)$	$Pr(3)$	$Pr(4)$	$Pr(5)$	$\#Spl$
f_1	0.0086	0.2135	0.5513	0.2177	0.0089	70292
f_2	0.0084	0.2164	0.5501	0.2167	0.0083	8474221
E_6	$1/120$	$26/120$	$66/120$	$26/120$	$1/120$	
	$= 0.0083$	$= 0.2167$	$= 0.5500$			

f_3, f_4 have a unique reduced degree 3.

f	$Pr(1)$	$Pr(2)$	$Pr(3)$	$Pr(4)$	$Pr(5)$	$\#Spl$
f_3	0	0.2497	0.4997	0.2506	0	705553
f_4	0	0.2506	0.4996	0.2498	0	706369
E_3^2	0	1/4	1/2	1/4	0	

2.5. $n = 7$

We put

$$f_1 = x^7 - x^5 - x^4 - x^3 - x^2 - x + 1,$$
$$f_2 = x^7 + x^6 - x^5 - x^4 - x^3 - x^2 - x + 1,$$
$$f_3 = x^7 + x^6 - 12x^5 - 7x^4 + 28x^3 + 14x^2 - 9x + 1.$$

f	$Pr(1)$	$Pr(2)$	$Pr(3)$	$Pr(4)$	$Pr(5)$	$Pr(6)$	$\#Spl$
f_1	0.0016	0.0823	0.4206	0.4113	0.0832	0.0011	10076
f_2	0.0017	0.0779	0.4189	0.4228	0.0775	0.0014	9994
f_3	0.0014	0.0790	0.4192	0.4198	0.0792	0.0014	7264359
E_7	1/6! $= 0.0014$	57/6! $= 0.0792$	302/6! $= 0.4194$	302/6!	57/6!	1/6!	

2.6. $n = 8$

We put

$$f_1 = x^8 + x + 2,$$
$$f_2 = x^8 + x^7 - 7x^6 - 6x^5 + 15x^4 + 10x^3 - 10x^2 - 4x + 1,$$
$$f_3 = y^2 + 1 \qquad (y = x^4 + x),$$
$$f_4 = y^2 + 2 \qquad (y = x^4 + x^2 + x).$$

f_1, f_2 are non-reduced. f_3, f_4 have a unique reduced degree 4.

f	$Pr(1)$	$Pr(2)$	$Pr(3)$	$Pr(4)$	$Pr(5)$	$Pr(6)$	$Pr(7)$	$\#Spl$
f_1	0	0.0204	0.2514	0.4686	0.2376	0.0220	0	1225
f_2	0.0002	0.0240	0.2364	0.4793	0.2361	0.0238	0.0002	6354766
E_8	0.0002	0.0238	0.2363	0.4794	0.2363	0.0238	0.0002	

f	$Pr(1)$	$Pr(2)$	$Pr(3)$	$Pr(4)$	$Pr(5)$	$Pr(6)$	$Pr(7)$	#Spl
f_3	0	0.0267	0.2203	0.5028	0.2227	0.0276	0	44089
f_4	0	0.0288	0.2221	0.5020	0.2200	0.0270	0	44112
E_4^2	0	1/36 $= 0.0278$	8/36 $= 0.222$	18/36 $= 0.5$	8/36	1/36	0	

2.7. $n = 9$

We put

$$f_1 = x^9 + x + 1,$$
$$f_2 = x^9 + x^8 - 8x^7 - 7x^6 + 21x^5 + 15x^4 - 20x^3 - 10x^2 + 5x + 1,$$
$$f_3 = y^3 + 2 \qquad (y = x^3 + x),$$
$$f_4 = y^3 + y^2 + 1 \qquad (y = x^3 + x).$$

f_1, f_2 are non-reduced. f_3, f_4 have a unique reduced degree 3.

	f_1	f_2	f_3	f_4
$Pr(1)$	0	0.0000	0	0
$Pr(2)$	0.0064	0.0061	0	0
$Pr(3)$	0.1026	0.1064	0.1267	0.1240
$Pr(4)$	0.3654	0.3871	0.3768	0.3763
$Pr(5)$	0.4295	0.3877	0.3758	0.3737
$Pr(6)$	0.0962	0.1065	0.1208	0.1259
$Pr(7)$	0	0.0061	0	0
$Pr(8)$	0	0.0000	0	0
#Spl	156	5649358	38912	38802

The following is the table of E_9.

k	1	2	3	4	5	6	7	8
	0.0000	0.0061	0.1065	0.3874	0.3874	0.1065	0.0061	0.0000

E_3^3 is

k	1	2	3	4	5	6	7	8
	0	0	0.125	0.375	0.375	0.125	0	0

2.8. $n = 10$

We put

$$f_1 = x^{10} + x + 1,$$
$$f_2 = (x^{11} - 1)/(x - 1),$$
$$f_3 = y^2 + y + 2 \qquad (y = x^5),$$
$$f_4 = y^2 + 3y + 3 \qquad (y = x^5).$$

f_1, f_2 are non-reduced, and 5 is the reduced degree of f_3, f_4.

	f_1	f_2	f_3	f_4
$Pr(1)$	0	0.0000	0	0
$Pr(2)$	0	0.0014	0.0018	0.0018
$Pr(3)$	0	0.0403	0.0387	0.0383
$Pr(4)$	0.1818	0.2432	0.2483	0.2477
$Pr(5)$	0.3636	0.4302	0.4235	0.4239
$Pr(6)$	0.3636	0.2432	0.2475	0.2483
$Pr(7)$	0.0909	0.0404	0.0385	0.0382
$Pr(8)$	0	0.0014	0.0017	0.0017
$Pr(9)$	0	0.0000	0	0
$\#Spl$	11	5084435	254385	1271165

The following is the table of E_{10}.

k	1	2	3	4	5	6	7	8	9
	0.0000	0.0014	0.0403	0.2431	0.4304	0.2431	0.0403	0.0014	0.0000

E_5^2 is

k	1	2	3	4	5	6	7	8	9
	0	0.0017	0.0382	0.2483	0.4236	0.2483	0.0382	0.0017	0

2.9. $n = 12$

We put

$$f_1 = (x^{13} - 1)/(x - 1),$$
$$f_2 = y^2 + y + 1 \qquad (y = x^6 + x),$$
$$f_3 = y^3 - 3y + 1 \qquad (y = x^4 + x),$$
$$f_4 = y^4 + y^3 + y^2 + y + 1 \quad (y = x^3 + x).$$

f_1 is non-reduced. The reduced degree of f_2, f_3, f_4 is $6, 4, 3$, respectively.

	f_1	f_2	f_3	f_4
$Pr(1)$	0.0000	0	0	0
$Pr(2)$	0.0001	0	0	0
$Pr(3)$	0.0038	0	0.0031	0
$Pr(4)$	0.0550	0.1296	0.0541	0.0605
$Pr(5)$	0.2444	0.2593	0.2085	0.2556
$Pr(6)$	0.3938	0.3333	0.4093	0.3707
$Pr(7)$	0.2438	0.2407	0.2556	0.2537
$Pr(8)$	0.0553	0.0370	0.0649	0.0594
$Pr(9)$	0.0038	0	0.0046	0
$Pr(10)$	0.0001	0	0	0
$Pr(11)$	0	0	0	0
$\#Spl$	4237228	54	1295	9862

The following is the table of E_{12} for $1 \le k \le 6$.

k	1	2	3	4	5	6
	0.0000	0.0001	0.0038	0.0552	0.2440	0.3939

E_6^2 is

k	1	2	3	4	5	6
	0	0.0001	0.0036	0.0561	0.2419	0.3965

E_4^3 is

k	1	2	3	4	5	6
	0	0	0.0046	0.0556	0.2361	0.4074

E_3^4 is

k	1	2	3	4	5	6
	0	0	0	0.0625	0.25	0.375

We give examples which have $3, 4$ only as reduced degrees. Polynomials with reduced degrees $3, 4$ only are of the form

$$x^{12} + 3cx^9 + 3c^2x^6 + c^3x^3 + d$$
$$= y^3 + d \qquad\qquad (y = x^4 + cx)$$
$$= y(y + c)^3 + d \qquad (y = x^3)$$

aside from the substitution of $x + a$ for x. For

$$f_5 = x^{12} - 9x^9 + 27x^6 - 27x^3 + 3 \quad (c = -3, d = 3),$$
$$f_6 = x^{12} - 3x^9 + 3x^6 - x^3 + 3 \qquad (c = -1, d = 3),$$

$[Pr(3, f), \cdots, Pr(8, f)]$ are

$$f_5 : [0, 0.0677, 0.2342, 0.3993, 0.2323, 0.0665] \quad (\#Spl = 353455),$$
$$f_6 : [0.0000, 0.0667, 0.2350, 0.3987, 0.2333, 0.0663] (\#Spl = 352211),$$

where 0.0000 does not mean zero but a very small number. They are approximated by a table of frequency distribution P defined by

$$[P[4], \cdots, P[8]]$$
$$= [1/15, 7/30, 2/5, 7/30, 1/15]$$
$$= [0.0667, 0.2333, 0.4000, 0.2333, 0.0667],$$

whose mean and variance are $n/2 = 6$ and $n/12 = 1$, respectively. If $n < 4$ or $n > 8$, then $P[n]$ should be 0.

Suppose that a polynomial $f(x)$ of degree 12 has reduced degrees $3, 6$, that is $f(x)$ is a quadratic polynomial of a polynomial $g(x)$ of degree 6 with $g(0) = 0$, and it is also equal to a quartic polynomial of a polynomial $h(x)$ of degree 3 with $h(0) = 0$; then by comparing coefficients, we have $g(x) = h(x)^2 + dh(x)$ for a constant d given by $f(x) = h(x)^4 + 2dh(x)^3 \cdots$. Here these polynomials above are monic. Therefore the reduced decomposition of reduced degree 6 is meaningless. Let us give an example.

$$f = y^4 + 2y^3 - y^2 - 2y + 5 \quad (y = x^3 + 2x^2 + 2x)$$
$$= z^2 - 2z + 5 \quad (z = x^6 + 4x^5 + 8x^4 + 9x^3 + 6x^2 + 2x = y^2 + y).$$

$Pr(k, f)$ $(4 \le k \le 8)$ is

$$[0.0629, 0.2500, 0.3576, 0.2599, 0.0695] \quad \#Spl = 604.$$

We may regard the above as being approximated by E_3^4 on considering $\#Spl = 604$ being small.

If a polynomial $f(x)$ of degree 12 has reduced degrees $4, 6$, then 2 is also a reduced degree of $f(x)$.

2.10. $n = 15$

We put

$$f_1 = x^{15} + x^{14} - 14x^{13} - 13x^{12} + 78x^{11} + 66x^{10} - 220x^9$$
$$-165x^8 + 330x^7 + 210x^6 - 252x^5 - 126x^4 + 84x^3$$
$$+28x^2 - 8x - 1,$$
$$f_2 = x^{15} - 3x^5 + 1,$$
$$f_3 = x^{15} + x^{10} - 2x^5 - 1,$$
$$f_4 = x^{15} + x^{12} - 4x^9 - 3x^6 + 3x^3 + 1,$$
$$f_5 = x^{15} + 2 = (x^3)^5 + 2 = (x^5)^3 + 2,$$

where f_1 is non-reduced, and the reduced degree of f_2, f_3 (resp. f_4) are 5 (resp. 3). The reduced degrees of f_5 are 3 and 5, and $\#Spl$ is $\#Spl(f_5, 10^{10})$.

	f_1	f_2	f_3	f_4	f_5
$Pr(1)$	0.0000	0	0	0	0
$Pr(2)$	0	0	0	0	0
$Pr(3)$	0.0001	0.0001	0.0001	0	0
$Pr(4)$	0.0023	0.0025	0.0024	0	0
$Pr(5)$	0.0295	0.0278	0.0280	0.0314	0.0313
$Pr(6)$	0.1472	0.1491	0.1495	0.1553	0.1562
$Pr(7)$	0.3206	0.3195	0.3200	0.3139	0.3125
$Pr(8)$	0.3212	0.3205	0.3190	0.3139	0.3127
$Pr(9)$	0.1474	0.1498	0.1500	0.1542	0.1562
$Pr(10)$	0.0294	0.0283	0.0284	0.0314	0.0311
$Pr(11)$	0.0023	0.0024	0.0026	0	0
$Pr(12))$	0.0000	0.0000	0.0001	0	0
$Pr(13)$	0	0	0	0	0
$Pr(14)$	0	0	0	0	0
$\#Spl$	3389785	169310	169660	62503	422912

The following is the table of E_{15} for $1 \le k \le 7$.

k	1	2	3	4	5	6	7
	0.0000	0.0000	0.0001	0.0023	0.0295	0.1473	0.3209

E_5^3 is

k	1	2	3	4	5	6	7
	0	0	0.0001	0.0024	0.0287	0.1490	0.3199

E_3^5 is

k	1	2	3	4	5	6	7
	0	0	0	0	0.0313	0.1563	0.3125

We give other examples whose reduced degrees are 3, 5.

$$
\begin{aligned}
f_6 &= y^5 - 2y^4 + y^3 + 3 & (y = x^3 - 3x^2 + 3x) \\
&= y^3 + 3 & (y = x^5 - 5x^4 + 10x^3 - 9x^2 + 3x), \\
f_7 &= y^5 + y^4 - 2y^3 - 2y^2 + y + 3 & (y = x^3 - 3x^2 + 3x) \\
&= y^3 + 3y^2 + 3y + 3 & (y = x^5 - 5x^4 + 10x^3 - 8x^2 + x), \\
f_8 &= y^5 - 3y^4 + 3y^3 - y^2 + 3 & (y = x^3) \\
&= y^3 + 3 & (y = x^5 - x^2), \\
f_9 &= y^5 + 3 & (y = x^3) \\
&= y^3 + 3 & (y = x^5).
\end{aligned}
$$

The following are $[Pr(3, f), \cdots, Pr(10, f)]$.

$f_6 : [0, 0, 0.0309, 0.1560, 0.3109, 0.3142, 0.1561, 0.0320]$ ($\#Spl = 70506$)

$f_7 : [0, 0, 0.0309, 0.1554, 0.3122, 0.3147, 0.1552, 0.0316]$ ($\#Spl = 70228$)

$f_8 : [0.0000, 0, 0.0301, 0.1575, 0.3110, 0.3126, 0.1573, 0.0315]$ ($\#Spl = 70729$)

$f_9 : [0, 0, 0.0316, 0.1557, 0.3113, 0.3128, 0.1571, 0.0314]$ ($\#Spl = 423207$).

These are approximated by E_3^5.

2.11. *Other examples*

For $f = x^{21} + 2$, $Pr(c, f, 10^9)$ is

$$[0, 0, 0, 0, 0, 0, 0.0075, 0.0555, 0.1657, 0.2727,$$
$$0.2719, 0.1642, 0.0546, 0.0080, 0, 0, 0, 0, 0]$$

and $\#Spl = 201543$.

$$E_3^7 = [0, 0, 0, 0, 0, 0, 0.0078, 0.0547, 0.1641, 0.2734,$$
$$0.2734, 0.1641, 0.0547, 0.0078, 0, 0, 0, 0, 0].$$

For $x^{35} + 2$, $Pr(c, f, 10^9)$ ($11 \leq c \leq 24$) is

$$[0.0002, 0.0011, 0.0069, 0.0292, 0.0806, 0.1594, 0.2220,$$
$$0.2252, 0.1582, 0.0798, 0.0289, 0.0072, 0.0011, 0.0001],$$

and $\#Spl = 60365$. $E_5^7(c)$ $(11 \le c \le 24)$ is

$$[0.0001, 0.0012, 0.0071, 0.0288, 0.0807, 0.1592, 0.2228,$$
$$0.2228, 0.1592, 0.0807, 0.0288, 0.0071, 0.0012, 0.0001],$$

For $f = x^{105} + 2$, $Pr(c, f, 10^9)$ $(42 \le c \le 65)$ is

$$[0.0004, 0.0003, 0.0012, 0.0060, 0.0115,$$
$$0.0242, 0.0435, 0.0683, 0.0963, 0.1183, 0.1348,$$
$$0.1303, 0.1145, 0.0944, 0.0663, 0.0455, 0.0254,$$
$$0.0107, 0.0060, 0.0012, 0.0004, 0.0002, 0, 0.0001],$$

and $\#Spl = 9954$. $E_3^{35}(c)$ $(42 \le c \le 65)$ is

$$[0.0002, 0.0007, 0.0021, 0.0053, 0.0121,$$
$$0.0243, 0.0430, 0.0675, 0.0945, 0.1182, 0.1321,$$
$$0.1321, 0.1182, 0.0945, 0.0675, 0.0430, 0.0243,$$
$$0.0121, 0.0053, 0.0021, 0.0007, 0.0002, 0, 0].$$

References

1. W. Duke, J.B. Friedlander and H. Iwaniec, Equidistribution of roots of a quadratic congruence to prime moduli, *Ann. of Math.* **141** (1995), 423-441.
2. D. Foata, *Distributions eulériennes et mahoniennes sur le groupe des permutations,* in "Higher Combinatorics, Proceedings of the NATO Advanced Study Institute, Berlin, West Germany, September 1-10, 1976" (M.Aigner, Ed), 27-49, Reidel, Dordrecht/Boston, 1977.
3. T. Hadano, Y. Kitaoka, T. Kubota, M. Nozaki, *Densities of sets of primes related to decimal expansion of rational numbers,* Number Theory: Tradition and Modernization, pp. 67-80, W. Zhang and Y. Tanigawa, eds. ©2006 Springer Science + Business Media,Inc.
4. Y. Kitaoka, *Introduction to Algebra (in Japanese),* Kin-en Shobo, 2004
5. Y. Kitaoka, A statistical relation of roots of a polynomial in different local fields, *Mathematics of Computation* **78** (2009), 523-536.
6. Á. Tóth, Roots of Quadratic congruences, *Internat. Math. Res. Notices* 2000, 719-739.

GENERALIZED MODULAR FUNCTIONS AND THEIR FOURIER COEFFICIENTS

WINFRIED KOHNEN

Mathematisches Institut der Universität Heidelberg,
INF 288, D-69120 Heidelberg, Germany
E-mail: winfried@mathi.uni-heidelberg.de

1. Introduction

The purpose of this short survey paper is to discuss certain arithmetical properties proved in [5] of Fourier coefficients of the so-called generalized modular functions. These are complex-valued functions on the complex upper half-plane $\mathcal{H}$ that satisfy the usual defining conditions of a classical modular function, *except* that the character need not be unitary. Many strange things may happen in this context, for example in general there exist non-constant generalized modular functions of weight zero with empty divisors, contrary to the classical situation.

Generalized modular functions used to be studied in the general framework of modular functions on "Grenzkreisgruppen" by Petersson (1937). An intensive study from a more modern point of view was initiated in 2003 by Knopp and Mason [3], pointing out in particular their connection with conformal field theory in physics. Indeed, generating functions of graded traces of automorphisms of certain vertex operator algebras appear as generalized modular functions, and the results of [5] also have some potential applications in this area.

2. Generalized modular functions, main features of the theory and examples

We start with the following formal

Definition 2.1. Let $\Gamma \subset \Gamma_1 := SL_2(\mathbf{Z})$ be a subgroup of finite index. A generalized modular function (GMF) (of weight zero) on Γ is a holomorphic function $f : \mathcal{H} \to \mathbf{C}$ such that

i) $f(\gamma \circ z) = \chi(\gamma) f(z) \ (\forall \gamma \in \Gamma)$

where $\chi : \Gamma \to \mathbf{C}^*$ is a (not necessarily unitary) character of Γ,

ii) f is meromorphic at the cusps of Γ, i.e. for all $\gamma \in \Gamma_1$ one has including the Fourier expansion

$$f(\gamma \circ z) = \sum_{n \geq n_0} a(n) q^{n/M}$$

for some $n_0 \in \mathbf{Z}$ and $M \in \mathbf{N}$ (depending on γ) convergent in a punctured neighborhood of $q = 0$. Here $q = e^{2\pi i z}$ for $z \in \mathcal{H}$ as usual.

We recall that χ is called unitary if $|\chi| = 1$. In this case we will call f *classical.*

More generally, one can define GMF's of integral or even real weight (admitting multiplier systems) as is done in [3]. Note that what we call here a GMF was called a PGMF in [3] (where the letter "P" stands for parabolic).

Remark 2.1. i) In the following we will always suppose that $f \not\equiv 0$.

ii) Since by our requirement ii) above f is invariant under parabolic elements of Γ, the character χ must be parabolic, i.e. $\chi(\gamma) = 1$ for $\forall \gamma \in \Gamma$ that are parabolic.

iii) It is a well-known fact that the abelianized group $\Gamma/[\Gamma, \Gamma]$ has rank $2g_\Gamma + t_\Gamma - 1$, where g_Γ is the genus of Γ and t_Γ is the number of cusps of Γ. Hence if $g_\Gamma > 0$, then there exist "many" non-unitary parabolic characters of Γ.

Let us now discuss *main features* of the theory and also give some *examples.*

Example 2.1. For each parabolic χ of Γ, there exists a GMF on Γ with character χ. *Indeed,* for a large positive even integer ℓ the Eisenstein series

$$E_{\ell,\chi}(z) := \sum_{\begin{pmatrix} \cdot & \cdot \\ c & d \end{pmatrix} \in \Gamma_\infty \backslash \Gamma} \chi(\gamma)^{-1}(cz+d)^{-\ell} \qquad (z \in \mathcal{H})$$

is well defined (since χ is parabolic) and abolutely locally uniformly convergent. The latter follows from a deeper result of Eichler (1965) on an estimate of the minimal length of a word in Γ. Now one constructs f as

$$f = \frac{E_{\ell,\chi}}{E_\ell} P(j)$$

where E_ℓ is the classical Eisenstein series on Γ_1 of weight ℓ and $P(j)$ is a polynomial in the j-invariant. For details we refer the reader to [3, §7].

Example 2.2. If f is a GMF on Γ, then its logarithmic derivative

$$g := \frac{f'}{f} \qquad (2.1)$$

is a meromorphic modular form of weight 2 on Γ with trivial character, such that

a) g has at most simple poles on $\mathcal{H}$ with residues equal to positive integers (in fact, $\mathrm{res}_z g = \mathrm{ord}_z f$);

b) g is meromorphic at the cusps of Γ, and the constant term in the expansion of g at the cusp s (essentially) is an integer (namely $\mathrm{ord}_z f$).

Conversely, if g is a meromorphic modular form of weight 2 on Γ with trivial character satisfying the conditions given in a) and b), then there is a GMF f on Γ (uniquely determined up to multiplication with non-zero complex numbers) satisfying (2.1).

In the context of Example 2, holomorphic modular forms g correspond to GMF's f with divisors supported at the cusps, and cusp forms g correspond to f with empty divisor.

Example 2 follows from more or less standard calculations. Note that for a given g, one sets

$$f := \exp G,$$

where G is an anti-derivative of g on $\mathcal{H}$ with the poles of g removed. The assumption that the poles of g are simple and the residues are positive integers together with the fact that $\mathcal{H}$ is simply connected then guarantee that $\exp G$ indeed is single-valued and extends holomorphically to the whole of $\mathcal{H}$.

Since the genus g_Γ is the dimension of the space of holomorphic cusp forms of weight 2 on Γ, we obtain as an immediate corollary that for $g_\Gamma > 0$ there exist non-constant GMF's f on Γ with empty divisor. This, of course is in contrast to the classical case $|\chi| = 1$.

iii) Let V be a rational, holomorphic, C_2-cofinite vertex operator algebra over $\mathbf{C}$ with a decomposition

$$V = \mathbf{C}1 \oplus V_1 \oplus V_2 \oplus \cdots$$

into finite-dimensional $L(0)$-eigenspaces V_n and with central charge c a positive integer divisible by 24, and let G be a finite group of automorphisms of V leaving each V_n stable. (For basic terminology about vertex operator algebras we refer to [2,5].) A famous example is the Frenkel-Lepowsky-Meurman Moonshine Module $V = V^\natural$, with $c = 24$ and $G = M$ the Monster simple group.

Let a_n be the character of G afforded on V_n, i.e.

$$a_n(h) := tr(h_{|V_n}) \qquad (h \in G).$$

For $h \in G$ let us define a formal q-series by

$$F(h; z) := q^{-c/24} \sum_{n \geq 0} a_n(h) q^n. \qquad (2.2)$$

This is called a "graded trace function". In particular

$$F(1; z) = q^{-c/24} \sum_{n \geq 0} (\dim V_n) q^n = q^{-c/24} + \cdots$$

is the "graded dimension" of V.

General theory implies that the coefficients $a_n(h)$ are always real algebraic integers in $\mathbf{Z}[\zeta_N]$, where ζ_N is a primitive N-th root of unity and N is the order of h in G.

Theorem 2.1. *i) (Zhu [6], 1996) The function $F(1; z)$ is a (classical) modular function of weight zero on Γ_1.*

ii) (Dong-Li-Mason [2], 2000) The function $F(h; z)$ is a GMF on $\Gamma_1(N)$ where N is the order of h in G. If $F(h; z)$ has integral Fourier coefficients, it is in fact on $\Gamma_0(N)$.

Recall that the groups $\Gamma_0(N)$ and $\Gamma_1(N)$ are defined as the subgroups of matrices $\begin{pmatrix} a & b \\ c & d \end{pmatrix}$ in Γ_1 such that $c \equiv 0 \pmod{N}$ and $c \equiv 0 \pmod{N}$, $a \equiv d \equiv 1 \pmod{N}$, respectively. Regarding statement i), recall that every character of the full modular group Γ_1 is unitary (in fact has order dividing 12).

Now one has the following general

Conjecture. *The function $F(h; z)$ is always a classical modular function on a congruence subgroup of Γ_1.*

In the case $V = V^\natural$, $G = M$ the conjecture was proved in a celebrated paper by Borcherds (1992), in which case the $F(h; z)$ indeed are "Hauptmoduln" (i.e. generators of function fields) of genus zero subgroups of $SL_2(\mathbf{R})$. In general the conjecture remains open. However, there is an approach by physicists that provides an argument, but it is not rigorous and relies on other seemingly even more difficult conjectures.

3. Fourier coefficients of GMF's

The general expectation is that the Fourier coefficients of a non-classical GMF are "badly" behaving. More precisely, let f be a non-classical GMF with rational Fourier coefficients $a(n)$ and non-zero leading coefficient equal to 1. Then the $a(n)$ should have unbounded denominators. (In the case of the group $\Gamma_0(N)$ this is stated as a formal conjecture in [5, §1].)

This expectation of course is theoretically supported by the conjecture in combination with Theorem 2.1 ii) stated in §2.

However, there are also numerical examples independent of the above that seem to support it, and the simplest is probably the following.

Example 3.1. Let

$$\eta(z) = e^{\pi i z/12} \prod_{n \geq 1} (1 - e^{2\pi i n z}) \qquad (z \in \mathcal{H})$$

be the Dedekind eta-function and put

$$g(z) := (\eta(z)\eta(11z))^2.$$

Then g is a cusp form of weight 2 on $\Gamma_0(11)$ whose Fourier expansion starts with q. Let us write (cf. (2.1))

$$2\pi i g = \frac{f'}{f} \tag{3.3}$$

and normalize the Fourier expansion of f to start with 1. Then f is a GMF of weight zero on $\Gamma_0(11)$ which is not classical (cf. §2) and (as can be seen inductively from (3.3)) has rational Fourier coefficients. Numerical calculations suggest that the $a(n)$ have unbounded denominators and in fact that infinitely many different primes p occur in the denominators.

Let us point out that to produce numerical examples as given above, one has to assume that g is a cusp form. Indeed, if p is a prime and one starts say with the Eisenstein series of weight 2 on $\Gamma_0(p)$ given by

$$g(z) := P(z) - pP(pz) \qquad (z \in \mathcal{H})$$

where

$$P(z) := 1 - 24 \sum_{n \geq 1} \sigma_1(n) q^n,$$

then one can obtain (3.3) with

$$f(z) = \frac{\Delta(z)}{\Delta(pz)} = q^{1-p} + \cdots,$$

and f is classical and has integral Fourier coefficients. Here

$$\Delta = \eta^{24} \qquad (3.4)$$

as usual denotes the unique normalized cusp form of weight 12 on Γ_1.

In the following, we shall give three results which under certain special and rather restrictive hypotheses show that the above expectation indeed is true. For complete proofs we refer the reader to [5].

Theorem 3.1 ([5]). *Let f be a GMF of weight zero on Γ with empty divisor. Suppose that $a(n)$ is integral for all $n \geq 0$ and $a(0) = 1$. Then $f = 1$ is a constant.*

Let us sketch the *proof.* For simplicity assume that $M = 1$ and write

$$2\pi i g \frac{f'}{f} = \sum_{n \geq 1} b(n) q^n$$

which is a cusp form of weight 2 on Γ (cf. §1). By general theory, the function f has a q-product expansion

$$f = \prod_{n \geq 1} (1 - q^n)^{c(n)} \qquad (|q| < 1)$$

where the exponents $c(n)$ are uniquely determined complex numbers and the general power of complex numbers as usual is defined in terms of the principal branch of the logarithm. The numbers $b(n)$ and $c(n)$ are related by the fomulas

$$b(n) = -\sum_{d|n} dc(d) \qquad (3.5)$$

or

$$nc(n) = -\sum_{d|n} \mu(d) b\left(\frac{n}{d}\right) \qquad (3.6)$$

with μ indicating the Möbius function (see for example [1]).

Since by hypothesis $a(n)$ is integral for all n, the same is true for $c(n)$ for all n, as is well-known and can directly be seen.

However, by a classical and well-known result of Rankin-Selberg (valid for any subgroup Γ of Γ_1 of finite index) one has

$$b(n) \ll_{\epsilon,g} n^{4/5+\epsilon} \qquad (\epsilon > 0).$$

Therefore by (3.6), $c(n) = 0$ for all n large enough, whence by (3.5) the sequence of the $b(n)$ is bounded. We thus deduce that the Rankin-Selberg zeta function

$$R_g(s) := \sum_{n \geq 1} b(n)^2 n^{-s} \qquad (\Re(s) > 2)$$

must converge for $\Re(s) > 1$. But the latter is known to have a pole at $s = 2$ with residue a non-zero absolute constant times the square of the Petersson norm of g. Therefore $g = 0$ and hence $f = 1$ as claimed.

One can prove a somewhat stronger result if one assumes that Γ is a congruence subgroup.

Theorem 3.2 ([5]). *Let f be a GMF on the congruence subgroup $\Gamma \subset \Gamma_1$ with empty divisor. Assume that $a(n)$ is rational for all $n \geq 0$ and in fact is p-integral for all $n \geq 0$, for all but a finite number of primes p. Then f is a constant.*

The proof uses similar ideas in the proof of Theorem 10, a bit harder, though for example, one uses the fact that the space of cusp forms of weight 2 on Γ has a basis of functions with integral Fourier coefficients.

Note that Theorem 3.2 implies that the GMF in Example 3.1 discussed above indeed has unbounded denominators.

Now finally let us look at the case $\Gamma = \Gamma_0(N)$. Recall that the cusps of $\Gamma_0(N)$ are represented by the numbers $\frac{a}{c}$ where $c | N$ and a runs through natural numbers $1 \leq a \leq N, (a, N) = 1$ that are incongruent modulo $(c, \frac{N}{c})$.

If f is a GMF on $\Gamma_0(N)$, then we say that f saitisfies condition (C) if for each $c | N$ the order $\mathrm{ord}_{\frac{a}{c}} f$ is independent of a.

Theorem 3.3 ([5]). *Let f be a GMF on $\Gamma_0(N)$ that satisfies condition (C). Assume that the divisor of f is supported at the cusps. Furthermore suppose that $a(n)$ is rational and p-integral for all n, for all but a finite number of primes p. Then f is an eta-quotient, i.e. there are integers $m \neq 0$ and m_t ($t | N$) and there is a non-zero complex number c such that*

$$f^m(z) = c \prod_{t | N} \Delta(tz)^{m_t}$$

where Δ is given by (3.4).

For the proof one applies a similar argument as given in [4] in the case of N squarefree, granting condition (C), to reduce to the case where f has empty divisor. One then applies Theorem 3.2.

One can show that the graded trace functions $F(h; z)$ discussed in §2 satisfy condition (C) [5]. Hence applying Theorem 3.3 one deduces

Theorem 3.4 ([5]). *Let V be a rational, holomorphic, C_2-cofinite vertex operator algebra and let $h \in G$ be an automorphism of V of order N. Suppose that $F(h; z)$ has integral Fourier coefficients and its poles and zeros are at the cusps. Then $F(h; z)$ is an eta-quotient. In particular it is a classical GMF on $\Gamma_0(N)$.*

One can show that if $V = V^\natural$ and $h \in M$ or if $V = V_L$ is the vertex operator algebra attached to an even positive definite self-dual lattice L and h arises from a fixed point free automorphism of L, then $F(h; z)$ has integral coefficients.

Unfortunately, it seems to be quite hard to verify *a priori* the condition imposed on the divisor of $F(h; z)$ in Theorem 3.4. In general, of course, this condition need not be true. In the case of $V = V^\natural$ it is known (*a posteriori*) that for a number of elements $h \in M$ there exist constants $\alpha_h \in \mathbf{C}$ such that $F(h; z) + \alpha_h$ satisfies the condition.

References

1. J.H. Bruinier, W. Kohnen and K. Ono, The arithmetic of the values of modular functions and the divisors of modular forms, *Compos. Math.* (3) **140** (2004), 552-566.
2. C. Dong, H. Li and G. Mason, Modular invariance of trace functions in orbifold theory and generalized moonshine, *Comm. Math. Phys.* **214** (2000), 1-56.
3. M. Knopp and G. Mason, Generalized modular forms, *J. Number Theory* **99** (2003), 1-18.
4. W. Kohnen, On a certain class of modular functions, *Proc. Amer. Math. Soc.* (1) **133** (2005), 65-70.
5. W. Kohnen and G. Mason, On generalized modular forms and their applications, *Nagoya Math. J.* **192** (2008), 119-136.
6. Y. Zhu, Modular invariance of characters of vertex operator algebras, *J. Amer. Math. Soc.* **9** (1996), 237-302.

FUNCTIONAL RELATIONS FOR ZETA-FUNCTIONS OF ROOT SYSTEMS

YASUSHI KOMORI

Graduate School of Mathematics, Nagoya University,
Chikusa-ku, Nagoya 464-8602, Japan
E-mail: komori@math.nagoya-u.ac.jp

KOHJI MATSUMOTO

Graduate School of Mathematics, Nagoya University,
Chikusa-ku, Nagoya 464-8602, Japan
E-mail: kohjimat@math.nagoya-u.ac.jp

HIROFUMI TSUMURA

Department of Mathematics and Information Sciences, Tokyo Metropolitan University,
1-1, Minami-Ohsawa, Hachioji, Tokyo 192-0397 Japan
E-mail: tsumura@tmu.ac.jp

We report on the theory of functional relations among zeta-functions of root systems, including known formulas for their special values. In the first part of this paper, we present known results on value-relations and functional relations for zeta-functions of root systems of A_2 type. Also, in view of the symmetry of underlying Weyl groups, we discuss a general framework of functional relations. In the second part of this paper, we prove several new results; we give a method for constructing functional relations systematically, and prove new functional relations among zeta-functions of root systems of types A_3, $C_2(\simeq B_2)$, B_3 and C_3, which include Witten's volume formulas as value-relations with explicit values of coefficients.

1. Introduction

Let $\mathbb{N}$ be the set of natural numbers, $\mathbb{N}_0$ the set of non-negative integers, $\mathbb{Z}$ the ring of rational integers, $\mathbb{Q}$ the field of rational numbers, $\mathbb{R}$ the field of real numbers and $\mathbb{C}$ the field of complex numbers.

The multiple zeta value (MZV) of depth r is defined by

$$\zeta(k_1, k_2, \ldots, k_r) = \sum_{m_1 > m_2 > \cdots > m_r \geq 1} \frac{1}{m_1^{k_1} m_2^{k_2} \cdots m_r^{k_r}} \qquad (1.1)$$

for $k_1, k_2, \ldots, k_r \in \mathbb{N}$ with $k_1 > 1$ (see Zagier [51] and Hoffman [12]). It was Euler who first studied the double zeta values and gave some relation formulas among them such as

$$\sum_{j=2}^{k-1} \zeta(j, k-j) = \zeta(k) \qquad (1.2)$$

for $k \in \mathbb{N}$ with $k \geq 3$, where $\zeta(s)$ is the Riemann zeta-function. Equation (1.2) is called the sum formula for double zeta values (see [9]). Research on MZVs has been conducted intensively in this decade (see the survey, [4,13,15]). A recent feature of studies on MZVs is to investigate the structure of the $\mathbb{Q}$-algebra generated by MZVs.

On the other hand, in the late 1990's, it was established that the multiple zeta-function $\zeta(s_1, s_2, ..., s_r)$ of complex variables can be continued meromorphically to the whole complex space $\mathbb{C}^r$ by, for example, Essouabri ([7,8]), Akiyama-Egami-Tanigawa ([1]), Arakawa-Kaneko ([3]), Zhao ([52]) and the second-named author ([23,25]).

Based on these researches, the second-named author raised the following problem several years ago (see, for example, [27]).

Problem. *Are the known relation formulas for multiple zeta values valid only at positive integers, or valid continuously also at other values?*

In other words, is it possible to find certain functional relations for multiple zeta-functions, which include some value-relations for MZVs? A classical example is the following formula which is often called the harmonic product relation:

$$\zeta(s_1)\zeta(s_2) = \zeta(s_1, s_2) + \zeta(s_2, s_1) + \zeta(s_1 + s_2).$$

As a related result, Bradley showed a certain class of functional relations called partition identities (see [5]). However, there are many kinds of relations among MZVs, so it is natural to expect that there will be many other classes of functional relations. For example, it seems interesting to prove certain functional relations which include sum formulas for MZVs. In order to give an "answer" to some specific cases of this Problem (the specification being clear from the context), we consider a wider class of multiple zeta-functions as follows.

Let $\mathfrak{g}$ be a complex semisimple Lie algebra with rank r. The Witten zeta-function associated with $\mathfrak{g}$ is defined by

$$\zeta_W(s; \mathfrak{g}) = \sum_{\varphi} (\dim \varphi)^{-s}, \qquad (1.3)$$

where the summation runs over all finite dimensional irreducible represen-tations φ of $\mathfrak{g}$.

Witten's motivation [50] for introducing the above zeta-function is to express the volumes of certain moduli spaces in terms of special values of (1.3). The expression is called Witten's volume formula, which especially implies that

$$\zeta_W(2k;\mathfrak{g}) = C_W(2k,\mathfrak{g})\pi^{2kn} \tag{1.4}$$

for any $k \in \mathbb{N}$, where n is the number of all positive roots of $\mathfrak{g}$ and $C_W(2k,\mathfrak{g}) \in \mathbb{Q}$ (Witten [50], Zagier [51]). In their work, the value of $C_W(2k,\mathfrak{g})$ is not explicitly given.

Let Δ be the set of all roots of $\mathfrak{g}$ in the vector space V equipped with an inner product $\langle \cdot, \cdot \rangle$, Δ_+ the set of all positive roots of $\mathfrak{g}$, $\Psi = \{\alpha_1, \ldots, \alpha_r\}$ the fundamental system of Δ, and $\alpha_j^\vee$ the coroot associated with α_j ($1 \leq j \leq r$). Let $\lambda_1, \ldots, \lambda_r$ be the fundamental weights satisfying $\langle \alpha_i^\vee, \lambda_j \rangle = \lambda_j(\alpha_i^\vee) = \delta_{ij}$ (Kronecker's delta). A more explicit form of $\zeta_W(s;\mathfrak{g})$ can be written down in terms of roots and weights by using Weyl's dimension formula (see (1.4) of [18]). Inspired by that form, we introduced in [18] the multi-variable version of Witten zeta-function

$$\zeta_r(\mathbf{s};\mathfrak{g}) = \sum_{m_1=1}^\infty \cdots \sum_{m_r=1}^\infty \prod_{\alpha \in \Delta_+} \langle \alpha^\vee, m_1\lambda_1 + \cdots + m_r\lambda_r \rangle^{-s_\alpha}, \tag{1.5}$$

where $\mathbf{s} = (s_\alpha)_{\alpha \in \Delta_+} \in \mathbb{C}^n$. In the case that $\mathfrak{g}$ is of type X_r, we call (1.5) the zeta-function of the root system of type X_r, and denote it by $\zeta_r(\mathbf{s}; X_r)$, where $X = A, B, C, D, E, F, G$. We also use the notation $\zeta_W(s; X_r)$ and $C_W(2k, X_r)$, instead of $\zeta_W(s;\mathfrak{g})$ and $C_W(2k,\mathfrak{g})$, respectively. Note that from (1.5) and [18, (1.7)], we have

$$\zeta_W(s; X_r) = K(X_r)^s \zeta_r(s, \ldots, s; X_r), \tag{1.6}$$

where

$$K(X_r) = \prod_{\alpha \in \Delta_+} \langle \alpha^\vee, \lambda_1 + \cdots + \lambda_r \rangle. \tag{1.7}$$

For example, $K(A_2) = 2$ and $K(C_2) = 6$ (see [18, (2.4) and (2.10)]).

More generally, in [18], we introduced multiple zeta-functions associated with sets of roots. In fact, we studied recursive structures in the family of those zeta-functions, which can be described in terms of Dynkin diagrams of underlying root systems. The meromorphic continuation of those zeta-functions is ensured as a special case of Essouabri's general theorem ([7,8]). It can also be proved by using the Mellin-Barnes integral formula (see [26]).

In [19], we established a general method for evaluating $\zeta_r(s, \ldots, s; X_r)$ at positive integers by considering generalizations of Bernoulli polynomials. In terms of those generalized Bernoulli polynomials, we gave a certain generalization of Witten's volume formula (1.4) with explicit determination of the constant $C_W(2k, X_r)$.

Several cases of zeta-functions of root systems had already been studied. A typical case is of A_2 type:

$$\zeta_2(s_1, s_2, s_3; A_2) = \sum_{m=1}^{\infty} \sum_{n=1}^{\infty} \frac{1}{m^{s_1} n^{s_2} (m+n)^{s_3}}. \qquad (1.8)$$

In the 1950's, Tornheim [39] first studied the value $\zeta_2(d_1, d_2, d_3; A_2)$ for $d_1, d_2, d_3 \in \mathbb{N}$, which is called the Tornheim double sum. Independently, Mordell [33] studied the value $\zeta_2(2d, 2d, 2d; A_2)$ $(d \in \mathbb{N})$ and proved, for example,

$$\zeta_2(2, 2, 2; A_2) = \frac{1}{2835} \pi^6. \qquad (1.9)$$

This determines the value of $C_W(2k, A_2)$ in (1.4). Following their works, several value-relations for $\zeta_2(s_1, s_2, s_3; A_2)$ were obtained by several authors (see [6,14,37,40,51]), and also those for its alternating analogues ([41,43,48]). On the other hand, from the analytic viewpoint, the second-named author [24] studied the multi-variable function $\zeta_2(s_1, s_2, s_3; A_2)$ for $s_1, s_2, s_3 \in \mathbb{C}$ which is also called the Mordell-Tornheim double zeta-function, denoted by $\zeta_{MT,2}(s_1, s_2, s_3)$.

Using $\zeta_2(s_1, s_2, s_3; A_2)$, we can give an "answer" to the Problem, that is, functional relations, for example,

$$\zeta(s+1, 1) - \zeta_2(s, 1, 1; A_2) + \zeta(s+2) = 0 \qquad (1.10)$$

which holds for all $s \in \mathbb{C}$ except for singularities of the three functions on the left-hand side. In fact, letting $s = k - 2$ for $k \geq 3$ in (1.10) and considering partial fraction decompositions, we can obtain the sum formula (1.2). This implies that (1.10) is an answer to the Problem. More generally the third-named author ([47]) proved functional relations for $\zeta_2(s_1, s_2, s_3; A_2)$ which include (1.10) (see Theorem 3.1), and for its alternating analogues ([45]), and its χ-analogues ([46]). A little later, Nakamura gave simple proofs of these results ([34,35]) whose method was inspired by Zagier's lecture.

As for the case of C_2 type, the second-named author defined $\zeta_2(s_1, s_2, s_3, s_4; C_2)$ and studied its analytic properties (see [26]). A little later, the third-named author gave some evaluation formulas for $\zeta_2(k_1, k_2, k_3, k_4; C_2)$ $(k_1, k_2, k_3, k_4 \in \mathbb{N})$ when $k_1 + k_2 + k_3 + k_4$ is odd ([44]).

As for the case of A_3 type, Gunnells and Sczech [10] gave explicit forms of Witten's volume formulas of this type. Recently the second and the third-named authors [30] studied $\zeta_3(\mathbf{s}; A_3)$, and gave certain functional relations for them.

Based on Zagier's work [51] and, in particular, on Nakamura's observation mentioned above, we found that the structural background of those functional relations is given by the symmetry with respect to Weyl groups. From this viewpoint, we considered this structure in [16–19]. In particular, in [19], we gave general forms of functional relations for zeta-functions of root systems. We will recall this result in Section 5.

In the first half of this paper, we summarize known results on functional relations for zeta-functions of root systems, which can be regarded as answers to the Problem. In Section 2, we recall a method of studying relations among Dirichlet series, which is called the 'u-method', introduced in [42]. In Section 3, we summarize known results on functional relations for $\zeta_2(\mathbf{s}; A_2)$. In Section 4, we introduce another method to construct functional relations for multiple Dirichlet series ([31]) which was inspired by Hardy's method of proving the functional equation for $\zeta(s)$ ([11]). In Section 5, we recall general forms of functional relations for $\zeta_r(\mathbf{s}; X_r)$ which we gave in [19]. This is the most general result stated in the present paper, but in general, from this theorem, it is not easy to deduce explicit forms of functional relations in each case. Therefore in the latter half of the paper we give a different method of constructing explicit functional relations. In Section 6, we prove a key lemma (Lemma 6.2) to give a certain procedure to construct functional relations systematically, which has the same flavour as the u-method. In Section 7, by using this lemma combined with a new idea of making use of polylogarithms, we give a functional relation for $\zeta_3(\mathbf{s}; A_3)$ which includes the explicit form of Witten's volume formula of A_3 type. (By "explicit form" we mean that the exact value of $C_W(2k, \mathfrak{g})$ is also determined.) In Sections 8 and 9, by a combination of the methods in Section 4 and in Section 7, we give functional relations for $\zeta_2(\mathbf{s}; C_2)$, for $\zeta_3(\mathbf{s}; B_3)$, and for $\zeta_3(\mathbf{s}; C_3)$ which include explicit forms of Witten's volume formulas.

2. A method to evaluate the Riemann zeta-function

In this section, we introduce a method for evaluating the (multiple) Dirichlet series at positive integers from the information of its trivial zeros, which is called the 'u-method'. In [42], the third-named author first established a method to prove Euler's formula for $\zeta(2k)$ ($k \in \mathbb{N}$). By applying this method to multiple series, several value-relations and functional relations for them have been given (see [40,44–47]). Here we briefly explain this method and recover Euler's formula for $\zeta(s)$:

$$\zeta(2m) = \frac{(-1)^{m-1}2^{2m-1}\pi^{2m}}{(2m)!}B_{2m} \quad (m \in \mathbb{N}), \qquad (2.11)$$

where B_n is the nth Bernoulli number defined by $t/(e^t - 1) = \sum_{n\geq 0} B_n t^n/n!$. For a small $\delta > 0$ and $u \in [1, 1+\delta]$, we let

$$F(t;u) := \frac{2e^t}{e^t + u} = \sum_{n=0}^{\infty} \mathcal{E}_n(u)\frac{t^n}{n!} \qquad (|t| < \pi), \qquad (2.12)$$

where each $\mathcal{E}_n(u)$ is a rational function in u and is continuous for $u \in [1, 1+\delta]$ because $(\partial^k/\partial t^k)F(t;u)$ is continuous for $(t, u) \in \{|t| < \pi\} \times [1, 1+\delta]$. Let $\gamma \in \mathbb{R}$ with $0 < \gamma < \pi$, and $\mathcal{C}_\gamma : z = \gamma e^{it}$ for $0 \leq t \leq 2\pi$, where $i = \sqrt{-1}$. From (2.12), we have

$$\int_{\mathcal{C}_\gamma} F(z;u)z^{-n-1}dz = \frac{(2\pi i)\mathcal{E}_n(u)}{n!} \quad (n \in \mathbb{N}_0). \qquad (2.13)$$

Let $M = M(\gamma) := \max |F(z,u)|$ for $(z, u) \in \mathcal{C}_\gamma \times [1, 1+\delta]$, which is independent of $u \in [1, 1+\delta]$. Then we obtain

$$\frac{|\mathcal{E}_n(u)|}{n!} \leq \frac{1}{2\pi}\int_{\mathcal{C}_\gamma} |F(z;u)|\,|z|^{-n-1}|dz| \leq \frac{M(\gamma)}{\gamma^n} \quad (n \in \mathbb{N}_0).$$

We let $\phi(s;u) = \sum_{n\geq 1}(-u)^{-n}n^{-s}$ for $s \in \mathbb{C}$. As is well known, $\phi(s;u)$ is convergent for $\Re s > 0$ when $u = 1$ and is convergent for any $s \in \mathbb{C}$ when $u > 1$. Furthermore, we see that $\phi(s;1) = (2^{1-s} - 1)\zeta(s)$. When $u > 1$, the second member of (2.12) can be expanded as $-2\sum_{n\geq 1}(-u)^{-n}e^{nt}$. Hence we have $\mathcal{E}_m(u) = -2\phi(-m;u)$ for $m \in \mathbb{N}_0$.

For any $k \in \mathbb{N}$ and $\theta \in (-\pi, \pi)$, we set

$$I_k(\theta;u) := i\sum_{n=1}^{\infty} \frac{(-u)^{-n}\sin(n\theta)}{n^{2k+1}}. \qquad (2.14)$$

Suppose $u \in (1, 1+\delta]$ and $\theta \in (-\pi, \pi)$, then

$$\begin{aligned}
&I_k(\theta;u)\\
&= \sum_{j=0}^{\infty} \phi(2k - 2j;u)\frac{(i\theta)^{2j+1}}{(2j+1)!}\\
&= \sum_{j=0}^{k-1} \phi(2k - 2j;u)\frac{(i\theta)^{2j+1}}{(2j+1)!} - \frac{1}{2}\sum_{m=0}^{\infty} \mathcal{E}_{2m}(u)\frac{(i\theta)^{2m+2k+1}}{(2m+2k+1)!}.
\end{aligned} \qquad (2.15)$$

If $|\theta| < \gamma < \pi$, we see that the right-hand side of (2.15) is uniformly convergent with respect to u on $[1, 1+\delta]$, so is continuous on $u \in [1, 1+\delta]$ (see Remark 2.1 (ii)). On the other hand, the left-hand side of (2.15) is also continuous on $u \in [1, 1+\delta]$ from the definition of $I_k(\theta; u)$. Hence we can let $u \to 1$ on both sides of (2.15).

Now we arrive at the crucial point of the argument. We use the fact that $\zeta(-2m) = 0$, that is, $\mathcal{E}_{2m}(1) = 0$ for $m \in \mathbb{N}$ and $\mathcal{E}_0(1) = 1$ (see Remark 2.1 (i)). Then, for $\theta \in (-\pi, \pi)$, we have

$$I_k(\theta; 1) = \sum_{j=0}^{k-1} \phi(2k - 2j; 1) \frac{(i\theta)^{2j+1}}{(2j+1)!} - \frac{(i\theta)^{2k+1}}{2(2k+1)!}. \tag{2.16}$$

Since $k \geq 1$, each side of (2.16) is continuous in $\theta \in [-\pi, \pi]$. Hence we can let $\theta \to \pi$ on both sides of (2.16) to obtain

$$0 = I_k(\pi; 1) = \sum_{j=0}^{k-1} \phi(2k - 2j; 1) \frac{(i\pi)^{2j+1}}{(2j+1)!} - \frac{(i\pi)^{2k+1}}{2(2k+1)!}.$$

For simplicity, we define

$$\mathcal{A}_{2m} = \phi(2m; 1) \frac{(2m)!}{(i\pi)^{2m}} = (2^{1-2m} - 1)\zeta(2m) \frac{(2m)!}{(i\pi)^{2m}} \quad (m \in \mathbb{N}_0), \tag{2.17}$$

and $\mathcal{A}_0 = -1/2$. Then (2.16) implies that

$$\sum_{j=0}^{k} \binom{2k+1}{2j+1} \mathcal{A}_{2k-2j} = 0$$

for $k \in \mathbb{N}$. Since $\mathcal{A}_0 = -1/2$, we obtain

$$-\frac{t}{2} = \sum_{k=0}^{\infty} \left(\sum_{j=0}^{k} \binom{2k+1}{2j+1} \mathcal{A}_{2k-2j} \right) \frac{t^{2k+1}}{(2k+1)!} = \left(\sum_{m=0}^{\infty} \mathcal{A}_{2m} \frac{t^{2m}}{(2m)!} \right) \frac{e^t - e^{-t}}{2}.$$

We can easily check that

$$\frac{2t}{e^t - e^{-t}} = \frac{2te^t}{e^{2t} - 1} = \sum_{m=0}^{\infty} (2 - 2^{2m}) B_{2m} \frac{t^{2m}}{(2m)!},$$

so we have $\mathcal{A}_{2m} = (2^{2m-1} - 1)B_{2m}$ for any nonnegative integer m. In view of (2.17), we obtain Euler's formula (2.11).

Remark 2.1. (i) It should be noted that the fact

$$\begin{aligned} &-2\phi(-2m; 1) \left(= -2 \left(2^{2m+1} - 1\right) \zeta(-2m)\right) \\ &= \mathcal{E}_{2m}(1) \left(= 0\right) \quad (m \in \mathbb{N}) \end{aligned} \tag{2.18}$$

(the trivial zeros of zeta-function!) plays a vital role in the above argument. In fact, equation (2.18) can be obtained by proving

$$\sum_{n=0}^{\infty} \{-2\phi(-n;1)\} \frac{t^n}{n!} = \sum_{n=0}^{\infty} \{-2\left(2^{n+1}-1\right)\zeta(-n)\} \frac{t^n}{n!}$$

$$= 1 + 2\sum_{n=1}^{\infty} \left(2^{n+1}-1\right) B_{n+1}\frac{t^n}{(n+1)!} \qquad (2.19)$$

$$= 2 + \frac{4}{e^{2t}-1} - \frac{2}{e^t-1} = \frac{2e^t}{e^t+1},$$

because $\zeta(1-k) = -B_k/k$ ($k \in \mathbb{N}$; $k \geq 2$) and $\zeta(0) = -1/2$. (ii) Also we note that $\phi(s;u)$ is continuous in u as $u \to 1+0$ for any $s \in \mathbb{C}$. In fact, similarly to the case of $\zeta(s)$, we can easily see that

$$\phi(s;u) = \frac{1}{(e^{2\pi is}-1)\Gamma(s)} \int_C \frac{e^t}{e^t+u} t^{s-1} dt, \qquad (2.20)$$

where C is the contour, that is, the path which starts at $+\infty$, passes through the real axis, goes around the origin counterclockwise and goes back to $+\infty$. From (2.20), we immediately obtain the desired continuity.

3. Functional relations for $\zeta_2(s_1, s_2, s_3; A_2)$

By applying the method introduced in Section 2 to $\zeta_2(s_1, s_2, s_3; A_2)$, the third-named author gave value-relation formulas for $\zeta_2(s_1, s_2, s_3; A_2)$ (see [40]). Moreover, applying the above method to the double series in complex variables, he gave functional relations for $\zeta_2(s_1, s_2, s_3; A_2)$ (see [47]). The original form in [47, Theorem 4.5] is a little complicated. By using a certain transformation formula (see Lemma 6.1, which is [28, Lemma 2.1]), we obtain the following simpler form.

Theorem 3.1. *For $k, l \in \mathbb{N}_0$,*

$$\zeta_2(k, l, s; A_2) + (-1)^k \zeta_2(k, s, l; A_2) + (-1)^l \zeta_2(l, s, k; A_2)$$

$$= 2\sum_{\rho=0}^{[k/2]} \binom{k+l-2\rho-1}{l-1} \zeta(2\rho)\zeta(s+k+l-2\rho)$$

$$+ 2\sum_{\rho=0}^{[l/2]} \binom{k+l-2\rho-1}{k-1} \zeta(2\rho)\zeta(s+k+l-2\rho) \qquad (3.21)$$

holds for all $s \in \mathbb{C}$ except for singularities of functions on both sides.

Proof. For $\theta, r, u \in \mathbb{R}$ with $r > 1$ and $u \in [1, 1 + \delta]$, and $k, p \in \mathbb{N}_0$, we let

$$\mathfrak{F}(i\theta; r; u) = \sum_{n=1}^{\infty} \frac{(-u)^{-n} e^{int}}{n^r},$$

$$\mathcal{J}_p(i\theta; k; u) = \frac{i^{p-1}}{2} \left\{ \mathfrak{F}(i\theta; k; u) + (-1)^{p-1} \mathfrak{F}(-i\theta; k; u) \right\}$$

$$- \sum_{j=0}^{k} \phi(k - j; u) \, \varepsilon_{p+1+j} \frac{(i\theta)^j}{j!},$$

where $\varepsilon_m = \{1 + (-1)^m\}/2$ for $m \in \mathbb{Z}$. Then, similarly to the proof of (2.16), we see that if $k \not\equiv p \pmod{2}$ and $\theta \in (-\pi, \pi)$ then $\mathcal{J}_p(i\theta; k; u) \to 0$ as $u \to 1$. Let

$$R(s_1, s_2; s_3; u) = \sum_{m,n=1}^{\infty} \frac{(-u)^{-2m-n}}{m^{s_1} n^{s_2} (m + n)^{s_3}},$$

$$S(s_1, s_2; s_3; u) = \sum_{m,n=1}^{\infty} \frac{(-u)^{-m-n}}{m^{s_1} n^{s_2} (m + n)^{s_3}},$$

which are double analogues of $\phi(s; u)$. Then, for $u \in (1, 1 + \delta]$,

$\mathcal{J}_p(i\theta; k; u) \mathfrak{F}(i\theta; r; u)$

$$= i^{p-1} \sum_{N=0}^{\infty} \frac{1}{2} \left\{ S(k, r; -N; u) + (-1)^{p-1} R(k, -N; r; u) \right.$$

$$\left. + (-1)^{p-1+N} R(r, -N; k; u) \right\} \frac{(i\theta)^N}{N!}$$

$$- \sum_{N=0}^{\infty} \sum_{j=0}^{k} \binom{N}{j} \phi(k - j; u) \phi(r + j - N; u) \varepsilon_{p+1+j} \frac{(i\theta)^N}{N!}$$

$$+ \frac{(-i)^{p-1}}{2} \sum_{m=1}^{\infty} \frac{u^{-2m}}{m^{k+r}}.$$

As noted above, the left-hand side tends to 0 as $u \to 1$ when $k \not\equiv p \pmod{2}$ and $\theta \in (-\pi, \pi)$. Therefore, similarly to the case of $\zeta(s)$, we can obtain the original form of the functional relation ([47, Lemma 4.5]):

$$\zeta_2(k, l, s; A_2) + (-1)^k \zeta_2(k, s, l; A_2) + (-1)^l \zeta_2(l, s, k; A_2)$$

$$= 2 \sum_{\substack{j=0 \\ j \equiv k \ (2)}}^{k} \left(2^{1-k+j} - 1\right) \zeta(k-j)$$

$$\times \sum_{\mu=0}^{[j/2]} \frac{(i\pi)^{2\mu}}{(2\mu)!} \binom{l-1+j-2\mu}{j-2\mu} \zeta(l+j+s-2\mu)$$

$$- 4 \sum_{\substack{j=0 \\ j \equiv k \ (2)}}^{k} \left(2^{1-k+j} - 1\right) \zeta(k-j) \sum_{\mu=0}^{[(j-1)/2]} \frac{(i\pi)^{2\mu}}{(2\mu+1)!} \sum_{\substack{\nu=0 \\ \nu \equiv l \ (2)}}^{l} \zeta(l-\nu)$$

$$\times \binom{\nu-1+j-2\mu}{j-2\mu-1} \zeta(\nu+j+s-2\mu)$$

holds for all $s \in \mathbb{C}$ except for singularities of functions on both sides, where $k, l \in \mathbb{N}$. Additionally, using a transformation formula in Lemma 6.1 below ([28, Lemma 2.1]), we obtain (3.21).   □

Example 3.1. Setting $(k, l) = (2, 2)$, $(3, 2)$ in (3.21), we have

$$\zeta_2(2, 2, s; A_2) + 2\zeta_2(2, s, 2; A_2) = 4\zeta(2)\zeta(s+2) - 6\zeta(s+4), \qquad (3.22)$$

$$\zeta_2(3, s, 2; A_2) - \zeta_2(3, 2, s; A_2) - \zeta_2(2, s, 3; A_2)$$

$$= 10\zeta(s+5) - 6\zeta(2)\zeta(s+3). \qquad (3.23)$$

Setting $s = 2$ in (3.22) and (3.23), we have (1.9) and

$$\zeta_2(2, 2, 3; A_2) = 6\zeta(2)\zeta(5) - 10\zeta(7), \qquad (3.24)$$

respectively, where (3.24) was given by Tornheim [39]. Note that $\zeta_2(k, 0, l; A_2) = \zeta(l, k)$. Then, setting $s = 0$ in (3.23), we have $\zeta(2, 3) - \zeta(3, 2) = 10\zeta(5) - 5\zeta(2)\zeta(3)$. On the other hand, it is well-known that $\zeta(3, 2) + \zeta(2, 3) = \zeta(2)\zeta(3) - \zeta(5)$. Combining these results, we obtain the known results

$$\zeta(2, 3) = \frac{9}{2}\zeta(5) - 2\zeta(2)\zeta(3); \quad \zeta(3, 2) = -\frac{11}{2}\zeta(5) + 3\zeta(2)\zeta(3),$$

which were originally obtained by double shuffle relations.

Remark 3.1. In [32], the second and the third-named authors generalized the result in Theorem 3.1 to the case of polylogarithmic analogues, that is,

$$
\sum_{m,n=1}^{\infty} \frac{x^n}{m^k n^l (m+n)^s} + (-1)^k \sum_{m,n=1}^{\infty} \frac{x^n}{m^k n^s (m+n)^l} + (-1)^l \sum_{m,n=1}^{\infty} \frac{x^{m+n}}{m^l n^s (m+n)^k}
$$

$$
= 2 \sum_{\rho=0}^{[k/2]} \binom{k+l-2\rho-1}{k-2\rho} \zeta(2\rho) \sum_{m=1}^{\infty} \frac{x^m}{m^{s+k+l-2\rho}}
$$

$$
+ 2 \sum_{\rho=0}^{[l/2]} \binom{k+l-2\rho-1}{l-2\rho} \zeta(2\rho) \sum_{m=1}^{\infty} \frac{x^m}{m^{s+k+l-2\rho}}, \tag{3.25}
$$

for $x \in \mathbb{C}$ with $|x| \leq 1$. The idea of this generalization gives an important key to construct functional relations for zeta-functions of the type of A_3, C_2, B_3 and C_3 (see Remark 7.1).

In [34], Nakamura gave an alternative simple proof of (3.21) whose method was inspired by Zagier's lecture. We explain this method. We denote by $\{B_n(x)\}$ the Bernoulli polynomials defined by $te^{xt}/(e^t - 1) = \sum_{n \geq 0} B_n(x) t^n / n!$ ($|t| < 2\pi$). It is known (see [2, p.266 - p.267]) that $B_{2j}(0) = B_{2j}$ for $j \in \mathbb{N}_0$ and

$$
B_j(x - [x]) = -\frac{j!}{(2\pi i)^j} \lim_{K \to \infty} \sum_{\substack{k=-K \\ k \neq 0}}^{K} \frac{e^{2\pi i k x}}{k^j} \qquad (j \in \mathbb{N}), \tag{3.26}
$$

where $[\cdot]$ is the integer part. For $k \in \mathbb{Z}$, $j \in \mathbb{N}$ we have

$$
\int_0^1 e^{-2\pi i k x} B_j(x)\, dx = \begin{cases} 0 & (k = 0), \\ -(2\pi i k)^{-j} j! & (k \neq 0), \end{cases} \tag{3.27}
$$

by (3.26). We further quote [2, p.276 19.(b)], for $p, q \geq 1$, which is

$$
B_p(x) B_q(x) = \sum_{k=0}^{\max(p,q)/2} \left\{ p \binom{q}{2k} + q \binom{p}{2k} \right\} \frac{B_{2k} B_{p+q-2k}(x)}{p+q-2k}
$$

$$
- (-1)^p \frac{p! q!}{(p+q)!} B_{p+q}. \tag{3.28}
$$

On the other hand, for $a, b \geq 2$, and $\Re(s) > 1$, we have

$$
\int_0^1 \sum_{l=1}^{\infty} \frac{e^{2\pi i l x}}{l^a} \sum_{m=1}^{\infty} \frac{e^{2\pi i m x}}{m^b} \sum_{n=1}^{\infty} \frac{e^{-2\pi i n x}}{n^s}\, dx = \zeta_2(a, b, s; A_2),
$$

$$\int_0^1 \sum_{l,m=1}^{\infty} \frac{e^{2\pi imx}}{(m+l)^a l^b} \sum_{n=1}^{\infty} \frac{e^{-2\pi inx}}{n^s} dx = \zeta_2(b,s,a;A_2),$$

$$\int_0^1 \sum_{l=1}^{\infty} \frac{e^{2\pi ilx}}{l^{a+b-j}} \sum_{m=1}^{\infty} \frac{e^{-2\pi inx}}{n^s} dx = \zeta(a+b+s-j).$$

Combining these relations and (3.26)-(3.28), we see that

$$\zeta_2(a,b,s;A_2) + (-1)^b \zeta_2(b,s,a;A_2) + (-1)^a \zeta_2(s,a,b;A_2)$$

$$= \frac{2}{a!b!} \sum_{k=0}^{\max(a,b)/2} \left\{ a \binom{b}{2k} + b \binom{a}{2k} \right\}$$

$$\times (a+b-2k-1)!(2k)!\zeta(2k)\zeta(a+b-s-2k), \qquad (3.29)$$

which coincides with (3.21). Nakamura also gave some more generalized formulas for double zeta and L-functions and triple zeta-functions of Mordell and Tornheim type ([35,36]). Furthermore triple zeta and L-functions were studied by Nakamura, Ochiai, and the second and the third-named authors ([28,29]). The aforementioned Lemma 6.1 first appeared in those studies.

4. Another method to construct functional relations for Dirichlet series

In this section, we introduce another method to study functional relations for Dirichlet series, whose basic idea was originally introduced by Hardy. Hardy gave an alternative proof of the functional equation for $\zeta(s)$ ([11], see also [38] Section 2.2). By generalizing this method, we can give functional relations for multiple zeta-functions, for example, (3.21) in Theorem 3.1.

First we consider a general Dirichlet series $Z(s) = \sum_{m=1}^{\infty} a_m m^{-s}$ where $\{a_n\} \subset \mathbb{C}$. Let $\Re s = \rho$ $(\rho \in \mathbb{R})$ be the abscissa of convergence of $Z(s)$. This means that if $\Re s > \rho$ then $Z(s)$ is convergent and if $\Re s < \rho$ then $Z(s)$ is not convergent. We further assume that $0 \le \rho < 1$.

Theorem 4.1 ([31], Theorem 3.1). *Assume that $\sum_{m=1}^{\infty} a_m \sin(mt) = 0$ is boundedly convergent for $t > 0$ and that, for $\rho < s < 1$,*

$$\lim_{\lambda \to \infty} \sum_{m=1}^{\infty} a_m \int_{\lambda}^{\infty} t^{s-1} \sin(mt) dt = 0. \qquad (4.30)$$

Then $Z(s)$ can be continued meromorphically to $\mathbb{C}$, and actually $Z(s) = 0$ for all $s \in \mathbb{C}$. The same conclusion holds if we assume the formulas similar to the above but "sin" (two places) is replaced by "cos".

We give a simple example showing how to apply this theorem. From (2.16) and the formula obtained by differentiating both sides of (2.16), we have

$$\sum_{l=1}^{\infty} \frac{(-1)^l \cos(l\theta)}{l^{2p}} = \sum_{\nu=0}^{p} \phi(2p - 2\nu) \frac{(-1)^\nu \theta^{2\nu}}{(2\nu)!}, \qquad (4.31)$$

$$\sum_{l=1}^{\infty} \frac{(-1)^l \sin(l\theta)}{l^{2q+1}} = \sum_{\nu=0}^{q} \phi(2q - 2\nu) \frac{(-1)^\nu \theta^{2\nu+1}}{(2\nu + 1)!} \qquad (4.32)$$

for $p \in \mathbb{N}$, $q \in \mathbb{N}_0$ and $\theta \in (-\pi, \pi)$. Note that the case $q = 0$ in (4.32) is a little delicate. To prove this case, we define $I_0(\theta; u)$ for $\theta \in (-\pi, \pi)$ and $u \in [1, 1 + \delta]$ by (2.14). Then equation (2.15) in the case $q = 0$ holds for $u \in (1, 1 + \delta]$. From [49, § 3.35] (see also [31, Lemma 4.1]) and Abel's theorem (see [49, § 3.71]), we can let $u \to 1$ in (2.15) for $I_0(\theta; u)$. Then, as well as (2.16), we obtain the case $q = 0$ in (4.32). Additionally we note that if $p, q \in \mathbb{N}$ then (4.31) and (4.32) hold for $\theta \in [-\pi, \pi]$ because both sides are continous for $\theta \in [-\pi, \pi]$.

Combining these results and putting $t = \theta + \pi$, we obtain, for $t \in \mathbb{R} \backslash 2\pi \mathbb{Z}$,

$$\sum_{m,n=1}^{\infty} \frac{\cos((m + n)t)}{m^2 n^2} + 2 \sum_{m,n=1}^{\infty} \frac{\cos(mt)}{n^2(m + n)^2}$$
$$+ 6 \sum_{m=1}^{\infty} \frac{\cos(mt)}{m^4} - 4\zeta(2) \sum_{m=1}^{\infty} \frac{\cos(mt)}{m^2} = 0 \qquad (4.33)$$

(see [31, Lemma 2.2]). We denote by $f(t)$ the left-hand side of (4.33). Note that each sum on the left-hand side of (4.33) is absolutely and uniformly convergent for $t \in \mathbb{R}$. Hence, for $s \in \mathbb{R}$ with $0 < s < 1$, we have

$$0 = \int_0^\infty t^{s-1} f(t) dx$$
$$= \int_0^\infty t^{s-1} \Bigg\{ \sum_{m,n=1}^{\infty} \frac{\cos((m + n)t)}{m^2 n^2} + 2 \sum_{m,n=1}^{\infty} \frac{\cos(mt)}{n^2(m + n)^2}$$
$$+ 6 \sum_{m=1}^{\infty} \frac{\cos(mt)}{m^4} - 4\zeta(2) \sum_{m=1}^{\infty} \frac{\cos(mt)}{m^2} \Bigg\} dt. \qquad (4.34)$$

By the same argument as in [38, Section 2.1], we have

$$\int_\lambda^\infty \frac{\cos(Nx)}{x^{1-s}} dx = \left[\frac{\sin(Nx)}{Nx^{1-s}} \right]_\lambda^\infty - \frac{s-1}{N} \int_\lambda^\infty \frac{\sin(Nx)}{x^{2-s}} dx = O\left(\frac{1}{N\lambda^{1-s}} \right)$$

for $N \in \mathbb{N}$. Using this result, we see that

$$\lim_{\lambda\to\infty} \sum_{m,n=1}^{\infty} \frac{1}{m^2 n^2} \int_{\lambda}^{\infty} \frac{\cos((m+n)x)}{x^{1-s}} dx = 0,$$

$$\lim_{\lambda\to\infty} \sum_{m,n=1}^{\infty} \frac{1}{n^2(m+n)^2} \int_{\lambda}^{\infty} \frac{\cos(mx)}{x^{1-s}} dx = 0,$$

$$\lim_{\lambda\to\infty} \sum_{m=1}^{\infty} \frac{1}{m^l} \int_{\lambda}^{\infty} \frac{\cos(mx)}{x^{1-s}} dx = 0 \quad (l = 2, 4)$$

hold for $0 < s < 1$. Hence we can justify term-by-term integration on the right-hand side of (4.34). Therefore it follows from Theorem 4.1 and the facts

$$\int_0^{\infty} \frac{\cos bx}{x^{1-s}} dx = \frac{\pi}{2} b^{-s} \frac{\sec(\pi(1-s)/2)}{\Gamma(1-s)},$$

$$\int_0^{\infty} \frac{\sin bx}{x^{1-s}} dx = \frac{\pi}{2} b^{-s} \frac{\mathrm{cosec}(\pi(1-s)/2)}{\Gamma(1-s)}$$

for $b > 0$ and $0 < s < 1$ (see [49, Chapter 12]) that the functional relation

$$\zeta_2(2, 2, s; A_2) + 2\zeta_2(s, 2, 2; A_2) + 6\zeta(s+4) - 4\zeta(2)\zeta(s+2) = 0 \quad (4.35)$$

holds for $0 < s < 1$ (and then for any $s \in \mathbb{C}$ by analytic continuation). This coincides with (3.22).

By using this method, we can give functional relations for more general types of multiple zeta-functions (see [31]).

5. A general form of functional relations

In the previous sections, we present various methods to obtain functional relations. However in those methods, it is not clear why these functional relations exist. From the viewpoint of Weyl group symmetry in the underlying Lie algebra structure, we can give a certain explanation of this phenomenon. In fact, in view of the Weyl group symmetry, we can show a general form of functional relations for zeta-functions of root systems. For the details, see [19,22].

First we prepare some notation. Let V be an r-dimensional real vector space equipped with an inner product $\langle \cdot, \cdot \rangle$. Let Δ be a finite reduced root system in V of X_r type and $\Psi = \{\alpha_1, \ldots, \alpha_r\}$ its fundamental system. Let Δ_+ and Δ_- be the set of all positive roots and negative roots respectively. Then we have a decomposition of the root system $\Delta = \Delta_+ \coprod \Delta_-$. Let $Q^\vee$

be the coroot lattice, P the weight lattice, P_+ the set of integral dominant weights and P_{++} the set of integral strongly dominant weights respectively defined by

$$Q^\vee = \bigoplus_{i=1}^{r} \mathbb{Z}\alpha_i^\vee, \quad P = \bigoplus_{i=1}^{r} \mathbb{Z}\lambda_i, \quad P_+ = \bigoplus_{i=1}^{r} \mathbb{N}_0\,\lambda_i, \quad P_{++} = \bigoplus_{i=1}^{r} \mathbb{N}\lambda_i, \quad (5.36)$$

where the fundamental weights $\{\lambda_j\}_{j=1}^r$ form a basis dual to $\Psi^\vee$ satisfying $\langle \alpha_i^\vee, \lambda_j \rangle = \delta_{ij}$. The reflection $\sigma_\alpha : V \to V$ with respect to a root $\alpha \in \Delta$ is defined by $\sigma_\alpha(v) = v - \langle \alpha^\vee, v \rangle \alpha$. For a subset $A \subset \Delta$, let $W(A)$ be the group generated by reflections σ_α for $\alpha \in A$. Let $W = W(\Delta)$ be the Weyl group. Then $\sigma_j = \sigma_{\alpha_j}$ $(1 \leq j \leq r)$ generates W. We denote the fundamental domain called the fundamental Weyl chamber by $C = \{v \in V \mid \langle \Psi^\vee, v \rangle \geq 0\}$, where $\langle \Psi^\vee, v \rangle$ means any of $\langle \alpha^\vee, v \rangle$ for $\alpha^\vee \in \Psi^\vee$. Then W acts on the set of Weyl chambers $WC = \{wC \mid w \in W\}$ simply transitively. Moreover if $wx = y$ for $x, y \in C$, then $x = y$ holds. The stabilizer W_x of a point $x \in V$ is generated by the reflections which stabilize x. We see that $P_+ = P \cap C$. For $w \in W$, we set $\Delta_w = \Delta_+ \cap w^{-1}\Delta_-$.

Let $I \subset \{1, \ldots, r\}$ and $\Psi_I = \{\alpha_i \mid i \in I\} \subset \Psi$. Let V_I be the linear subspace spanned by Ψ_I. Then $\Delta_I = \Delta \cap V_I$ is a root system in V_I whose fundamental system is Ψ_I. For the root system Δ_I, we denote the corresponding coroot lattice (resp. weight lattice etc.) by $Q_I^\vee = \bigoplus_{i \in I} \mathbb{Z}\alpha_i^\vee$ (resp. $P_I = \bigoplus_{i \in I} \mathbb{Z}\lambda_i$ etc.). Let $\Delta_+^\vee$ be the set of all positive coroots, and $W^I = \{w \in W \mid \Delta_{I+}^\vee \subset w\Delta_+^\vee\}$.

Let $\mathbf{y} \in V$ and $\mathbf{s} = (s_\alpha)_{\alpha \in \overline{\Delta}} \in \mathbb{C}^{|\Delta_+|}$, where $\overline{\Delta}$ is the quotient of Δ obtained by identifying α and $-\alpha$. Define an action of W to $\mathbf{s}$ by $(w\mathbf{s})_\alpha = s_{w^{-1}\alpha}$. Now we introduce the "twisted" multiple zeta-function of the form

$$\zeta_r(\mathbf{s}, \mathbf{y}; \Delta) = \sum_{\lambda \in P_{++}} e^{2\pi\sqrt{-1}\langle \mathbf{y}, \lambda \rangle} \prod_{\alpha \in \Delta_+} \frac{1}{\langle \alpha^\vee, \lambda \rangle^{s_\alpha}}. \quad (5.37)$$

A motivation of introducing such a generalized form with exponential factors is to study multiple L-functions of root systems (see [17,21]). When $\mathbf{y} = 0$ in (5.37), the function $\zeta_r(\mathbf{s}; \Delta) = \zeta_r(\mathbf{s}, 0; \Delta)$ coincides with the zeta-function of the root system Δ, defined by (1.5).

For $s \in \mathbb{C}$, $\Re s > 1$ and $x, c \in \mathbb{R}$, let

$$\mathcal{L}_s(x, c) = -\frac{\Gamma(s+1)}{(2\pi\sqrt{-1})^s} \sum_{\substack{n \in \mathbb{Z} \\ n+c \neq 0}} \frac{e^{2\pi\sqrt{-1}(n+c)x}}{(n+c)^s}. \quad (5.38)$$

Then we obtain the following general form of functional relations for zeta-functions of root systems.

Theorem 5.1 ([19], Theorem 4.3). *When $I \neq \emptyset$, for $\mathbf{s} \in S$ and $\mathbf{y} \in V$, we have*

$$S(\mathbf{s}, \mathbf{y}; I; \Delta) \tag{5.39}$$

$$:= \sum_{w \in W^I} \left(\prod_{\alpha \in \Delta_{w^{-1}}} (-1)^{-s_\alpha} \right) \zeta_r(w^{-1}\mathbf{s}, w^{-1}\mathbf{y}; \Delta)$$

$$= (-1)^{|\Delta_+ \backslash \Delta_{I+}|} \left(\prod_{\alpha \in \Delta_+ \backslash \Delta_{I+}} \frac{(2\pi\sqrt{-1})^{s_\alpha}}{\Gamma(s_\alpha + 1)} \right) \sum_{\lambda \in P_{I++}} e^{2\pi\sqrt{-1}\langle \mathbf{y}, \lambda \rangle} \prod_{\alpha \in \Delta_{I+}} \frac{1}{\langle \alpha^\vee, \lambda \rangle^{s_\alpha}}$$

$$\times \int_0^1 \cdots \int_0^1 \exp\left(-2\pi\sqrt{-1} \sum_{\alpha \in \Delta_+ \backslash (\Delta_{I+} \cup \Psi)} x_\alpha \langle \alpha^\vee, \lambda \rangle \right)$$

$$\times \left(\prod_{\alpha \in \Delta_+ \backslash (\Delta_{I+} \cup \Psi)} \mathcal{L}_{s_\alpha}(x_\alpha, 0) \right)$$

$$\times \left(\prod_{i \in I^c} \mathcal{L}_{s_{\alpha_i}} \left(\langle \mathbf{y}, \lambda_i \rangle - \sum_{\alpha \in \Delta_+ \backslash (\Delta_{I+} \cup \Psi)} x_\alpha \langle \alpha^\vee, \lambda_i \rangle, 0 \right) \right) \prod_{\alpha \in \Delta_+ \backslash (\Delta_{I+} \cup \Psi)} dx_\alpha.$$

Remark 5.1. We also studied the case $I = \emptyset$ and gave an integral expression of $S(\mathbf{s}, \mathbf{y}; \emptyset; \Delta)$ similar to (5.39) (see [19, Theorem 4.4]).

Example 5.1. Here we give an alternative proof of (3.23). Set $\Delta_+ = \Delta_+(A_2) = \{\alpha_1, \alpha_2, \alpha_1 + \alpha_2\}$, and $\mathbf{y} = 0$, $\mathbf{s} = (2, s, 3)$ for $s \in \mathbb{C}$ with $\Re s > 1$, $I = \{2\}$, that is, $\Delta_{I+} = \{\alpha_2\}$. Then we see that the left-hand side of (5.39) is

$$S(\mathbf{s}, \mathbf{y}; I; \Delta) = \sum_{m,n=1}^{\infty} \frac{1}{m^2 n^s (m+n)^3} - \sum_{\substack{m,n=1 \\ m \neq n}}^{\infty} \frac{1}{m^2 n^s (-m+n)^3}$$

$$= \zeta_2(2, s, 3; A_2) - \zeta_2(3, 2, s; A_2) + \zeta_2(3, s, 2; A_2).$$

On the other hand, the right-hand side of (5.39) is

$$\left(\frac{(2\pi\sqrt{-1})^2}{2!} \right) \left(\frac{(2\pi\sqrt{-1})^3}{3!} \right) \sum_{m=1}^{\infty} \frac{1}{m^s} \int_0^1 e^{-2\pi\sqrt{-1}mx} \mathcal{L}_2(x, 0) \mathcal{L}_3(-x, 0) dx$$

$$= \left(\frac{(2\pi\sqrt{-1})^2}{2!} \right) \left(\frac{(2\pi\sqrt{-1})^3}{3!} \right) \sum_{m=1}^{\infty} \frac{1}{m^s} \int_0^1 e^{-2\pi\sqrt{-1}mx} B_2(x) B_3(1-x) dx,$$

by $\mathcal{L}_k(x, 0) = B_k(x - [x])$ for $x \in \mathbb{R}$ (see (3.26)). Hence, by using (3.27) and (3.28), we obtain (3.23).

6. Some lemmas for explicit construction of functional relations

From the general form of functional relations in Theorem 5.1, it is possible to deduce explicit formulas of functional relations for zeta-functions of root systems, e.g., as in Example 5.1. However, if a rank of the root system is high, then it seems quite hard to give explicit forms directly from Theorem 5.1. Therefore now we introduce a different procedure to construct explicit functional relations. For this aim, we give some general preparatory lemmas. We first quote the following lemma from our previous paper. Let $\phi(s) := \sum_{n \geq 1}(-1)^n n^{-s} = (2^{1-s} - 1)\zeta(s)$, and $\varepsilon_\nu := (1 + (-1)^\nu)/2$ $(\nu \in \mathbb{Z})$.

Lemma 6.1 ([28] Lemma 2.1). *Let $f, g : \mathbb{N}_0 \to \mathbb{C}$ be arbitrary functions. Then, for $a \in \mathbb{N}$, we have*

$$\sum_{k=0}^{a} \phi(a-k)\varepsilon_{a-k} \sum_{\mu=0}^{[k/2]} f(k-2\mu)\frac{(-1)^\mu \pi^{2\mu}}{(2\mu)!} = \sum_{\xi=0}^{[a/2]} \zeta(2\xi)f(a-2\xi), \quad (6.40)$$

and

$$\sum_{k=1}^{a} \phi(a-k)\varepsilon_{a-k} \sum_{\mu=0}^{[(k-1)/2]} g(k-2\mu)\frac{(-1)^\mu \pi^{2\mu}}{(2\mu+1)!} = -\frac{1}{2}g(a). \quad (6.41)$$

Corollary 6.1. *With the same notation as in Lemma 6.1, put*

$$h(d) := \sum_{\mu=0}^{[d/2]} g(d-2\mu)\frac{(-1)^\mu \pi^{2\mu}}{(2\mu+1)!} \quad (d \in \mathbb{N}_0).$$

Then we have

$$g(d) = -2\sum_{\mu=0}^{d} \phi(d-\mu)\varepsilon_{d-\mu}h(\mu) \quad (d \in \mathbb{N}_0).$$

Proof. In (6.41), we replace $g(x)$ by $g(x-1)$. Then (6.41) implies that

$$\sum_{k=1}^{a} \phi(a-k)\varepsilon_{a-k}h(k-1) = -\frac{1}{2}g(a-1)$$

for $a \in \mathbb{N}$. Replacing a by $d+1$ and $k-1$ by μ, respectively, we obtain the desired assertion. □

Using Lemma 6.1, we prove the following lemma which is a key to construct functional relations. Let $h \in \mathbb{N}$, and

$$\mathfrak{C} := \{C(l) \in \mathbb{C} \mid l \in \mathbb{Z},\ l \neq 0\},$$
$$\mathfrak{D} := \{D(N;m;\eta) \in \mathbb{R} \mid N,m,\eta \in \mathbb{Z},\ N \neq 0,\ m \geq 0,\ 1 \leq \eta \leq h\},$$
$$\mathfrak{A} := \{a_\eta \in \mathbb{N} \mid 1 \leq \eta \leq h\}$$

be sets of numbers indexed by integers. We let

$$\binom{x}{k} := \begin{cases} \dfrac{x(x-1)\cdots(x-k+1)}{k!} & (k \in \mathbb{N}), \\ 1 & (k = 0). \end{cases}$$

Lemma 6.2. *With the above notation, we assume that the infinite series appearing in*

$$\sum_{\substack{N \in \mathbb{Z} \\ N \neq 0}} (-1)^N C(N) e^{iN\theta} - 2 \sum_{\eta=1}^{h} \sum_{k=0}^{a_\eta} \phi(a_\eta - k) \varepsilon_{a_\eta - k}$$

$$\times \sum_{\xi=0}^{k} \left\{ \sum_{\substack{N \in \mathbb{Z} \\ N \neq 0}} (-1)^N D(N; k - \xi; \eta) e^{iN\theta} \right\} \frac{(i\theta)^\xi}{\xi!} \qquad (6.42)$$

are absolutely convergent for $\theta \in [-\pi, \pi]$, and that (6.42) is a constant function for $\theta \in [-\pi, \pi]$. Then, for $d \in \mathbb{N}_0$,

$$\sum_{\substack{N \in \mathbb{Z} \\ N \neq 0}} \frac{(-1)^N C(N) e^{iN\theta}}{N^d} = 2 \sum_{\eta=1}^{h} \sum_{k=0}^{a_\eta} \phi(a_\eta - k) \varepsilon_{a_\eta - k}$$

$$\times \sum_{\xi=0}^{k} \left\{ \sum_{\omega=0}^{k-\xi} \binom{\omega + d - 1}{\omega} (-1)^\omega \sum_{\substack{m \in \mathbb{Z} \\ m \neq 0}} \frac{(-1)^m D(m; k - \xi - \omega; \eta) e^{im\theta}}{m^{d+\omega}} \right\} \frac{(i\theta)^\xi}{\xi!}$$

$$- 2 \sum_{k=0}^{d} \phi(d-k) \varepsilon_{d-k} \sum_{\xi=0}^{k} \left\{ \sum_{\eta=1}^{h} \sum_{\omega=0}^{a_\eta - 1} \binom{\omega + k - \xi}{\omega} (-1)^\omega \right.$$

$$\left. \times \sum_{\substack{m \in \mathbb{Z} \\ m \neq 0}} \frac{D(m; a_\eta - 1 - \omega; \eta)}{m^{k-\xi+\omega+1}} \right\} \frac{(i\theta)^\xi}{\xi!} \qquad (6.43)$$

holds for $\theta \in [-\pi, \pi]$, where the infinite series appearing on both sides of (6.43) are absolutely convergent for $\theta \in [-\pi, \pi]$.

Proof. For $d \in \mathbb{N}_0$, put

$$G_d(\theta; \mathfrak{C}; \mathfrak{D}; \mathfrak{A}) \tag{6.44}$$

$$:= \frac{1}{i^d} \left[\sum_{\substack{l \in \mathbb{Z} \\ l \neq 0}} \frac{(-1)^l C(l) e^{il\theta}}{l^d} - 2 \sum_{\eta=1}^{h} \sum_{k=0}^{a_\eta} \phi(a_\eta - k) \varepsilon_{a_\eta - k} \right.$$

$$\left. \times \sum_{\xi=0}^{k} \left\{ \sum_{\nu=0}^{\xi} \binom{d-1+\xi-\nu}{\xi-\nu} (-1)^\xi \sum_{\substack{m \in \mathbb{Z} \\ m \neq 0}} \frac{(-1)^m D(m; k-\xi; \eta) e^{im\theta}}{m^{d+\xi-\nu}} \frac{(-i\theta)^\nu}{\nu!} \right\} \right]$$

$$= \frac{1}{i^d} \left[\sum_{\substack{l \in \mathbb{Z} \\ l \neq 0}} \frac{(-1)^l C(l) e^{il\theta}}{l^d} - 2 \sum_{\eta=1}^{h} \sum_{k=0}^{a_\eta} \phi(a_\eta - k) \varepsilon_{a_\eta - k} \right.$$

$$\left. \times \sum_{\nu=0}^{k} \left\{ \sum_{\omega=0}^{k-\nu} \binom{d-1+\omega}{\omega} (-1)^\omega \sum_{\substack{m \in \mathbb{Z} \\ m \neq 0}} \frac{(-1)^m D(m; k-\nu-\omega; \eta) e^{im\theta}}{m^{d+\omega}} \right\} \frac{(i\theta)^\nu}{\nu!} \right].$$

Note that the second equality of (6.44) follows by putting $\omega = \xi - \nu$. Then the assumption of (6.42) implies that

$$G_0(\theta; \mathfrak{C}; \mathfrak{D}; \mathfrak{A}) = R_0 \quad (\theta \in [-\pi, \pi]), \tag{6.45}$$

where there exists a constant $R_0 = R_0(\mathfrak{C}; \mathfrak{D}; \mathfrak{A}) \in \mathbb{C}$, because $\binom{-1+\xi-\nu}{\xi-\nu} = 0$ if $\xi > \nu$. Furthermore we can check that $G_d(\theta; \mathfrak{C}; \mathfrak{D}; \mathfrak{A})$ is absolutely convergent with respect to $\theta \in [-\pi, \pi]$ and that

$$\frac{d}{d\theta} G_d(\theta; \mathfrak{C}; \mathfrak{D}; \mathfrak{A}) = G_{d-1}(\theta; \mathfrak{C}; \mathfrak{D}; \mathfrak{A}) \quad (d \in \mathbb{N}). \tag{6.46}$$

In fact, if we differentiate the second member of (6.44) with respect to θ, then we have

$$\frac{d}{d\theta} G_d(\theta; \mathfrak{C}; \mathfrak{D}; \mathfrak{A})$$

$$= \frac{1}{i^{d-1}} \left[\sum_{\substack{l \in \mathbb{Z} \\ l \neq 0}} \frac{(-1)^l C(l) e^{il\theta}}{l^{d-1}} - 2 \sum_{\eta=1}^{h} \sum_{k=0}^{a_\eta} \phi(a_\eta - k) \varepsilon_{a_\eta - k} \right.$$

$$\times \sum_{\xi=0}^{k} \left\{ \sum_{\nu=0}^{\xi} \binom{d-1+\xi-\nu}{\xi-\nu} (-1)^\xi \sum_{\substack{m \in \mathbb{Z} \\ m \neq 0}} \frac{(-1)^m D(m; k-\xi; \eta) e^{im\theta}}{m^{d+\xi-\nu-1}} \frac{(-i\theta)^\nu}{\nu!} \right.$$

$$\left. \left. - \sum_{\nu=1}^{\xi} \binom{d-1+\xi-\nu}{\xi-\nu} (-1)^\xi \sum_{\substack{m \in \mathbb{Z} \\ m \neq 0}} \frac{(-1)^m D(m; k-\xi; \eta) e^{im\theta}}{m^{d+\xi-\nu}} \frac{(-i\theta)^{\nu-1}}{(\nu-1)!} \right\} \right].$$

Replacing $\nu - 1$ by μ in the last member of the above summation, and using the well-known relation

$$-\binom{m-1}{l-1} + \binom{m}{l} = \binom{m-1}{l} \quad (l, m \in \mathbb{N}),$$

we obtain (6.46).

By integrating both sides of (6.45) and multiplying by i on both sides, we have $iG_1(\theta; \mathfrak{C}; \mathfrak{D}; \mathfrak{A}) = R_0(i\theta) + R_1$ with some constant $R_1 = R_1(\mathfrak{C}; \mathfrak{D}; \mathfrak{A})$. Repeating this operation, and by (6.46), we obtain

$$i^d G_d(\theta; \mathfrak{C}; \mathfrak{D}; \mathfrak{A}) = \sum_{k=0}^{d} R_{d-k} \frac{(i\theta)^k}{k!}, \tag{6.47}$$

where there exist constants $R_k = R_k(\mathfrak{C}; \mathfrak{D}; \mathfrak{A})$ $(0 \leq k \leq d)$. We can explicitly determine $\{R_k\}$ as follows. Putting $\theta = \pm\pi$ in (6.47) with $d+1$ $(d \in \mathbb{N}_0)$, we have

$$\frac{i^{d+1}}{2(i\pi)} \{G_{d+1}(\pi; \mathfrak{C}; \mathfrak{D}; \mathfrak{A}) - G_{d+1}(-\pi; \mathfrak{C}; \mathfrak{D}; \mathfrak{A})\}$$

$$= \sum_{\mu=0}^{[d/2]} R_{d-2\mu} \frac{(i\pi)^{2\mu}}{(2\mu+1)!}. \tag{6.48}$$

It follows from (6.44) that the left-hand side of (6.48) is equal to

$$-2 \sum_{\eta=1}^{h} \sum_{k=0}^{a_\eta} \phi(a_\eta - k) \varepsilon_{a_\eta - k} \tag{6.49}$$

$$\times \sum_{\tau=0}^{[(k-1)/2]} \left\{ \sum_{\omega=0}^{k-2\tau-1} \binom{d+\omega}{\omega} (-1)^\omega \sum_{\substack{m \in \mathbb{Z} \\ m \neq 0}} \frac{D(m; k - 2\tau - 1 - \omega; \eta)}{m^{d+\omega+1}} \right\} \frac{(i\pi)^{2\tau}}{(2\tau+1)!}.$$

Applying (6.41) with

$$g(x) = \sum_{\omega=0}^{x-1} \binom{d+\omega}{\omega} (-1)^\omega \sum_{\substack{m \in \mathbb{Z} \\ m \neq 0}} \frac{D(m; x - 1 - \omega; \eta)}{m^{d+\omega+1}},$$

we can rewrite (6.48) as

$$\sum_{\eta=1}^{h} \sum_{\omega=0}^{a_\eta-1} \binom{d+\omega}{\omega} (-1)^\omega \sum_{\substack{m \in \mathbb{Z} \\ m \neq 0}} \frac{D(m; a_\eta - 1 - \omega; \eta)}{m^{d+\omega+1}}$$

$$= \sum_{\nu=0}^{[d/2]} R_{d-2\nu} \frac{(i\pi)^{2\nu}}{(2\nu+1)!}. \tag{6.50}$$

Hence, by Corollary 6.1, we have

$$R_d = R_d(\mathfrak{C}; \mathfrak{D}; \mathfrak{A}) \tag{6.51}$$

$$= -2 \sum_{\nu=0}^{d} \phi(d - \nu)\varepsilon_{d-\nu} \sum_{\eta=1}^{h} \sum_{\omega=0}^{a_\eta - 1} \binom{\nu + \omega}{\omega} (-1)^\omega \sum_{\substack{m \in \mathbb{Z} \\ m \neq 0}} \frac{D(m; a_\eta - 1 - \omega; \eta)}{m^{\nu+\omega+1}}$$

for $d \in \mathbb{N}_0$. Therefore, combining (6.44),(6.47) and (6.51), we have

$$\sum_{\substack{l \in \mathbb{Z} \\ l \neq 0}} \frac{(-1)^l C(l) e^{il\theta}}{l^d} - 2 \sum_{\eta=1}^{h} \sum_{k=0}^{a_\eta} \phi(a_\eta - k)\varepsilon_{a_\eta - k} \tag{6.52}$$

$$\times \sum_{\nu=0}^{k} \left\{ \sum_{\omega=0}^{k-\nu} \binom{d - 1 + \omega}{\omega} (-1)^\omega \sum_{\substack{m \in \mathbb{Z} \\ m \neq 0}} \frac{(-1)^m D(m; k - \nu - \omega; \eta) e^{im\theta}}{m^{d+\omega}} \right\} \frac{(i\theta)^\nu}{\nu!}$$

$$= -2 \sum_{\mu=0}^{d} \sum_{\nu=0}^{d-\mu} \phi(d - \mu - \nu)\varepsilon_{d-\mu-\nu}$$

$$\times \sum_{\eta=1}^{h} \sum_{\omega=0}^{a_\eta - 1} \binom{\nu + \omega}{\omega} (-1)^\omega \sum_{\substack{m \in \mathbb{Z} \\ m \neq 0}} \frac{D(m; a_\eta - 1 - \omega; \eta)}{m^{\nu+\omega+1}} \frac{(i\theta)^\mu}{\mu!}.$$

Changing the running indices (μ, ν) into (k, ξ) with $k = \mu + \nu$ and $\xi = \mu \leq k$, we find that the right-hand side of (6.52) is equal to the second term on the right-hand side of (6.43). $\square$

7. Functional relations for $\zeta_3(s; A_3)$

In the rest of this paper, we will give explicit forms of functional relations for zeta-functions of root systems by using lemmas proved in Section 6. In this section, we consider the case of A_r type.

Fix $p \in \mathbb{N}$ and $s \in \mathbb{R}$ with $s > 1$ and $x \in \mathbb{C}$ with $|x| = 1$. From (4.31), we have

$$\left(\sum_{\substack{l \in \mathbb{Z} \\ l \neq 0}} \frac{(-1)^l e^{il\theta}}{l^{2p}} - 2 \sum_{j=0}^{p} \phi(2p - 2j) \frac{(i\theta)^{2j}}{(2j)!} \right) \sum_{m=1}^{\infty} \frac{(-1)^m x^m e^{im\theta}}{m^s} = 0 \tag{7.53}$$

for $\theta \in [-\pi, \pi]$. Hence we have

$$
\sum_{\substack{l \in \mathbb{Z}, \; l \neq 0 \\ m \geq 1 \\ l+m \neq 0}} \frac{(-1)^{l+m} x^m e^{i(l+m)\theta}}{l^{2p} m^s}
$$

$$
- 2 \sum_{j=0}^{p} \phi(2p - 2j) \left\{ \sum_{m=1}^{\infty} \frac{(-1)^m x^m e^{im\theta}}{m^s} \right\} \frac{(-1)^j \theta^{2j}}{(2j)!}
$$

$$
= - \sum_{m=1}^{\infty} \frac{x^m}{m^{s+2p}} \tag{7.54}
$$

for $\theta \in [-\pi, \pi]$. Now we use Lemma 6.2 with $h = 1$, $a_1 = 2p$,

$$
C(N) = \sum_{\substack{l \neq 0, \; m \geq 1 \\ l+m=N}} \frac{x^m}{l^{2p} m^s} \quad (N \in \mathbb{Z}, \; N \neq 0),
$$

and $D(N; \mu; 1) = x^N N^{-s}$ (if $\mu = 0$ and $N \geq 1$), or $= 0$ (otherwise). Under these choices, we see that the left-hand side of (7.54) is of the form (6.42) because $\varepsilon_{2p-k} = 1$ $(0 \leq k \leq 2p)$ implies $k = 2j$ $(0 \leq j \leq p)$. Furthermore the right-hand side of (7.54) is a constant, because we fix s, x and p. Therefore we can apply Lemma 6.2 with $d = 2q$ for $q \in \mathbb{N}$. Then (6.43) gives that

$$
0 = \sum_{\substack{l \neq 0, \; m \geq 1 \\ l+m \neq 0}} \frac{(-1)^{l+m} x^m e^{i(l+m)\theta}}{l^{2p} m^s (l+m)^{2q}}
$$

$$
- 2 \sum_{j=0}^{p} \phi(2p - 2j) \sum_{\xi=0}^{2j} \binom{2q - 1 + 2j - \xi}{2q - 1} (-1)^{2j - \xi}
$$

$$
\times \sum_{m=1}^{\infty} \frac{(-1)^m x^m e^{im\theta}}{m^{s+2q+2j-\xi}} \frac{(i\theta)^\xi}{\xi!}
$$

$$
+ 2 \sum_{j=0}^{q} \phi(2q - 2j) \sum_{\xi=0}^{2j} \binom{2p - 1 + 2j - \xi}{2p - 1} (-1)^{2p-1}
$$

$$
\times \sum_{m=1}^{\infty} \frac{x^m}{m^{s+2p+2j-\xi}} \frac{(i\theta)^\xi}{\xi!} = 0 \tag{7.55}
$$

for $\theta \in [-\pi, \pi]$, where we replace k by $2j$ in (6.43) because $(a_1, d) = (2p, 2q)$ as mentioned above. This relation will play an important role in the next section. Here we apply Lemma 6.1 to the real part of (7.55) in the case $\theta = \pi$ and $x = 1$. Then we have the following.

Proposition 7.1. *For $p, q \in \mathbb{N}$,*

$$\sum_{\substack{l \in \mathbb{Z},\, l \neq 0 \\ m \geq 1 \\ l+m \neq 0}} \frac{1}{l^{2p} m^s (l+m)^{2q}}$$

$$= 2 \sum_{\nu=0}^{p} \binom{2p + 2q - 2\nu - 1}{2q - 1} \zeta(2\nu) \zeta(s + 2p + 2q - 2\nu)$$

$$+ 2 \sum_{\nu=0}^{q} \binom{2p + 2q - 2\nu - 1}{2p - 1} \zeta(2\nu) \zeta(s + 2p + 2q - 2\nu) \tag{7.56}$$

holds for $s \in \mathbb{C}$ except for singularities of functions on both sides.

Note that (7.56) essentially coincides with (3.21) in the case $(k, l) = (2p, 2q)$, because the left-hand side of (7.56) can be easily transformed to that of (3.21) in the case $(k, l) = (2p, 2q)$. This implies that, from relation (7.53) which is given by multiplying two quantities of A_1 type, we can obtain relation (7.56) for zeta-functions of A_2 and A_1 type. From the view point of Dynkin diagrams, we may say that (7.53) corresponds to two vertices, and the above procedure of applying Lemma 6.2 to obtain (7.56) corresponds to the fact that the Dynkin diagram of A_2 can be produced by joining those two vertices. Based on this observation, instead of (7.53), we next combine a quantity of A_2 type and a quantity of A_1 type to get a relation for zeta-functions of A_3 and of A_2 type. From (1.5), we see that the zeta-function of root system of A_3 type is defined by

$$\zeta_3(s_1, s_2, s_3, s_4, s_5, s_6; A_3)$$

$$= \sum_{l=1}^{\infty} \sum_{m=1}^{\infty} \sum_{n=1}^{\infty} \frac{1}{l^{s_1} m^{s_2} n^{s_3} (l+m)^{s_4} (m+n)^{s_5} (l+m+n)^{s_6}}. \tag{7.57}$$

Fix $p, q, b \in \mathbb{N}$ and $s \in \mathbb{R}$ with $s > 1$ and $x \in \mathbb{C}$ with $|x| = 1$. From (4.31), we have

$$\left(\sum_{\substack{l \in \mathbb{Z} \\ l \neq 0}} \frac{(-1)^l e^{il\theta}}{l^{2p}} - 2 \sum_{j=0}^{p} \phi(2p - 2j) \frac{(i\theta)^{2j}}{(2j)!} \right)$$

$$\times \sum_{\substack{m \in \mathbb{Z},\, m \neq 0 \\ n \geq 1 \\ m+n \neq 0}} \frac{(-1)^{m+n} x^n e^{i(m+n)\theta}}{m^{2q} n^s (m+n)^{2b}} = 0 \tag{7.58}$$

for $\theta \in [-\pi, \pi]$. This formula corresponds to a diagram of A_2 and another vertex. Next we use Lemma 6.2, which gives the procedure of joining these two figures to obtain the diagram of A_3. First, by separating the terms corresponding to $l + m + n = 0$, we have

$$\sum_{\substack{l,m\neq 0,\, n\geq 1 \\ m+n\neq 0,\, l+m+n\neq 0}} \frac{(-1)^{l+m+n}x^n e^{i(l+m+n)\theta}}{l^{2p}m^{2q}n^s(m+n)^{2b}}$$

$$-2\sum_{j=0}^{p}\phi(2p-2j)\left\{\sum_{\substack{m\neq 0 \\ n\geq 1 \\ m+n\neq 0}} \frac{(-1)^{m+n}x^n e^{i(m+n)\theta}}{m^{2q}n^s(m+n)^{2b}}\right\}\frac{(-1)^j\theta^{2j}}{(2j)!}$$

$$=-\sum_{\substack{m\neq 0 \\ n\geq 1 \\ m+n\neq 0}} \frac{x^n}{m^{2q}n^s(m+n)^{2p+2b}} \qquad (7.59)$$

for $\theta \in [-\pi,\pi]$. We can apply Lemma 6.2 with $h=1$, $a_1 = 2p$,

$$C(N) = \sum_{\substack{l,m\neq 0 \\ n\geq 1 \\ m+n\neq 0 \\ l+m+n=N}} \frac{x^n}{l^{2p}m^{2q}n^s(m+n)^{2b}},$$

$$D(N;\mu;1) = \begin{cases} \displaystyle\sum_{\substack{m\neq 0 \\ n\geq 1 \\ m+n=N}} \frac{x^n}{m^{2q}n^s(m+n)^{2b}} & (\nu=0,\, N\neq 0), \\[2mm] 0 & (\nu\geq 1,\, N\neq 0), \end{cases}$$

and $d = 2c$ ($c \in \mathbb{N}$). Formula (7.59) implies that the assumptions of Lemma 6.2 are satisfied, so consequently we have

$$\sum_{\substack{l,m\neq 0,\, n\geq 1 \\ m+n\neq 0,\, l+m+n\neq 0}} \frac{(-1)^{l+m+n}x^n e^{i(l+m+n)\theta}}{l^{2p}m^{2q}n^s(m+n)^{2b}(l+m+n)^{2c}}$$

$$=2\sum_{j=0}^{p}\phi(2p-2j)\sum_{\xi=0}^{2j}\binom{2j-\xi+2c-1}{2c-1}(-1)^{2j-\xi}$$

$$\times \sum_{\substack{m\neq 0 \\ n\geq 1 \\ m+n\neq 0}} \frac{(-1)^{m+n}x^n e^{i(m+n)\theta}}{m^{2q}n^s(m+n)^{2b+2c+2j-\xi}}\frac{(i\theta)^\xi}{\xi!}$$

$$-2\sum_{j=0}^{c}\phi(2c-2j)\sum_{\xi=0}^{2j}\binom{2j-\xi+2p-1}{2p-1}(-1)^{2p-1}$$

$$\times \sum_{\substack{m\neq 0 \\ n\geq 1 \\ m+n\neq 0}} \frac{x^n}{m^{2q}n^s(m+n)^{2p+2b+2j-\xi}}\frac{(i\theta)^\xi}{\xi!}.$$

Putting $x = -e^{-i\theta}$ ($\theta \in \mathbb{R}$) on both sides and separating the terms corresponding to $l + m = 0$, we have

$$\sum_{\substack{l,m \neq 0,\, n \geq 1 \\ l+m \neq 0,\, m+n \neq 0 \\ l+m+n \neq 0}} \frac{(-1)^{l+m} e^{i(l+m)\theta}}{l^{2p} m^{2q} n^s (m+n)^{2b} (l+m+n)^{2c}}$$

$$-2 \sum_{j=0}^{p} \phi(2p-2j) \sum_{\xi=0}^{2j} \binom{2j-\xi+2c-1}{2c-1}(-1)^{2j-\xi}$$

$$\times \sum_{\substack{m \neq 0 \\ n \geq 1 \\ m+n \neq 0}} \frac{(-1)^m e^{im\theta}}{m^{2q} n^s (m+n)^{2b+2c+2j-\xi}} \frac{(i\theta)^\xi}{\xi!}$$

$$+2 \sum_{j=0}^{c} \phi(2c-2j) \sum_{\xi=0}^{2j} \binom{2j-\xi+2p-1}{2p-1}(-1)^{2p-1}$$

$$\times \sum_{\substack{m \neq 0 \\ n \geq 1 \\ m+n \neq 0}} \frac{(-1)^n e^{-in\theta}}{m^{2q} n^s (m+n)^{2p+2b+2j-\xi}} \frac{(i\theta)^\xi}{\xi!}$$

$$= - \sum_{\substack{m \neq 0 \\ n \geq 1 \\ m+n \neq 0}} \frac{1}{m^{2p+2q} n^{s+2c} (m+n)^{2b}}.$$

Again we apply Lemma 6.2 with $h = 2$, $a_1 = 2p$, $a_2 = 2c$ and $d = 2a$ for $a \in \mathbb{N}$. Then

$$\sum_{\substack{l,m \neq 0,\, n \geq 1 \\ l+m \neq 0,\, m+n \neq 0 \\ l+m+n \neq 0}} \frac{(-1)^{l+m} e^{i(l+m)\theta}}{l^{2p} m^{2q} n^s (l+m)^{2a} (m+n)^{2b} (l+m+n)^{2c}}$$

$$= 2 \sum_{j=0}^{p} \phi(2p-2j) \sum_{\xi=0}^{2j} \sum_{\omega=0}^{2j-\xi} \binom{\omega+2a-1}{\omega}(-1)^{\omega} \binom{2j-\xi-\omega+2c-1}{2c-1}$$

$$\times (-1)^{2j-\xi-\omega} \sum_{\substack{m \neq 0 \\ n \geq 1 \\ m+n \neq 0}} \frac{(-1)^m e^{im\theta}}{m^{2q+2a+\omega} n^s (m+n)^{2b+2c+2j-\xi-\omega}} \frac{(i\theta)^\xi}{\xi!}$$

$$-2 \sum_{j=0}^{c} \phi(2c-2j) \sum_{\xi=0}^{2j} \sum_{\omega=0}^{2j-\xi} \binom{\omega+2a-1}{\omega}(-1)^{\omega} \binom{2j-\xi-\omega+2p-1}{2p-1}$$

$$\times (-1)^{2p-1}(-1)^{2a+\omega} \sum_{\substack{m \neq 0 \\ n \geq 1 \\ m+n \neq 0}} \frac{(-1)^n e^{-in\theta}}{m^{2q} n^{s+2a+\omega} (m+n)^{2p+2b+2j-\xi-\omega}} \frac{(i\theta)^\xi}{\xi!}$$

$$-2\sum_{j=0}^{a}\phi(2a-2j)\sum_{\xi=0}^{2j}\sum_{\omega=0}^{2p-1}\binom{\omega+2j-\xi}{\omega}(-1)^{\omega}\binom{2p+2c-2-\omega}{2c-1}$$

$$\times(-1)^{2p-1-\omega}\sum_{\substack{m\neq 0\\ n\geq 1\\ m+n\neq 0}}\frac{1}{m^{2q+2j-\xi+\omega+1}n^{s}(m+n)^{2p+2b+2c-1-\omega}}\frac{(i\theta)^{\xi}}{\xi!}$$

$$+2\sum_{j=0}^{a}\phi(2a-2j)\sum_{\xi=0}^{2j}\sum_{\omega=0}^{2c-1}\binom{\omega+2j-\xi}{\omega}(-1)^{\omega}\binom{2p+2c-2-\omega}{2p-1}$$

$$\times(-1)^{2p-1}(-1)^{2j-\xi+\omega+1}\sum_{\substack{m\neq 0\\ n\geq 1\\ m+n\neq 0}}\frac{1}{m^{2q}n^{s+2j-\xi+\omega+1}(m+n)^{2p+2b+2c-1-\omega}}\frac{(i\theta)^{\xi}}{\xi!}$$

holds for $\theta\in[-\pi,\pi]$. Now we put $\theta=\pi$ in this equation and take its real part. For simplicity, we denote the obtained equation by $J_1=J_2+J_3+J_4+J_5$. First we consider J_1. This can be divided into the following:

$$\sum_{\substack{l\geq 1\\ m\geq 1\\ n\geq 1}}+\sum_{\substack{l\leq -1\\ m\geq 1\\ n\geq 1\\ l+m\neq 0\\ l+m+n\neq 0}}+\sum_{\substack{l\geq 1\\ m\leq -1\\ n\geq 1\\ l+m\neq 0\\ m+n\neq 0\\ l+m+n\neq 0}}+\sum_{\substack{l\leq -1\\ m\leq -1\\ n\geq 1\\ m+n\neq 0\\ l+m+n\neq 0}},$$

which we denote by $J_{11}+J_{12}+J_{13}+J_{14}$. We can immediately see that $J_{11}=\zeta_3(2p,2q,s,2a,2b,2c;A_3)$. For J_{12}, replacing l by $-l$, we have

$$J_{12}=\sum_{\substack{l\geq 1,\, m\geq 1\\ n\geq 1,\, l\neq m\\ l\neq m+n}}\frac{1}{(-l)^{2p}m^{2q}n^{s}(-l+m)^{2a}(m+n)^{2b}(-l+m+n)^{2c}}.$$

Here, putting $j=-l+m$ if $l<m$ and $k=l-m$ if $l>m$, respectively, we have

$$J_{12}=\sum_{\substack{l\geq 1,\, j\geq 1\\ n\geq 1}}\frac{1}{l^{2p}(l+j)^{2q}n^{s}j^{2a}(l+j+n)^{2b}(j+n)^{2c}}$$

$$+\sum_{\substack{k\geq 1,\, m\geq 1\\ n\geq 1,\, k\neq n}}\frac{1}{(k+m)^{2p}m^{2q}n^{s}k^{2a}(m+n)^{2b}(-k+n)^{2c}},$$

where the first term on the right-hand side is $\zeta_3(2p,2a,s,2q,2c,2b;A_3)$. Furthermore, putting $j'=-k+n$ if $k<n$ and $k'=k-n$ if $k>n$, respectively, in the second term on the right-hand side, we can obtain

$$J_{12}=\zeta_3(2p,2a,s,2q,2c,2b;A_3)+\zeta_3(2q,2a,2c,2p,s,2b;A_3)$$
$$+\zeta_3(2q,s,2c,2b,2a,2p;A_3).$$

Similarly we can express J_{13} and J_{14} as sums of values of the zeta-function of A_3 type. Therefore J_1 can be transformed to the left-hand side of the following theorem. On the other hand, if we apply (6.40) to $J_2 + J_3 + J_4 + J_5$, then it can be transformed to the right-hand side of the following theorem with

$$T(2d, s, 2e) = \sum_{\substack{m \neq 0,\ n \geq 1 \\ m+n \neq 0}} \frac{1}{m^{2d} n^s (m+n)^{2e}}$$

for $d, e \in \mathbb{N}$. From Proposition 7.1, we see that $T(2d, s, 2e)$ can be written as (7.60) below.

Theorem 7.1. *For* $p, q, a, b, c \in \mathbb{N}$,

$$\zeta_3(2p, 2q, s, 2a, 2b, 2c; A_3) + \zeta_3(2p, 2a, s, 2q, 2c, 2b; A_3)$$

$$+ \zeta_3(2q, 2a, 2c, 2p, s, 2b; A_3) + \zeta_3(2q, s, 2c, 2b, 2a, 2p; A_3)$$

$$+ \zeta_3(2a, 2p, 2b, 2q, 2c, s; A_3) + \zeta_3(2a, 2c, 2b, s, 2p, 2q; A_3)$$

$$+ \zeta_3(s, 2c, 2p, 2a, 2b, 2q; A_3) + \zeta_3(2b, 2q, 2a, s, 2p, 2c; A_3)$$

$$+ \zeta_3(2b, s, 2a, 2q, 2c, 2p; A_3) + \zeta_3(2p, 2b, s, 2c, 2q, 2a; A_3)$$

$$+ \zeta_3(2c, 2p, 2q, 2b, 2a, s; A_3) + \zeta_3(2c, 2b, 2q, 2p, s, 2a; A_3)$$

$$= 2 \sum_{\xi=0}^{p} \zeta(2\xi) \sum_{\omega=0}^{2p-2\xi} \binom{\omega + 2a - 1}{\omega} \binom{2p + 2c - 2\xi - \omega - 1}{2c - 1}$$

$$\times T(2q + 2a + \omega, s, 2p + 2b + 2c - 2\xi - \omega)$$

$$+ 2 \sum_{\xi=0}^{c} \zeta(2\xi) \sum_{\omega=0}^{2c-2\xi} \binom{\omega + 2a - 1}{\omega} \binom{2p + 2c - 2\xi - \omega - 1}{2p - 1}$$

$$\times T(2q, s + 2a + \omega, 2p + 2b + 2c - 2\xi - \omega)$$

$$+ 2 \sum_{\xi=0}^{a} \zeta(2\xi) \sum_{\omega=0}^{2p-1} \binom{\omega + 2a - 2\xi}{\omega} \binom{2p + 2c - 2 - \omega}{2c - 1}$$

$$\times T(2q + 2a - 2\xi + \omega + 1, s, 2p + 2b + 2c - 1 - \omega)$$

$$+ 2 \sum_{\xi=0}^{a} \zeta(2\xi) \sum_{\omega=0}^{2c-1} \binom{\omega + 2a - 2\xi}{\omega} \binom{2p + 2c - \omega - 2}{2p - 1}$$

$$\times T(2q, s + 2a - 2\xi + \omega + 1, 2p + 2b + 2c - \omega - 1)$$

holds for $s \in \mathbb{C}$ except for singularities of functions on both sides, where

$$T(2d, s, 2e) = 2 \sum_{\nu=0}^{d} \binom{2d + 2e - 2\nu - 1}{2e - 1} \zeta(2\nu)\zeta(s + 2d + 2e - 2\nu)$$
$$+ 2 \sum_{\nu=0}^{e} \binom{2d + 2e - 2\nu - 1}{2d - 1} \zeta(2\nu)\zeta(s + 2d + 2e - 2\nu). \tag{7.60}$$

Example 7.1. In the case when $(p, q, a, b, c) = (k, k, k, k, k)$ and $s = 2k$ for $k \in \mathbb{N}$ in Theorem 7.1, we recover the explicit expression for Witten's volume formula of A_3 type, which has been proved by Gunnells and Sczech ([10, Proposition 8.5]). For example, in the case when $(p, q, a, b, c) = (1, 1, 1, 1, 1)$, we obtain

$$4\zeta_3(2, 2, s, 2, 2, 2; A_3) + 2\zeta_3(2, s, 2, 2, 2, 2; A_3)$$
$$+ 4\zeta_3(2, 2, 2, s, 2, 2; A_3) + 2\zeta_3(2, 2, 2, 2, 2, s; A_3)$$
$$= 678\zeta(s + 10) - 512\zeta(2)\zeta(s + 8) \tag{7.61}$$
$$+ 148\zeta(4)\zeta(s + 6) + 4\zeta(6)\zeta(s + 4),$$

because $\zeta_3(s_1, s_2, s_3, s_4, s_5, s_6; A_3) = \zeta_3(s_3, s_2, s_1, s_5, s_4, s_6; A_3)$. In particular when $s = 2$, we obtain the explicit value of $C_W(2, A_3)$, that is,

$$\zeta_3(2, 2, 2, 2, 2, 2; A_3) = \frac{23}{2554051500}\pi^{12}. \tag{7.62}$$

In our previous work [16, Theorem 3.4], we already obtained the functional relation between $\zeta_3(\mathbf{s}; A_3)$ and $\zeta_2(\mathbf{s}; A_2)$, and checked that the functional relation implicitly implies (7.62), by using the properties of $\zeta_2(\mathbf{s}; A_2)$. On the other hand, we can see that the above formula in Theorem 7.1 itself includes the explicit form of Witten's volume formula of A_3 type.

Example 7.2. By the same method as above, we can obtain the following formulas ([30]):

$$\zeta_3(1, 1, 1, 2, 1, 2; A_3) = -\frac{29}{175}\zeta(2)^4 + \zeta(3)\zeta(5) - \frac{1}{2}\zeta(6, 2),$$

$$\zeta_3(1, 1, 2, 1, 2, 1; A_3) = \frac{2683}{1050}\zeta(2)^4 + \frac{1}{2}\zeta(2)\zeta(3)^2 - 16\zeta(3)\zeta(5) + \frac{29}{4}\zeta(6, 2),$$

$$\zeta_3(1, 1, 1, 2, 1, 3; A_3) = \frac{2}{5}\zeta(2)^2\zeta(5) + 10\zeta(2)\zeta(7) - \frac{53}{3}\zeta(9).$$

Remark 7.1. Here we summarize the method developed in this section. The starting point is the simple identity (4.31) (and (4.32)), which is based on the fact $\zeta(-2n) = 0$ ($n \in \mathbb{N}$). One basic idea is to multiply (4.31) by an infinite series (see (7.53))) to obtain a new identity (see (7.54)). Then we apply the argument of repeated integration, embodied in Lemma 6.2, to deduce the functional relations. This procedure is the essence of the "u-method" mentioned in Sections 2 and 3, though the parameter $u > 1$ does not appear in this section.

However, the original u-method (developed, for instance, in [30]) is unsatisfactory because it only produces functional relations in which some of the variables should be equal to 0. In order to remove this restriction, we introduce the idea of considering the infinite series of polylogarithm type (that is, with an additional parameter x in the numerators). This idea, inspired by the method in [32] (see Remark 3.3), was first successfully used in [16] under the name of the "polylogarithm technique". This additional flexibility enables us to deduce more general type of functional relations such as Theorem 7.1. We will also use this technique in the following sections.

We may proceed further. Next we combine a quantity of A_3 type and a quantity of A_1 type to obtain

$$\left(\sum_{\substack{k \in \mathbb{Z} \\ k \neq 0}} \frac{(-1)^k e^{ik\theta}}{k^{2p}} - 2 \sum_{j=0}^{p} \phi(2p - 2j) \frac{(i\theta)^{2j}}{(2j)!} \right)$$

$$\times \sum_{\substack{l,m \in \mathbb{Z},\, n \geq 1, \\ l,m \neq 0,\, l+m \neq 0 \\ m+n \neq 0,\, l+m+n \neq 0}} \frac{(-1)^{l+m+n} x^m y^n e^{i(l+m+n)\theta}}{l^{2q} m^{2r} n^s (l+m)^{2a} (m+n)^{2b} (l+m+n)^{2c}} = 0$$

for $p, q, r, a, b, c \in \mathbb{N}$ and $x, y \in \mathbb{C}$ with $|x| = 1$ and $|y| = 1$. Again, by using Lemma 6.2 repeatedly, we will be able to obtain the functional relation for zeta-functions of A_4 and A_3 type. Then, by using the result in Theorem 7.1, we will be able to obtain functional relations for zeta-functions of A_4 and A_1 type, which include explicit forms of Witten's volume formulas of A_4 type, for example,

$$\zeta_4(2, 2, 2, 2, 2, 2, 2, 2, 2, 2; A_4) = \frac{1}{650970015609375} \pi^{20}. \tag{7.63}$$

By continuing this procedure inductively, it seems to be possible to obtain functional relations which include explicit forms of Witten's volume formulas of A_r type for any $r \in \mathbb{N}$.

8. Functional relations for $\zeta_2(\mathbf{s}; C_2)$

In this section, we study

$$\zeta_2(s_1, s_2, s_3, s_4; C_2) = \sum_{m=1}^{\infty} \sum_{n=1}^{\infty} \frac{1}{m^{s_1} n^{s_2} (m+n)^{s_3} (m+2n)^{s_4}} \qquad (8.64)$$

(see [18, (6.1)], also [19, Example 7.3]). As noted in [18, Section 2], we know that

$$\zeta_2(s_1, s_2, s_3, s_4; B_2) = \sum_{m=1}^{\infty} \sum_{n=1}^{\infty} \frac{1}{m^{s_1} n^{s_2} (m+n)^{s_3} (2m+n)^{s_4}} \qquad (8.65)$$

(see [18, (2.11)]), which coincides with $\zeta_2(s_2, s_1, s_3, s_4; C_2)$. This fact is the natural consequence of the isomorphism $B_2 \simeq C_2$.

Here we consider $\zeta_2(\mathbf{s}; C_2)$ and construct explicit functional relations which include explicit forms of Witten's volume formulas of C_2 type.

As we mentioned in the previous section, the procedure of producing a functional relation for $\zeta_2(\mathbf{s}; A_2)$ corresponds to the fact that the Dynkin diagram of A_2 can be produced by adding one edge which joins two vertices. From this viewpoint, we should step on the procedure corresponding to adding another edge to the Dynkin diagram of A_2 to obtain the diagram of C_2, by using Lemma 6.2.

Replacing x by $-xe^{i\theta}$ on the left-hand side of (7.55), we have

$$\sum_{\substack{l \in \mathbb{Z}, \, l \neq 0 \\ m \geq 1, \, l+m \neq 0}} \frac{(-1)^l x^m e^{i(l+2m)\theta}}{l^{2p} m^s (l+m)^{2q}}$$

$$- 2 \sum_{j=0}^{p} \phi(2p-2j) \sum_{\xi=0}^{2j} \binom{2q-1+2j-\xi}{2q-1} (-1)^{2j-\xi} \sum_{m=1}^{\infty} \frac{x^m e^{2im\theta}}{m^{s+2q+2j-\xi}} \frac{(i\theta)^\xi}{\xi!}$$

$$- 2 \sum_{j=0}^{q} \phi(2q-2j) \sum_{\xi=0}^{2j} \binom{2p-1+2j-\xi}{2p-1} \sum_{m=1}^{\infty} \frac{(-1)^m x^m e^{im\theta}}{m^{s+2p+2j-\xi}} \frac{(i\theta)^\xi}{\xi!} = 0$$

for $\theta \in [-\pi, \pi]$. From the first sum, we separate the terms corresponding to the condition $l + 2m = 0$ and move them to the right-hand side. Then, as well as in the case of (7.54), applying Lemma 6.2 with $(h, a_1, a_2, d) =$

$(2, 2p, 2q, 2r)$ for $r \in \mathbb{N}$, we obtain

$$\sum_{\substack{l \in \mathbb{Z},\, l \neq 0 \\ m \geq 1,\, l+m \neq 0 \\ l+2m \neq 0}} \frac{(-1)^l x^m e^{i(l+2m)\theta}}{l^{2p} m^s (l+m)^{2q} (l+2m)^{2r}}$$

$$- 2\sum_{j=0}^{p} \phi(2p-2j) \sum_{\xi=0}^{2j} \sum_{\omega=0}^{2j-\xi} \binom{\omega+2r-1}{\omega}(-1)^{\omega}$$

$$\times \binom{2q-1+2j-\xi-\omega}{2q-1}(-1)^{2j-\xi-\omega} \frac{1}{2^{2r+\omega}} \sum_{m=1}^{\infty} \frac{x^m e^{2im\theta}}{m^{s+2q+2j-\xi+2r}} \frac{(i\theta)^{\xi}}{\xi!}$$

$$- 2\sum_{j=0}^{q} \phi(2q-2j) \sum_{\xi=0}^{2j} \sum_{\omega=0}^{2j-\xi} \binom{\omega+2r-1}{\omega}(-1)^{\omega}$$

$$\times \binom{2p-1+2j-\xi-\omega}{2p-1} \sum_{m=1}^{\infty} \frac{(-1)^m x^m e^{im\theta}}{m^{s+2p+2j-\xi+2r}} \frac{(i\theta)^{\xi}}{\xi!}$$

$$+ 2\sum_{j=0}^{r} \phi(2r-2j) \sum_{\xi=0}^{2j} \sum_{\omega=0}^{2p-1} \binom{\omega+2j-\xi}{\omega}(-1)^{\omega}$$

$$\times \binom{2p+2q-2-\omega}{2q-1}(-1)^{2p-1-\omega} \frac{1}{2^{2j-\xi+\omega+1}} \sum_{m=1}^{\infty} \frac{x^m}{m^{s+2q+2j-\xi+2p}} \frac{(i\theta)^{\xi}}{\xi!}$$

$$+ 2\sum_{j=0}^{r} \phi(2r-2j) \sum_{\xi=0}^{2j} \sum_{\omega=0}^{2q-1} \binom{\omega+2j-\xi}{\omega}(-1)^{\omega}$$

$$\times \binom{2p+2q-2-\omega}{2p-1} \sum_{m=1}^{\infty} \frac{x^m}{m^{s+2p+2j-\xi+2q}} \frac{(i\theta)^{\xi}}{\xi!} = 0 \qquad (8.66)$$

for $\theta \in [-\pi, \pi]$. Then, putting $(x, \theta) = (1, \pi)$ in (8.66) and applying Lemma 6.1 to the real part of this equation, we obtain the following relation which holds for $s > 1$, and furthermore for $s \in \mathbb{C}$ except for singularities by the meromorphic continuation of $\zeta_2(s; C_2)$.

Theorem 8.1. *For $p, q, r \in \mathbb{N}$,*

$$\zeta_2(2p, s, 2q, 2r; C_2) + \zeta_2(2p, 2q, s, 2r; C_2)$$

$$+ \zeta_2(2r, 2q, s, 2p; C_2) + \zeta_2(2r, s, 2q, 2p; C_2)$$

$$= 2 \sum_{\nu=0}^{p} \zeta(2\nu)\zeta(2p + 2q + 2r - 2\nu + s)$$

$$\times \sum_{\mu=0}^{2p-2\nu} \frac{1}{2^{2r+\mu}} \binom{2p + 2q - 2\nu - \mu - 1}{2q - 1} \binom{2r - 1 + \mu}{2r - 1}$$

$$+ 2 \sum_{\nu=0}^{q} \zeta(2\nu)\zeta(2p + 2q + 2r - 2\nu + s)$$

$$\times \sum_{\mu=0}^{2q-2\nu} (-1)^{\mu} \binom{2p + 2q - 2\nu - \mu - 1}{2p - 1} \binom{2r - 1 + \mu}{2r - 1}$$

$$+ 2 \sum_{\nu=0}^{r} \zeta(2\nu)\zeta(2p + 2q + 2r - 2\nu + s)$$

$$\times \sum_{\mu=0}^{2p-1} \frac{1}{2^{2r-2\nu+\mu+1}} \binom{2p + 2q - \mu - 2}{2q - 1} \binom{2r - 2\nu + \mu}{2r - 2\nu}$$

$$+ 2 \sum_{\nu=0}^{r} \zeta(2\nu)\zeta(2p + 2q + 2r - 2\nu + s)$$

$$\times \sum_{\mu=0}^{2q-1} (-1)^{\mu+1} \binom{2p + 2q - \mu - 2}{2p - 1} \binom{2r - 2\nu + \mu}{2r - 2\nu}$$

holds for all $s \in \mathbb{C}$ except for singularities of functions on both sides. Note that singularities of $\zeta_2(s, C_2)$ have been determined in [18, Theorem 6.2].

Example 8.1. Putting $(p, q, r) = (1, 1, 1)$ in Theorem 8.1, we can obtain

$$\zeta_2(2, s, 2, 2; C_2) + \zeta_2(2, 2, s, 2; C_2) = -\frac{39}{16}\zeta(s + 6) + \frac{3}{2}\zeta(2)\zeta(s + 4). \quad (8.67)$$

In particular when $s = 2$, we obtain

$$\zeta_2(2, 2, 2, 2; C_2) = \frac{\pi^8}{302400}, \quad (8.68)$$

which have already been obtained in [19, (7.24)]. It should be noted that Equations (8.67) and (8.70) mentioned below coincide with Equations (2.6) and (2.7) in [16], respectively. Note that, in [16], we used the notation $\zeta_2(s_1, s_2, s_3, s_4; B_2)$ defined by

$$\sum_{m=1}^{\infty} \sum_{n=1}^{\infty} \frac{1}{(2m+n)^{s_1} n^{s_2} m^{s_3} (m+n)^{s_4}},$$

different from (8.65) (see [18, Section 2]).

Remark 8.1. Using the same method as in the proof of Theorem 8.1, we can prove that

$$\zeta_2(p, s, q, r; C_2) + (-1)^p \zeta_2(p, q, s, r; C_2) + (-1)^{p+q} \zeta_2(r, q, s, p; C_2)$$
$$+ (-1)^{p+q+r} \zeta_2(r, s, q, p; C_2)$$

$$= 2(-1)^p \Bigg\{ \sum_{\nu=0}^{[p/2]} \zeta(2\nu)\zeta(p+q+r-2\nu+s)$$

$$\times \sum_{\mu=0}^{p-2\nu} \frac{1}{2^{r+\mu}} \binom{p+q-2\nu-\mu-1}{q-1} \binom{r-1+\mu}{r-1}$$

$$+ \sum_{\nu=0}^{[q/2]} \zeta(2\nu)\zeta(p+q+r-2\nu+s)$$

$$\times \sum_{\mu=0}^{q-2\nu} (-1)^\mu \binom{p+q-2\nu-\mu-1}{p-1} \binom{r-1+\mu}{r-1}$$

$$+ \sum_{\nu=0}^{[r/2]} \zeta(2\nu)\zeta(p+q+r-2\nu+s)$$

$$\times \sum_{\mu=0}^{p-1} \frac{1}{2^{r-2\nu+\mu+1}} \binom{p+q-\mu-2}{q-1} \binom{r-2\nu+\mu}{r-2\nu}$$

$$+ \sum_{\nu=0}^{[r/2]} \zeta(2\nu)\zeta(p+q+r-2\nu+s)$$

$$\times \sum_{\mu=0}^{q-1} (-1)^{\mu+1} \binom{p+q-\mu-2}{p-1} \binom{r-2\nu+\mu}{r-2\nu} \Bigg\} \tag{8.69}$$

holds for all $s \in \mathbb{C}$ except for singularities of functions on both sides, where $p, q, r \in \mathbb{N}$. For example, we have

$$\zeta_2(2, s, 2, 1; C_2) + \zeta_2(2, 2, s, 1; C_2) + \zeta_2(1, 2, s, 2; C_2) - \zeta_2(1, s, 2, 2; C_2)$$

$$= 3\zeta(2)\zeta(s+3) - \frac{39}{8}\zeta(s+5). \tag{8.70}$$

In particular, putting $s = 2$ in (8.70), we have

$$\zeta_2(2,2,2,1;C_2) = \frac{3}{2}\zeta(2)\zeta(5) - \frac{39}{16}\zeta(7),$$

which coincides with our previous result in [44, Example in §3]. Note that the left-hand side of (8.69) is equal to $S(\mathbf{s},\mathbf{y};I;\Delta)$ for $\Delta = \Delta(C_2)$, $\mathbf{s} = (p,s,q,r)$, $\mathbf{y} = 0$ and $I = \{2\}$. Therefore we can see that Theorem 8.1 corresponds to the case C_2 of Theorem 5.1.

9. Functional relations for $\zeta_3(\mathbf{s};B_3)$ and for $\zeta_3(\mathbf{s};C_3)$

In this section, we consider $\zeta_3(\mathbf{s};B_3)$ and $\zeta_3(\mathbf{s};C_3)$ defined by

$$\zeta_3(s_1,s_2,s_3,s_4,s_5,s_6,s_7,s_8,s_9;B_3)$$

$$= \sum_{m_1=1}^{\infty}\sum_{m_2=1}^{\infty}\sum_{m_3=1}^{\infty} m_1^{-s_1}m_2^{-s_2}m_3^{-s_3}(m_1+m_2)^{-s_4}(m_2+m_3)^{-s_5}$$

$$\times (2m_2+m_3)^{-s_6}(m_1+m_2+m_3)^{-s_7}(m_1+2m_2+m_3)^{-s_8}$$

$$\times (2m_1+2m_2+m_3)^{-s_9}, \tag{9.71}$$

and

$$\zeta_3(s_1,s_2,s_3,s_4,s_5,s_6,s_7,s_8,s_9;C_3)$$

$$= \sum_{m_1=1}^{\infty}\sum_{m_2=1}^{\infty}\sum_{m_3=1}^{\infty} m_1^{-s_1}m_2^{-s_2}m_3^{-s_3}(m_1+m_2)^{-s_4}(m_2+m_3)^{-s_5}$$

$$\times (m_2+2m_3)^{-s_6}(m_1+m_2+m_3)^{-s_7}(m_1+m_2+2m_3)^{-s_8}$$

$$\times (m_1+2m_2+2m_3)^{-s_9}, \tag{9.72}$$

which have been continued meromorphically to the whole space whose possible singularities have been determined in [18, Theorems 6.1 and 6.3]. Note that $\zeta_3(\mathbf{s};D_3)$ essentially coincides with $\zeta_3(\mathbf{s};A_3)$ which has been considered in [16,30].

We aim to prove functional relations for these functions, namely generalize the result in Theorems 7.1 and 8.1 to the cases B_3 and C_3. However, it seems too complicated to treat these cases in full generality. Hence we study some special cases as follows.

First we prove the following functional relation for $\zeta_3(\mathbf{s};C_3)$. The basic structure of the proof, based on Lemma 6.2, is similar to that in the proof of Theorem 7.1 for $\zeta_3(\mathbf{s};A_3)$. A novel point here is that we will also use the result described in Section 4.

Theorem 9.1. *The functional relation*

$$8\zeta_3(2,2,s,2,2,2,2,2,2;C_3) + 8\zeta_3(2,2,2,2,2,s,2,2,2,2;C_3)$$

$$+ 8\zeta_3(2,2,2,2,2,2,s,2,2;C_3)$$

$$= \frac{184775}{512}\zeta(s+16) - \frac{16875}{64}\zeta(2)\zeta(s+14) + \frac{513}{8}\zeta(4)\zeta(s+12) \qquad (9.73)$$

$$+ \frac{25}{8}\zeta(6)\zeta(s+10) + \frac{1}{4}\zeta(8)\zeta(s+8)$$

holds for $s \in \mathbb{C}$ except for singularities of functions on both sides. In particular when $s = 2$,

$$\zeta_3(2,2,2,2,2,2,2,2,2;C_3) = \frac{19}{8403115488768000}\pi^{18}, \qquad (9.74)$$

hence $C_W(2,C_3) = 19/16209713520$ in Witten's volume formula (1.4).

Proof. Instead of (7.53) or (7.58), we start the same argument as in the proof of Proposition 7.1 or Theorem 7.1 from the relation

$$\{G(\theta;2,2,2;x) + G(-\theta;2,2,2;x^{-1})\} \sum_{n=1}^{\infty} \frac{(-1)^n y^n e^{in\theta}}{n^s} = 0 \qquad (9.75)$$

for $s > 1$, where we denote by $G(\theta;2,2,2;x)$ the left-hand side of (7.55) in the case $(2p,2q,s) = (2,2,2)$. Then, by replacing $-m$ $(m \geq 1)$ by m $(\dot{m} \leq -1)$ on the left-hand side of (9.75), we can rewrite (9.75) to

$$\sum_{\substack{l\neq 0,\, m\neq 0 \\ n\geq 1,\, l+m\neq 0}} \frac{(-1)^{l+m+n} x^m y^n e^{i(l+m+n)\theta}}{l^2 m^2 n^s (l+m)^2}$$

$$-2\sum_{j=0}^{1} \phi(2-2j) \sum_{\xi=0}^{2j} \binom{1+2j-\xi}{1}(-1)^{2j-\xi}$$

$$\times \sum_{\substack{m\neq 0 \\ n\geq 1}} \frac{(-1)^{m+n} x^m y^n e^{i(m+n)\theta}}{m^{4+2j-\xi} n^s} \frac{(i\theta)^\xi}{\xi!}$$

$$+2\sum_{j=0}^{1} \phi(2-2j) \sum_{\xi=0}^{2j} \binom{1+2j-\xi}{1}(-1) \sum_{\substack{m\neq 0 \\ n\geq 1}} \frac{(-1)^n x^m y^n e^{in\theta}}{m^{4+2j-\xi} n^s} \frac{(i\theta)^\xi}{\xi!} = 0.$$

As well as in the proof of Theorem 7.1, separate the constant terms corresponding to $l + m + n = 0$ in the first term and to $m + n = 0$ in the second term on the left-hand side, move them to the right-hand side, and apply

Lemma 6.2 with $d = 2$. Then we obtain

$$\sum_{\substack{l \neq 0,\, m \neq 0 \\ n \geq 1,\, l+m \neq 0 \\ l+m+n \neq 0}} \frac{(-1)^{l+m+n} x^m y^n e^{i(l+m+n)\theta}}{l^2 m^2 n^s (l+m)^2 (l+m+n)^2}$$

$$- 2 \sum_{j=0}^{1} \phi(2-2j) \sum_{\xi=0}^{2j} \sum_{\omega=0}^{2j-\xi} \binom{\omega+1}{\omega} (-1)^\omega \binom{1+2j-\xi-\omega}{1} (-1)^{2j-\xi-\omega}$$

$$\times \sum_{\substack{m \neq 0 \\ n \geq 1}} \frac{(-1)^{m+n} x^m y^n e^{i(m+n)\theta}}{m^{4+2j-\xi-\omega} n^s (m+n)^{2+\omega}} \frac{(i\theta)^\xi}{\xi!}$$

$$+ \cdots = 0,$$

where we omit three terms on the left-hand side, which are of the form similar to the second term on the left-hand side. Note that each of their denominators is of the form of A_2 type. Next we replace y by $-ye^{i\theta}$, move the constant terms to the right-hand side and apply Lemma 6.2 with $d = 2$. Then we have

$$\sum_{\substack{l \neq 0,\, m \neq 0 \\ n \geq 1,\, l+m \neq 0 \\ l+m+n \neq 0 \\ l+m+2n \neq 0}} \frac{(-1)^{l+m} x^m y^n e^{i(l+m+2n)\theta}}{l^2 m^2 n^s (l+m)^2 (l+m+n)^2 (l+m+2n)^2}$$

$$- 2 \sum_{j=0}^{1} \phi(2-2j) \sum_{\xi=0}^{2j} \sum_{\sigma=0}^{2j-\sigma} \binom{\sigma+1}{\sigma} (-1)^\sigma \sum_{\omega=0}^{2j-\xi-\sigma} \binom{\omega+1}{\omega} (-1)^{2j-\xi-\sigma}$$

$$\times \binom{1+2j-\xi-\sigma-\omega}{1} \sum_{\substack{m \neq 0 \\ n \geq 1}} \frac{(-1)^m x^m y^n e^{i(m+2n)\theta}}{m^{4+2j-\xi-\omega} n^s (m+n)^{2+\omega} (m+2n)^{2+\sigma}} \frac{(i\theta)^\xi}{\xi!}$$

$$+ \cdots = 0,$$

where we omit seven terms of the forms similar to the second term. Each denominator of these terms is of the form of C_2 type. Replacing x by $-xe^{i\theta}$, applying Lemma 6.2 with $d = 2$, and putting $\theta = \pi$, we obtain

$$\sum_{\substack{l \neq 0,\, m \neq 0 \\ n \geq 1,\, l+m \neq 0 \\ l+m+n \neq 0 \\ l+m+2n \neq 0 \\ l+2m+2n \neq 0}} \frac{x^m y^n}{l^2 m^2 n^s (l+m)^2 (l+m+n)^2 (l+m+2n)^2 (l+2m+2n)^2}$$

$$+ \cdots = 0, \tag{9.76}$$

where the omitted terms are of the form of C_2 type. Next, we replace (x, y) by $(xe^{i\theta}, ye^{i\theta})$ and $(xe^{-i\theta}, ye^{-i\theta})$ respectively, and subtract these terms.

Then we have

$$\sum_{\substack{l\neq 0,\, m\neq 0 \\ n\geq 1,\, l+m\neq 0 \\ m+n\neq 0 \\ l+m+n\neq 0 \\ l+m+2n\neq 0 \\ l+2m+2n\neq 0}} \frac{x^m y^n \sin((m+n)\theta)}{l^2 m^2 n^s (l+m)^2(l+m+n)^2(l+m+2n)^2(l+2m+2n)^2}$$

$$+ \cdots = 0, \tag{9.77}$$

where the omitted terms are double series of similar forms.

Finally, in order to complete the proof of this theorem, we need to apply Theorem 4.1 to each term on the left-hand side of (9.77) with $s = 2$. Then we consequently obtain

$$\sum_{\substack{l\neq 0,\, m\neq 0 \\ n\geq 1,\, l+m\neq 0 \\ m+n\neq 0 \\ l+m+n\neq 0 \\ l+m+2n\neq 0 \\ l+2m+2n\neq 0}} \frac{x^m y^n}{l^2 m^2 n^s (l+m)^2(m+n)^2(l+m+n)^2(l+m+2n)^2(l+2m+2n)^2}$$

$$+ \cdots = 0. \tag{9.78}$$

We further replace (x, y) by $\left(e^{i\theta}, e^{2i\theta}\right)$ and $\left(e^{-i\theta}, e^{-2i\theta}\right)$ respectively, subtract these terms, and apply Theorem 4.1 with $s = 2$. Then we obtain

$$\sum_{\substack{l\neq 0,\, m\neq 0 \\ n\geq 1,\, l+m\neq 0 \\ m+n\neq 0,\, m+2n\neq 0 \\ l+m+n\neq 0 \\ l+m+2n\neq 0 \\ l+2m+2n\neq 0}} \frac{1}{l^2 m^2 n^s (l+m)^2(m+n)^2(m+2n)^2(l+m+n)^2}$$

$$\times \frac{1}{(l+m+2n)^2(l+2m+2n)^2}$$

$$+ \cdots = 0, \tag{9.79}$$

where the omitted terms are finite sums of zeta values of C_2 type. Though we omit their explicit forms, we can also apply Theorem 8.1 to these terms, and can express them as the right-hand side of (9.73). On the other hand, similarly to the case of Theorem 7.1, we can transform the first term on the left-hand side of (9.79) to the left-hand side of (9.73).

Moreover, from (2.16) in [18], we can easily check that $K(C_3) = 720$ (see definition (1.7)). Hence, combining (1.6) and (9.74), we obtain the value of $C_W(2, C_3)$. □

Similarly we can obtain the following formula in the case B_3.

Theorem 9.2. *The functional relation*

$$4\zeta_3(2, s, 2, 2, 2, 2, 2, 2, 2; B_3) + 4\zeta_3(s, 2, 2, 2, 2, 2, 2, 2, 2; B_3) \tag{9.80}$$
$$+ 4\zeta_3(2, 2, 2, s, 2, 2, 2, 2, 2; B_3) + 4\zeta_3(2, 2, 2, 2, s, 2, 2, 2, 2; B_3)$$
$$+ 4\zeta_3(2, 2, 2, 2, 2, 2, s, 2, 2; B_3) + 4\zeta_3(2, 2, 2, 2, 2, 2, 2, s, 2; B_3)$$
$$= \left(9 \cdot 2^{-s-6} + \frac{5626955}{256}\right)\zeta(s+16) + \left(5 \cdot 2^{-s-5} - \frac{59131}{4}\right)\zeta(2)\zeta(s+14)$$
$$+ \left(5 \cdot 2^{-s-5} + \frac{17155}{8}\right)\zeta(4)\zeta(s+12) + \frac{241}{16}\zeta(6)\zeta(s+10) + \frac{1}{8}\zeta(8)\zeta(s+8)$$

holds for $s \in \mathbb{C}$ except for singularities of functions on both sides. In particular when $s = 2$, we have

$$\zeta_3(2, 2, 2, 2, 2, 2, 2, 2, 2; B_3) = \frac{19}{8403115488768000}\pi^{18}, \tag{9.81}$$

hence $C_W(2, B_3) = 19/16209713520$ in Witten's volume formula (1.4).

Proof. The argument is similar to that in the proof of Theorem 9.1. In fact, instead of (9.75), we start the same argument from the relation

$$\{H(\theta; 2, 2, 2, 2; x) + H(-\theta; 2, 2, 2, 2; x)\}\sum_{n=1}^{\infty}\frac{(-1)^n y^n e^{in\theta}}{n^s} = 0 \tag{9.82}$$

for $s > 1$, where we denote by $H(\theta; 2, 2, 2, 2; x)$ the left-hand side of (8.66) in the case $(2p, 2q, 2r, s) = (2, 2, 2, 2)$. Repeating the same procedure as in the proof of Theorem 9.1, we can describe the left-hand side of (9.80) as a finite sum of the forms of the left-hand side of (8.69). Hence, by using (8.69), we can obtain (9.80). The value of $C_W(2, B_3)$ can be calculated from (1.6), (9.81), and the fact $K(B_3) = 720$. □

Remark 9.1. Comparing the above two theorems, we see that $C_W(2, B_3) = C_W(2, C_3)$. However it does not always hold that $C_W(2k, B_3) = C_W(2k, C_3)$, that is, $\zeta_W(2k; B_3) = \zeta_W(2k; C_3)$ for $k \geq 2$. In fact, we can compute that

$$\zeta_W(4; B_3) = 1.00066856607695295\cdots,$$

$$\zeta_W(4; C_3) = 1.00082905650461486\cdots,$$

hence $C_W(4, B_3) \neq C_W(4, C_3)$.

Note that the left-hand side of (9.73) corresponds to $S(\mathbf{s}, \mathbf{y}; I; \Delta)$ for $\Delta = \Delta(C_3), \mathbf{s} = (2, s, 2, 2, 2, 2, 2, 2, 2), \mathbf{y} = 0$ and $I = \{3\}$ in the terminology of Section 5. Next we prove the following result which is corresponding to the case $\Delta = \Delta(C_3)$, $\mathbf{s} = (2, s, t, 2, u, v, 2, 2, 2)$, $\mathbf{y} = 0$ and $I = \{2, 3\}$.

Theorem 9.3. *The functional relation*

$$\zeta_3(2, s, t, 2, u, v, 2, 2, 2; C_3) + 2\zeta_3(2, 2, t, s, 2, 2, u, v, 2; C_3)$$
$$+ 2\zeta_3(s, 2, 2, 2, t, 2, u, 2, v; C_3)$$
$$= \sum_{\xi=0,1} \zeta(2\xi) \sum_{\tau=0}^{2-2\xi} (\tau+1) \sum_{\nu=0}^{2-2\xi-\tau} (\nu+1)$$
$$\times \left\{ \sum_{\omega=0}^{2-2\xi-\tau-\nu} (\omega+1)(3-2\xi-\tau-\nu-\omega) \right.$$
$$\times \frac{1}{2^{\tau+2}} \zeta_2(s+2+\omega, t, u+6-2\xi-\nu-\omega, v+2+\nu; C_2)$$
$$+ (-1)^{\tau+\nu} \sum_{\omega=0}^{2-2\xi-\tau-\nu} (\omega+1)(3-2\xi-\tau-\nu-\omega)$$
$$\times \zeta_2(s, t+4+\omega+\nu, u+6-2\xi-\omega-\nu, v; C_2)$$
$$+ (-1)^{\tau+\nu} \sum_{\omega=0,1} \binom{\omega+2-2\xi-\tau-\nu}{\omega} (2-\omega)$$
$$\times \frac{1}{2^{\nu+2}} \zeta_2(s+3-2\xi-\tau-\nu+\omega, t+2+\nu, u+3-\omega, v+2+\tau; C_2)$$
$$+ \sum_{\omega=0,1} \binom{\omega+2-2\xi-\tau-\nu}{\omega} (2-\omega)$$
$$\left. \times \frac{1}{2^{\nu+2}} \zeta_2(s, t+5-2\xi-\tau+\omega, u+3-\omega, v+2+\tau; C_2) \right\}$$
$$- \sum_{\xi=0,1} \zeta(2\xi) \sum_{\tau=0}^{2-2\xi} (\tau+1) \sum_{\nu=0,1} \binom{\nu+2-2\xi-\tau}{\nu}$$
$$\times \left\{ (-1)^{\tau+1} \sum_{\omega=0}^{1-\nu} (\omega+1)(2-\nu-\omega) \right.$$
$$\times \zeta_2(q, r+3-\nu-\omega, v+3-2\xi-\tau+\nu, p+4+\tau+\omega; C_2)$$
$$+ (-1)^{\tau+\nu} \sum_{\omega=0}^{1-\nu} (\omega+1)(2-\nu-\omega)$$
$$\times \zeta_2(s+2+\tau, t+5-2\xi-\tau+\nu+\omega, u+3-\nu-\omega, v; C_2)$$

$$+ (-1)^{\tau + \nu} \sum_{\omega = 0,1} \binom{\omega + 1 - \nu}{\omega} (2 - \omega)$$

$$\times \frac{1}{2^{3 - 2\xi - \tau + \nu}} \zeta_2(s + 4 + \tau - \nu + \omega, t + 3 - 2\xi - \tau + \nu, u + 3 - \omega, v; C_2)$$

$$+ (-1)^{\tau + 1} \sum_{\omega = 0,1} \binom{\omega + 1 - \nu}{\omega} (2 - \omega)$$

$$\times \frac{1}{2^{3 - 2\xi - \tau + \nu}} \zeta_2(s + 2 + \tau, t + 5 - 2\xi - \tau + \omega, u + 3 - \omega, v; C_2) \bigg\}$$

$$- \sum_{\xi = 0,1} \zeta(2\xi) \sum_{\tau = 0,1} \binom{\tau + 2 - 2\xi}{\tau} \sum_{\nu = 0}^{1 - \tau} (\nu + 1)$$

$$\times \bigg\{ - \sum_{\omega = 0}^{1 - \tau - \nu} (\omega + 1)(2 - \tau - \nu - \omega)$$

$$\times \frac{1}{2^{3 - 2\xi + \tau}} \zeta_2(s + 2 + \omega, t, u + 6 - 2\xi - \nu - \omega, v + 2 + \nu, ; C_2)$$

$$+ (-1)^{\tau + \nu} \sum_{\omega = 0}^{1 - \tau - \nu} (\omega + 1)(2 - \tau - \nu - \omega)$$

$$\times \zeta_2(s, t + 4 + \nu + \omega, u + 6 - 2\xi - \nu - \omega, v; C_2)$$

$$+ (-1)^{\tau + \nu} \sum_{\omega = 0,1} \binom{\omega + 1 - \tau - \nu}{\omega} (2 - \omega)$$

$$\times \frac{1}{2^{\nu + 2}} \zeta_2(s + 2 - \tau - \nu + \omega, t + 2 + \nu, u + 3 - \omega, v + 3 - 2\xi + \tau; C_2)$$

$$- \sum_{\omega = 0,1} \binom{\omega + 1 - \tau - \nu}{\omega} (2 - \omega)$$

$$\times \frac{1}{2^{\nu + 2}} \zeta_2(s, t + 4 - \tau + \omega, u + 3 - \omega, v + 3 - 2\xi + \tau; C_2) \bigg\}$$

$$+ \sum_{\xi = 0,1} \zeta(2\xi) \sum_{\tau = 0,1} \binom{\tau + 2 - 2\xi}{\tau} \sum_{\nu = 0,1} \binom{\nu + 1 - \tau}{\nu}$$

$$\times \bigg\{ (-1)^{\tau + 1} \sum_{\omega = 0}^{1 - \nu} (\omega + 1)(2 - \nu - \omega)$$

$$\times \zeta_2(s + 5 - 2\xi + \tau + \omega, t, u + 3 - \nu - \omega, v + 2 - \tau + \nu; C_2)$$

$$+ (-1)^{\tau + \nu} \sum_{\omega = 0}^{1 - \nu} (\omega + 1)(2 - \nu - \omega)$$

$$\times \zeta_2(s + 3 - 2\xi + \tau, t + 4 + \omega - \tau + \nu, u + 3 - \nu - \omega, v; C_2)$$

$$+ (-1)^{\tau + \nu} \sum_{\omega = 0,1} \binom{\omega + 1 - \nu}{\omega} (2 - \omega)$$

$$\times \frac{1}{2^{\nu + 2 - \tau}} \zeta_2(s + 5 - 2\xi + \tau - \nu + \omega, t + 2 - \tau + \nu, u + 3 - \omega, v; C_2)$$

$$+ (-1)^{\tau + 1} \sum_{\omega = 0,1} \binom{\omega + 1 - \nu}{\omega} (2 - \omega)$$

$$\times \frac{1}{2^{\nu + 2 - \tau}} \zeta_2(s + 3 - 2\xi + \tau, t + 4 - \tau + \omega, u + 3 - \omega, v; C_2) \Big\}$$

holds for $s, t, u, v \in \mathbb{C}$ except for singularities of functions on both sides.

Proof. As well as (7.53) and (7.58), we begin by combining quantities of type A_1 and of type C_2, that is,

$$\left(\sum_{l=1}^{\infty} \frac{(-1)^l (e^{il\theta} + e^{-il\theta})}{l^2} - 2 \sum_{j=0,1} \phi(2 - 2j) \frac{(i\theta)^{2j}}{(2j)!} \right)$$

$$\times \sum_{m,n=1}^{\infty} \frac{(-1)^{m+n} x^m y^n e^{i(m+n)\theta}}{m^s n^t (m+n)^u (m+2n)^v} = 0 \qquad (9.83)$$

for $\theta \in [-\pi, \pi]$, where we fix $s, t, u, v \in \{z \in \mathbb{R} \mid z > 1\}$ and $x, y \in \{z \in \mathbb{R} \mid |z| = 1\}$. Then

$$\sum_{l,m,n=1}^{\infty} \frac{(-1)^{l+m+n} x^m y^n e^{i(l+m+n)\theta}}{l^2 m^s n^t (m+n)^u (m+2n)^v} + \sum_{\substack{l,m,n=1 \\ l \neq m+n}}^{\infty} \frac{(-1)^{l+m+n} x^m y^n e^{i(-l+m+n)\theta}}{l^2 m^s n^t (m+n)^u (m+2n)^v}$$

$$- 2 \sum_{j=0,1} \phi(2 - 2j) \frac{(-1)^j \theta^{2j}}{(2j)!} \sum_{m,n=1}^{\infty} \frac{(-1)^{m+n} x^m y^n e^{i(m+n)\theta}}{m^s n^t (m+n)^u (m+2n)^v}$$

$$= - \sum_{m,n=1}^{\infty} \frac{x^m y^n}{m^s n^t (m+n)^{u+2} (m+2n)^v} \qquad (9.84)$$

for $\theta \in [-\pi, \pi]$. Applying Lemma 6.2 with $d = 2$, we have

$$\sum_{l,m,n=1}^{\infty} \frac{(-1)^{l+m+n} x^m y^n e^{i(l+m+n)\theta}}{l^2 m^s n^t (m+n)^u (m+2n)^v (l+m+n)^2}$$

$$+ \sum_{\substack{l,m,n=1 \\ l \neq m+n}}^{\infty} \frac{(-1)^{l+m+n} x^m y^n e^{i(-l+m+n)\theta}}{l^2 m^s n^t (m+n)^u (m+2n)^v (-l+m+n)^2}$$

$$-2\sum_{j=0,1}\phi(2-2j)\sum_{\xi=0}^{2j}\binom{2j-\xi+1}{2j-\xi}(-1)^{2j-\xi}\frac{(i\theta)^\xi}{\xi!}$$

$$\times\sum_{m,n=1}^{\infty}\frac{(-1)^{m+n}x^my^ne^{i(m+n)\theta}}{m^sn^t(m+n)^{u+2+2j-\xi}(m+2n)^v}$$

$$-2\sum_{j=0,1}\phi(2-2j)\sum_{\xi=0}^{2j}\binom{1+2j-\xi}{1}\frac{(i\theta)^\xi}{\xi!}$$

$$\times\sum_{m,n=1}^{\infty}\frac{x^my^n}{m^sn^t(m+n)^{u+2+2j-\xi}(m+2n)^v}=0\quad(\theta\in[-\pi,\pi]).\quad(9.85)$$

For simplicity, we denote the sum of the third and the fourth terms on the left-hand side of (9.85) by

$$-2\sum_{j=0,1}\phi(2-2j)\sum_{\xi=0}^{2j}\frac{(i\theta)^\xi}{\xi!}$$

$$\times\left[\sum_{m,n=1}^{\infty}\left\{(-1)^{m+n}\mathcal{D}_1(m,n;2j-\xi)e^{i(m+n)\theta}+\mathcal{D}_2(m,n;2j-\xi)\right\}x^my^n\right],$$

where $\mathcal{D}_j(m,n;\nu)\in\mathbb{R}$ $(j=1,2)$. Since (9.85) holds for $y\in\mathbb{C}$ with $|y|=1$, we replace y by $-ye^{-i\theta}$ with $y\in\mathbb{C}$ $(|y|=1)$. Then we have

$$\sum_{l,m,n=1}^{\infty}\frac{(-1)^{l+m}x^my^ne^{i(l+m)\theta}}{l^2m^sn^t(m+n)^u(m+2n)^v(l+m+n)^2}$$

$$+\sum_{\substack{l,m,n=1\\l\neq m+n,\,l\neq m}}^{\infty}\frac{(-1)^{l+m}x^my^ne^{i(-l+m)\theta}}{l^2m^sn^t(m+n)^u(m+2n)^v(-l+m+n)^2}$$

$$-2\sum_{j=0,1}\phi(2-2j)\sum_{\xi=0}^{2j}\sum_{m,n=1}^{\infty}\left\{(-1)^m\mathcal{D}_1(m,n;2j-\xi)e^{im\theta}\right.$$

$$\left.+(-1)^n\mathcal{D}_2(m,n;2j-\xi)e^{-in\theta}\right\}x^my^n\frac{(i\theta)^\xi}{\xi!}$$

$$=-\sum_{m,n=1}^{\infty}\frac{x^my^n}{m^{s+2}n^{t+2}(m+n)^u(m+2n)^v}\quad(\theta\in[-\pi,\pi]).\quad(9.86)$$

Therefore, applying Lemma 6.2 with $d = 2$, we have

$$\sum_{l,m,n=1}^{\infty} \frac{(-1)^{l+m} x^m y^n e^{i(l+m)\theta}}{l^2 m^s n^t (l+m)^2 (m+n)^u (m+2n)^v (l+m+n)^2} \tag{9.87}$$

$$= - \sum_{\substack{l,m,n=1 \\ l \neq m+n,\, l \neq m}}^{\infty} \frac{(-1)^{l+m} x^m y^n e^{i(-l+m)\theta}}{l^2 m^s n^t (-l+m)^2 (m+n)^u (m+2n)^v (-l+m+n)^2}$$

$$+ 2 \sum_{j=0,1} \phi(2-2j) \sum_{\xi=0}^{2j} \sum_{m,n=1}^{\infty} \left\{ (-1)^m \mathcal{D'}_1(m,n;2j-\xi) e^{im\theta} \right.$$

$$\left. + (-1)^n \mathcal{D'}_2(m,n;2j-\xi) e^{-in\theta} + \mathcal{D'}_3(m,n;2j-\xi) \right\} x^m y^n \frac{(i\theta)^\xi}{\xi!}$$

for $\theta \in [-\pi, \pi]$ with some $\mathcal{D'}_j(m,n;\nu) \in \mathbb{R}$ ($j = 1,2,3$).

Now we repeat this procedure. Namely, replace y by $ye^{2i\theta}$ and apply Lemma 6.2 with $d = 2$. Furthermore, replace x by $-xe^{i\theta}$ and apply Lemma 6.2 with $d = 2$. Then we can obtain the equation

$$\sum_{l,m,n=1}^{\infty} \frac{(-1)^l x^m y^n e^{i(l+2m+2n)\theta}}{l^2 m^s n^t (l+m)^2 (m+n)^u (m+2n)^v (l+m+n)^2}$$

$$\times \frac{1}{(l+m+2n)^2 (l+2m+2n)^2}$$

$$= - \sum_{\substack{l,m,n=1 \\ l \neq m,\, l \neq m+n \\ l \neq m+2n,\, l \neq 2m+2n}}^{\infty} \frac{1}{l^2 m^s n^t (-l+m)^2 (m+n)^u (m+2n)^v (-l+m+n)^2}$$

$$\times \frac{(-1)^l x^m y^n e^{i(-l+2m+2n)\theta}}{(-l+m+2n)^2 (-l+2m+2n)^2}$$

$$+ 2 \sum_{j=0,1} \phi(2-2j) \sum_{\xi=0}^{2j} \sum_{m,n=1}^{\infty} \left\{ \widetilde{\mathcal{D}}_1(m,n;2j-\xi) e^{i(2m+2n)\theta} \right.$$

$$+ (-1)^{m+n} \widetilde{\mathcal{D}}_2(m,n;2j-\xi) e^{i(m+n)\theta} + (-1)^m \widetilde{\mathcal{D}}_3(m,n;2j-\xi) e^{i(m+2n)\theta}$$

$$\left. + (-1)^m \widetilde{\mathcal{D}}_4(m,n;2j-\xi) e^{im\theta} + \widetilde{\mathcal{D}}_5(m,n;2j-\xi) \right\} x^m y^n \frac{(i\theta)^\xi}{\xi!} \tag{9.88}$$

for $\theta \in [-\pi, \pi]$ with some $\widetilde{\mathcal{D}}_j(m,n;\nu) \in \mathbb{R}$ ($j = 1,2,\ldots,5$). Put $\theta = \pi$ and

$(x, y) = (1, 1)$, and consider the real part. Then we have

$$
\sum_{l,m,n=1}^{\infty} \frac{1}{l^2 m^s n^t (l+m)^2 (m+n)^u (m+2n)^v (l+m+n)^2}
$$

$$
\times \frac{1}{(l+m+2n)^2 (l+2m+2n)^2}
$$

$$
+ \sum_{\substack{l,m,n=1 \\ l \neq m, \ l \neq m+n \\ l \neq m+2n, \ l \neq 2m+2n}}^{\infty} \frac{1}{l^2 m^s n^t (-l+m)^2 (m+n)^u (m+2n)^v (-l+m+n)^2}
$$

$$
\times \frac{1}{(-l+m+2n)^2 (-l+2m+2n)^2}
$$

$$
- 2 \sum_{j=0,1} \phi(2-2j) \sum_{\tau=0}^{j} \sum_{m,n=1}^{\infty} \left\{ \widetilde{D}_1(m,n;2j-2\tau) + \widetilde{D}_2(m,n;2j-2\tau) \right.
$$

$$
\left. + \widetilde{D}_3(m,n;2j-2\tau) + \widetilde{D}_4(m,n;2j-2\tau) + \widetilde{D}_5(m,n;2j-2\tau) \right\}
$$

$$
\times \frac{(-1)^\tau \pi^{2\tau}}{(2\tau)!} = 0. \tag{9.89}
$$

By (9.72), we see that the first term on the left-hand side of (9.89) coincides with

$$
\zeta_3(2, s, t, 2, u, v, 2, 2, 2; C_3).
$$

For the second term on the left-hand side of (9.89), change the running indices of summation corresponding to the conditions $l \neq m$, $l \neq m+n$, $l \neq m+2n$, $l \neq 2m+2n$. Then we can see that the second term on the left-hand side of (9.89) coincides with

$$
\zeta_3(2, s, t, 2, u, v, 2, 2, 2; C_3) + 2\zeta_3(2, 2, t, s, 2, 2, u, v, 2; C_3)
$$

$$
+ 2\zeta_3(s, 2, 2, 2, t, 2, u, 2, v; C_3).
$$

Furthermore, using Lemma 6.1, we can rewrite the third term on the left-hand side of (9.89) as

$$
- 2 \sum_{\xi=0,1} \zeta(2\xi) \sum_{m,n=1}^{\infty} \left\{ \widetilde{D}_1(m,n;2-2\xi) + \widetilde{D}_2(m,n;2-2\xi) \right.
$$

$$
\left. + \widetilde{D}_3(m,n;2-2\xi) + \widetilde{D}_4(m,n;2-2\xi) + \widetilde{D}_5(m,n;2-2\xi) \right\}.
$$

We can concretely calculate the value $\widetilde{D}_j(m,n;\nu)$ in terms of $\zeta_2(\mathbf{s};C_2)$ and $\zeta(s)$. Combining these results, we obtain the assertion. $\square$

Example 9.1. Putting $(s, t, u, v) = (2, 2, 2, 2)$, we obtain

$$3\,\zeta_3(2, 2, 2, 2, 2, 2, 2, 2, 2; C_3)$$

$$= \zeta_2(5, 5, 4, 2; C_2)\zeta(2) - \frac{3}{2}\zeta_2(5, 5, 6, 2; C_2) + \frac{1}{2}\zeta_2(6, 4, 4, 2; C_2)\zeta(2)$$

$$- \frac{3}{4}\zeta_2(6, 4, 6, 2; C_2) + \zeta_2(6, 5, 3, 2; C_2)\zeta(2) - \frac{3}{2}\zeta_2(6, 5, 5, 2; C_2)$$

$$- \frac{3}{2}\zeta_2(6, 8, 2, 2; C_2) + \frac{3}{4}\zeta_2(7, 4, 3, 2; C_2)\zeta(2) - \frac{9}{8}\zeta_2(7, 4, 5, 2; C_2)$$

$$- \frac{11}{16}\zeta_2(7, 5, 4, 2; C_2) - \frac{23}{32}\zeta_2(8, 4, 4, 2; C_2) - \zeta_2(8, 6, 2, 2; C_2)$$

$$+ \zeta_2(4, 5, 4, 3; C_2)\zeta(2) - \frac{3}{2}\zeta_2(4, 5, 6, 3; C_2) + \frac{1}{2}\zeta_2(5, 5, 3, 3; C_2)\zeta(2)$$

$$- \frac{3}{4}\zeta_2(5, 5, 5, 3; C_2) + \zeta_2(6, 5, 2, 3; C_2)\zeta(2) - \frac{3}{16}\zeta_2(6, 5, 4, 3; C_2)$$

$$- \frac{3}{4}\zeta_2(7, 4, 2, 3; C_2)\zeta(2) + \zeta_2(2, 5, 5, 4; C_2)\zeta(2) + \frac{3}{2}\zeta_2(2, 6, 4, 4; C_2)\zeta(2)$$

$$- \frac{3}{8}\zeta_2(2, 6, 6, 4; C_2) - \frac{7}{8}\zeta_2(2, 7, 5, 4; C_2) - \frac{15}{16}\zeta_2(2, 8, 4, 4; C_2)$$

$$+ \frac{1}{2}\zeta_2(4, 4, 4, 4; C_2)\zeta(2) - \frac{3}{4}\zeta_2(4, 4, 6, 4; C_2) - \frac{1}{2}\zeta_2(4, 5, 3, 4; C_2)\zeta(2)$$

$$+ \frac{3}{4}\zeta_2(4, 5, 5, 4; C_2) + \frac{1}{4}\zeta_2(5, 4, 3, 4; C_2)\zeta(2) - \frac{3}{8}\zeta_2(5, 4, 5, 4; C_2)$$

$$- \frac{3}{2}\zeta_2(5, 5, 2, 4; C_2)\zeta(2) + \frac{1}{4}\zeta_2(5, 5, 4, 4; C_2) + \frac{1}{2}\zeta_2(6, 4, 2, 4; C_2)\zeta(2)$$

$$- \frac{3}{32}\zeta_2(6, 4, 4, 4; C_2) + \frac{11}{16}\zeta_2(7, 5, 2, 4; C_2) - \frac{23}{32}\zeta_2(8, 4, 2, 4; C_2)$$

$$- 2\zeta_2(2, 4, 5, 5; C_2)\zeta(2) - \zeta_2(2, 5, 4, 5; C_2)\zeta(2) - \frac{1}{2}\zeta_2(2, 6, 5, 5; C_2)$$

$$- \frac{7}{8}\zeta_2(2, 7, 4, 5; C_2) - \frac{1}{2}\zeta_2(4, 4, 3, 5; C_2)\zeta(2) + \frac{3}{4}\zeta_2(4, 4, 5, 5; C_2)$$

$$+ \frac{3}{2}\zeta_2(4, 5, 2, 5; C_2)\zeta(2) - \frac{1}{4}\zeta_2(4, 5, 4, 5; C_2) - \frac{1}{4}\zeta_2(5, 4, 2, 5; C_2)\zeta(2)$$

$$+ \frac{1}{4}\zeta_2(5, 4, 4, 5; C_2) - \frac{27}{16}\zeta_2(6, 5, 2, 5; C_2) + \frac{9}{8}\zeta_2(7, 4, 2, 5; C_2)$$

$$- \frac{3}{2}\zeta_2(2, 4, 6, 6; C_2) + 4\zeta_2(2, 5, 3, 6; C_2)\zeta(2) - \zeta_2(2, 5, 5, 6; C_2)$$

$$- \frac{3}{8}\zeta_2(2, 6, 4, 6; C_2) - 2\zeta_2(3, 5, 2, 6; C_2)\zeta(2) + \zeta_2(4, 4, 2, 6; C_2)\zeta(2)$$

$$- \frac{3}{8}\zeta_2(4, 4, 4, 6; C_2) + \frac{19}{8}\zeta_2(5, 5, 2, 6; C_2) - \frac{27}{32}\zeta_2(6, 4, 2, 6; C_2)$$

$$+ 4\zeta_2(2, 4, 3, 7; C_2)\zeta(2) + 2\zeta_2(2, 4, 5, 7; C_2) + 3\zeta_2(2, 5, 4, 7; C_2)$$

$$- 2\zeta_2(3, 4, 2, 7; C_2)\zeta(2) - \frac{9}{4}\zeta_2(4, 5, 2, 7; C_2) + \frac{1}{2}\zeta_2(5, 4, 2, 7; C_2)$$

$$- 6\zeta_2(2, 5, 3, 8; C_2) + 3\zeta_2(3, 5, 2, 8; C_2) - \frac{3}{2}\zeta_2(4, 4, 2, 8; C_2)$$

$$- 6\zeta_2(2, 4, 3, 9; C_2) + 3\zeta_2(3, 4, 2, 9; C_2).$$

The authors also checked this equation numerically by using definitions (8.64) and (9.72).

Remark 9.2. From these considerations, we can see that our method may be applied to much wider class of multiple zeta-functions. As another example, we will consider the zeta-function $\zeta_2(\mathbf{s}; G_2)$ associated with the exceptional Lie algebra of type G_2 and will give certain functional relations including explicit forms of Witten's volume formulas of type G_2 in a forthcoming paper [20].

Acknowledgements. The authors greatly thank Dr. Takuya Okamoto for his pointing out a mistake in this paper.

References

1. S. Akiyama, S. Egami, and Y. Tanigawa, Analytic continuation of multiple zeta-functions and their values at non-positive integers, *Acta Arith.* **98** (2001), 107-116.
2. T. M. Apostol, *Introduction to Analytic Number Theory*, Springer, 1976.
3. T. Arakawa and M. Kaneko, Multiple zeta values, poly-Bernoulli numbers, and related zeta functions, *Nagoya Math. J.* **153** (1999), 189-209.
4. D. Bowman and D. M. Bradley, Multiple polylogarithms: a brief survey, in *'Conference on q-Series with Applications to Combinatorics, Number Theory, and Physics' (Urbana, IL, 2000)*, Contemp. Math. **291**, B. C. Berndt and K. Ono (eds.), Amer. Math. Soc., Providence, RI, 2001, 71-92.
5. D. M. Bradley, Partition identities for the multiple zeta function, in *'Zeta Functions, Topology and Quantum Physics', Developments in Mathematics 14*, T. Aoki et al. (eds.), Springer, New York, 2005, 19-29.
6. O. Espinosa and V. H. Moll, The evaluation of Tornheim double sums, Part I, *J. Number Theory* **116** (2006), 200-229.
7. D. Essouabri, *Singularités des séries de Dirichlet associées à des polynômes de plusieurs variables et applications à la théorie analytique des nombres*, Thése, Univ. Henri Poincaré - Nancy I, 1995.
8. D. Essouabri, Singularités des séries de Dirichlet associées à des polynômes de plusieurs variables et applications en théorie analytique des nombres, *Ann. Inst. Fourier* **47** (1997), 429-483.
9. L. Euler, Meditationes circa singulare serierum genus, *Novi Comm. Acad. Sci. Petropol.* **20** (1775), 140-186. Reprinted in *Opera Omnia, ser. I*, vol. 15, B. G. Teubner, Berlin, 1927, 217-267.

10. P. E. Gunnells and R. Sczech, Evaluation of Dedekind sums, Eisenstein cocycles, and special values of *L*-functions, *Duke Math. J.* **118** (2003), 229-260.

11. G. H. Hardy, Notes on some points in the integral calculus LV, On the integration of Fourier series, *Messenger of Math.* **51** (1922), 186-192. Reprinted in Collected papers of G. H. Hardy (including joint papers with J. E. Littlewood and others), Vol. III, Clarendon Press, Oxford, 1969, 506-512.

12. M. E. Hoffman, Multiple harmonic series, *Pacific J. Math.* **152** (1992), 275-290.

13. M. E. Hoffman, Algebraic aspects of multiple zeta values, in *'Zeta Functions, Topology and Quantum Physics'*, Developments in Mathematics 14, T. Aoki et al. (eds.), Springer, New York, 2005, 51-74.

14. J. G. Huard, K. S. Williams and N.-Y. Zhang, On Tornheim's double series, *Acta Arith.* **75** (1996), 105-117.

15. M. Kaneko, Multiple zeta values, *Sugaku Expositions* **18** (2005), 221-232.

16. Y. Komori, K. Matsumoto and H. Tsumura, Zeta-functions of root systems, in *'Proceedings of the Conference on L-functions' (Fukuoka, 2006)*, L. Weng and M. Kaneko (eds), World Scientific, 2007, 115-140.

17. Y. Komori, K. Matsumoto and H. Tsumura, Zeta and *L*-functions and Bernoulli polynomials of root systems, *Proc. Japan Acad., Ser. A* **84** (2008), 57-62.

18. Y. Komori, K. Matsumoto and H. Tsumura, On Witten multiple zeta-functions associated with semisimple Lie algebras II, to appear in *J. Math. Soc. Japan*.

19. Y. Komori, K. Matsumoto and H. Tsumura, On Witten multiple zeta-functions associated with semisimple Lie algebras III, preprint, arXiv:math/0907.0955.

20. Y. Komori, K. Matsumoto and H. Tsumura, On Witten multiple zeta-functions associated with semisimple Lie algebras IV, preprint, arXiv:math/0907.0972.

21. Y. Komori, K. Matsumoto and H. Tsumura, On multiple Bernoulli polynomials and multiple *L*-functions of root systems, to appear in *Proc. London Math. Soc.*

22. Y. Komori, K. Matsumoto and H. Tsumura, An introduction to the theory of zeta-functions of root systems, preprint.

23. K. Matsumoto, Asymptotic expansions of double zeta-functions of Barnes, of Shintani, and Eisenstein series, *Nagoya Math J.*, **172** (2003), 59-102.

24. K. Matsumoto, On the analytic continuation of various multiple zeta-functions, in *'Number Theory for the Millennium II, Proc. Millennial Conference on Number Theory'*, M. A. Bennett et al. (eds.), A K Peters, 2002, 417-440.

25. K. Matsumoto, The analytic continuation and the asymptotic behaviour of certain multiple zeta-functions I, *J. Number Theory*, **101** (2003), 223-243.

26. K. Matsumoto, On Mordell-Tornheim and other multiple zeta-functions, in *'Proceedings of the Session in analytic number theory and Diophantine equations' (Bonn, January-June 2002)*, D. R. Heath-Brown and B. Z. Moroz

(eds.), Bonner Mathematische Schriften Nr. 360, Bonn 2003, n.25, 17pp.

27. K. Matsumoto, Analytic properties of multiple zeta-functions in several variables, in *'Number Theory: Tradition and Modernization'*, W. Zhang and Y. Tanigawa (eds.), Springer, 2006, 153-173.

28. K. Matsumoto, T. Nakamura, H. Ochiai and H. Tsumura, On value-relations, functional relations and singularities of Mordell-Tornheim and related triple zeta-functions, *Acta Arith.* **132** (2008), 99-125.

29. K. Matsumoto, T. Nakamura and H. Tsumura, Functional relations and special values of Mordell-Tornheim triple zeta and *L*-functions, *Proc. Amer. Math. Soc.* **136** (2008), 2135-2145.

30. K. Matsumoto and H. Tsumura, On Witten multiple zeta-functions associated with semisimple Lie algebras I, *Ann. Inst. Fourier* **56** (2006), 1457-1504.

31. K. Matsumoto and H. Tsumura, A new method of producing functional relations among multiple zeta-functions, *Quart. J. Math.* (Oxford) **59** (2008), 55-83.

32. K. Matsumoto and H. Tsumura, Functional relations among certain double polylogarithms and their character analogues, *Šiauliai Math. Semin.* (11) **3** (2008), 189-205.

33. L. J. Mordell, On the evaluation of some multiple series, *J. London Math. Soc.* **33** (1958), 368-371.

34. T. Nakamura, A functional relation for the Tornheim double zeta function, *Acta Arith.* **125** (2006), 257-263.

35. T. Nakamura, Double Lerch series and their functional relations, *Aequationes Math.* **75** (2008), 251-259.

36. T. Nakamura, Double Lerch value relations and functional relations for Witten zeta functions, *Tokyo J. Math.* **31** (2008), 551-574.

37. M. V. Subbarao and R. Sitaramachandrarao, On some infinite series of L. J. Mordell and their analogues, *Pacific J. Math.* **119** (1985), 245-255.

38. E. C. Titchmarsh, *The Theory of the Riemann Zeta-function*, 2nd ed. (revised by D. R. Heath-Brown), Oxford University Press, 1986.

39. L. Tornheim, Harmonic double series, *Amer. J. Math.* **72** (1950), 303-314.

40. H. Tsumura, On some combinatorial relations for Tornheim's double series, *Acta Arith.* **105** (2002), 239-252.

41. H. Tsumura, On alternating analogues of Tornheim's double series, *Proc. Amer. Math. Soc.* **131** (2003), 3633-3641.

42. H. Tsumura, An elementary proof of Euler's formula for $\zeta(2m)$, *Amer. Math. Monthly* 111 (2004), 430-431.

43. H. Tsumura, Evaluation formulas for Tornheim's type of alternating double series, *Math. Comp.* **73** (2004), 251-258.

44. H. Tsumura, On Witten's type of zeta values attached to $SO(5)$, *Arch. Math. (Basel)* **84** (2004), 147-152.

45. H. Tsumura, Certain functional relations for the double harmonic series related to the double Euler numbers, *J. Austral. Math. Soc., Ser. A.* **79** (2005), 317-333.

46. H. Tsumura, On some functional relations between Mordell-Tornheim dou-

ble *L*-functions and Dirichlet *L*-functions, *J. Number Theory* **120** (2006), 161-178.

47. H. Tsumura, On functional relations between the Mordell-Tornheim double zeta functions and the Riemann zeta function, *Math. Proc. Cambridge Philos. Soc.* **142** (2007), 395-405.

48. H. Tsumura, On alternating analogues of Tornheim's double series II, *Ramanujan J.* **18** (2009), 81-90.

49. E. T. Whittaker and G. N. Watson, *A Course of Modern Analysis*, 4th ed., Cambridge University Press, 1958.

50. E. Witten, On quantum gauge theories in two dimensions, *Comm. Math. Phys.* **141** (1991), 153-209.

51. D. Zagier, Values of zeta functions and their applications, in *'First European Congress of Mathematics'* Vol. II, A. Joseph et al. (eds.), Progr. Math. **120**, Birkhäuser, 1994, 497-512.

52. J. Zhao, Analytic continuation of multiple zeta functions, *Proc. Amer. Math. Soc.* **128** (2000), 1275-1283.

A QUICK INTRODUCTION TO MAASS FORMS

JIANYA LIU

School of Mathematics, Shandong University,
Jinan, Shandong 250100, China
E-mail: jyliu@sdu.edu.cn

We first introduce the spectral theory of Maass forms for the full modular group. Then we state, without proof, some basic properties of Maass forms for the congruence subgroup $\Gamma_0(N)$. Finally, as an application of these materials, we treat a Linnik-type problem for Maass cusp forms, that is the occurrence of the first negative coefficient in the Fourier expansion of Maass new forms.

1. Introduction

1.1. *The aim of the paper*

For simplicity, we shall mainly treat the theory for the full modular group

$$\Gamma = SL_2(\mathbb{Z}) = \left\{ \begin{pmatrix} a & b \\ c & d \end{pmatrix} : a,b,c,d \in \mathbb{Z}, \ ad-bc = 1 \right\}. \qquad (1.1)$$

Even for the simplest case (1.1), there are already a large amount of materials in the literature. To explain these materials in limited number of pages, I have to omit most of the proofs, and content myself just with illustrations and explanations. The general theory for congruence subgroups $\Gamma_0(N)$ is similar in philosophy but more involved in details, and therefore we just state their basic properties. Finally, as an application of these materials, we treat a Linnik-type problem for Maass cusp forms, i.e. the first negative coefficient in the Fourier expansion of Maass new forms.

For detailed treatment of Maass forms in book form, the reader is referred to e.g. Borel [1], Bump [2], Iwaniec [5], and Ye [25].

The contents are as follows:

§2. Maass forms for $SL_2(\mathbb{Z})$
§3. Fourier expansion for Maass forms
§4. Spectral decomposition of non-Euclidean Laplacian Δ
§5. Hecke theory for Maass forms

1.2. *Notation*

Throughout the notes, Γ is always $SL_2(\mathbb{Z})$, as in (1.1). We use $s = \sigma + it$ and $z = x + iy$ to denote complex variables and $\Re$ and $\Im$ for their real and the imaginary parts. The Vinogradov symbol $A \ll B$ means that $A = O(B)$, and as usual $e(x) = e^{2\pi i x}$. Let $\mathbb{H}$ denote the upper half plane $\{z \in \mathbb{C} \mid \Im z > 0\}$. The group Γ acts on $\mathbb{H}$ through

$$gz = \frac{az + b}{cz + d}, \quad g = \begin{pmatrix} a & b \\ c & d \end{pmatrix} \in \Gamma.$$

Then, since

$$\Im gz = \frac{y}{|cz + d|^2}, \quad g = \begin{pmatrix} * & * \\ c & d \end{pmatrix} \in \Gamma, \tag{1.2}$$

$\mathbb{H}$ is stable under the action of Γ. Note that $-1 \in \Gamma$ acts trivially on $\mathbb{H}$, and so we sometimes refer to the group $PSL_2(\mathbb{Z}) = SL_2(\mathbb{Z})/\{\pm 1\}$.

In this notes, we shall make a conventional use of the associative law for composition of functions or operators whenever possible: $(f \circ g) \circ h = f \circ (g \circ h)$ to the effect that these are operated successively.

We recall that $\Gamma = SL_2(\mathbb{Z})$ is generated by $S = \begin{pmatrix} 0 & -1 \\ 1 & 0 \end{pmatrix}$ and $T = \begin{pmatrix} 1 & 1 \\ 0 & 1 \end{pmatrix}$, where S is for Spiegelung and T for translation, and that

$$\Gamma \backslash \mathbb{H} = \left\{ z \in \mathbb{H} : |z| > 1, -\frac{1}{2} \le \Re z < \frac{1}{2} \right\}$$

$$\cup \left\{ z \in \mathbb{H} : |z| = 1, \frac{\pi}{2} \le \arg z < \frac{2\pi}{3} \right\}$$

is a fundamental domain for the action of Γ on the upper half-plane $\mathbb{H}$.

2. Maass forms for $SL_2(\mathbb{Z})$

The standard Euclidean Laplace operator on the complex plane $\mathbb{C}$ is defined by

$$\Delta^e = \frac{\partial^2}{\partial x^2} + \frac{\partial^2}{\partial y^2},$$

while on the upper half-plane $\mathbb{H}$, the non-Euclidean Laplace operator is given by

$$\Delta = -y^2\left(\frac{\partial^2}{\partial x^2} + \frac{\partial^2}{\partial y^2}\right) = -y^2\Delta^e.$$

By the above convention, we may understand

$$(\Delta f) \circ g = \Delta(f \circ g), \qquad (2.1)$$

where the latter means that

$$\Delta(f \circ g)(z) = \Delta((f \circ (gz))),$$

for any smooth function f on $\mathbb{H}$, any element $g = \left(\begin{smallmatrix} a & b \\ c & d \end{smallmatrix}\right) \in SL_2(\mathbb{R})$, and any $z \in \mathbb{H}$. We say that the operator Δ is invariant under the action of Γ.

In order to show (2.1), it is sufficient to check it for the generators.

Definition 2.1. A smooth function $f \neq 0$ on $\mathbb{H}$ is called a Maass form for the group Γ if it satisfies the following conditions:

(i) Strict automorphy: for all $g \in \Gamma$ and all $z \in \mathbb{H}$,

$$f(gz) = f(z);$$

(ii) Eigentlichkeit: f is an eigenfunction of the non-Euclidean Laplacian,

$$\Delta f = \lambda f,$$

where λ is an eigenvalue of Δ;

(iii) Growth condition: there exists a positive integer N such that

$$f(z) \ll y^N, \quad y \to +\infty.$$

Definition 2.2. A Maass form f is said to be a cusp form if

$$\int_0^1 f\left(\left(\begin{smallmatrix} 1 & b \\ 0 & 1 \end{smallmatrix}\right)z\right)db = \int_0^1 f(z+b)db = 0$$

holds for all $z \in \mathbb{H}$.

Note that Definition 2.2 entails the identical vanishing of the constant term in the Fourier expansion, i.e. it is of the form of (3.7) below.

As an example of Maass forms, we consider the non-analytic Eisenstein series

$$E^*(z,s) = \frac{\Gamma(s)}{2\pi^s} \sum_{\substack{m,n\in\mathbb{Z} \\ (m,n)\neq(0,0)}} \frac{y^s}{|mz+n|^{2s}}. \qquad (2.2)$$

This is the completed Epstein zeta-function, cf. Siegel [24].

Theorem 2.1. *Let $s = \sigma + it$. Then*

(i) *The Eisenstein series $E^*(z,s)$ is absolutely convergent for $\sigma > 1$.*

(ii) *Although $E^*(z,s)$ is not analytic as a function of $z \in \mathbb{H}$, it is strictly automorphic, i.e. for $\sigma > 1$,*

$$E^*(gz,s) = E^*(z,s), \quad g \in \Gamma. \tag{2.3}$$

(iii) *For $\sigma > 1$, the Eisenstein series $E^*(s,z)$ is an eigenfunction of Δ,*

$$\Delta E^*(z,s) = s(1-s)E^*(z,s). \tag{2.4}$$

Proof. (i) We have

$$\frac{y^s}{|mz+n|^{2s}} \ll \frac{|y|^\sigma}{\min(|my|,|n|)^{2\sigma}}.$$

Therefore, for $\sigma > 1$,

$$\sum_{\substack{m,n\in\mathbb{Z}\\(m,n)\neq(0,0)}} \frac{y^s}{|mz+n|^{2s}} \ll |y|^\sigma + |y|^\sigma \sum_{m\geq 1}\sum_{n\leq m|y|} \frac{1}{n^{2\sigma}} + \frac{1}{|y|^\sigma}\sum_{n\geq 1}\sum_{m\leq n/|y|} \frac{1}{m^{2\sigma}}$$

$$\ll |y|^\sigma + \frac{|y|^{1-\sigma}}{2\sigma-2} + \frac{|y|^{\sigma-1}}{2\sigma-2}.$$

Hence the series in (2.2) is absolutely convergent in $\sigma > 1$.

(ii) Since the imaginary part is separated in (2.1), $E^*(z,s)$ is not analytic as a function of $z \in \mathbb{H}$.

To see that $E^*(s,z)$ is strictly automorphic, we introduce a non-analytic Eisenstein series $E(z,s)$ by the formula

$$E^*(z,s) = E(z,s)\xi(2s), \tag{2.5}$$

where

$$\xi(s) = \pi^{-s/2}\Gamma(s/2)\zeta(s),$$

and $\zeta(s)$ is the Riemann zeta-function. By (2.2) and (1.2) we have

$$E(z,s) = y^s + \frac{1}{2}\sum_{\substack{(c,d)=1\\c\neq 0}} \frac{y^s}{|cz+d|^{2s}} = \frac{1}{2}\sum_{g\in\Gamma_\infty\backslash\Gamma} (\Im gz)^s, \tag{2.6}$$

where Γ_∞ is the subgroup of Γ generated by $S = \left(\begin{smallmatrix}1&1\\0&1\end{smallmatrix}\right)$. From (2.6), it is easy to see that $E(z,s)$, and *a fortiori* $E^*(z,s)$, is invariant under Γ, whence (2.3).

(iii) That $E(z,s)$ is an eigenfunction of Δ in $\sigma > 1$ can be checked by applying Δ term by term in the expansion (2.7). In fact, we have

$$\Delta y^s = -y^2\left(\frac{\partial^2}{\partial x^2} + \frac{\partial^2}{\partial y^2}\right)(y^s) = s(1-s)y^s,$$

so that the first term y^s in (2.6) is an eigenfunction of Δ with eigenvalue $\lambda = s(1-s)$. The general term $\frac{y^s}{|cz+d|^{2s}}$, which is $(\Im gz)^s$ by (1.2), is also an eigenfunction of Δ with eigenvalue $\lambda = s(1-s)$, since Δ is invariant under Γ. Therefore,

$$\Delta E(z,s) = s(1-s)E(z,s), \tag{2.7}$$

and hence (2.4) follows. □

We remark that Theorem 2.1, holding only for $\sigma > 1$, does not prove that $E^*(z,s)$ is a Maass form, since the growth condition in Definition 2.1 is to be checked. This will be done via the analytic continuation of $E^*(z,s)$ in §3.2.

3. Analytic continuation of Eisenstein series

3.1. *Fourier expansion of Maass forms*

We will first give the Fourier expansion of a Maass form f in the following theorem.

Theorem 3.1. *Suppose $f(z)$ satisfies the conditions (i) and (ii) in Definition 2.1. Then it has the Fourier expansion*

$$f(z) = a_0 y^{1/2+i\nu} + b_0 y^{1/2-i\nu}$$
$$+ \sum_{n \neq 0} \{ a_n \sqrt{y} K_{i\nu}(2\pi|n|y) + b_n \sqrt{y} I_{i\nu}(2\pi|n|y) \} e(nx), \tag{3.1}$$

where $\lambda = 1/4 + \nu^2$, and $K_{i\nu}$ and $I_{i\nu}$ are the modified Bessel functions of the third kind and the first kind, respectively. Here and throughout, a summation over $n \neq 0$ means that n runs over all non-zero integers.

Proof. Since

$$f(Tz) = f(z), \quad \text{for } T = \begin{pmatrix} 1 & 1 \\ 0 & 1 \end{pmatrix} \in \Gamma,$$

we have

$$f(x+1+iy) = f(z+1) = f(z) = f(x+iy),$$

so that $f(z)$, being a smooth function, must have the Fourier expansion in x:

$$f(z) = \sum_{n \in \mathbb{Z}} c_n(y) e(nx).$$

It remains to determine the coefficients $c_n(y)$.

By Definition 2.1, we have $\Delta f = \lambda f$, and hence $c_n(y)$ satisfies

$$-y^2 c_n''(y) + 4\pi^2 n^2 y^2 c_n(y) = \lambda c_n(y). \tag{3.2}$$

If $n \neq 0$, this is a modified Bessel equation, which is known to have the general solution

$$c_n(y) = a_n \sqrt{y} K_{i\nu}(2\pi|n|y) + b_n \sqrt{y} I_{i\nu}(2\pi|n|y). \tag{3.3}$$

For $n = 0$, (3.2) takes the form

$$c_0''(y) + \lambda y^{-2} c_0(y) = 0, \tag{3.4}$$

which has the solution

$$c_0(y) = a_0 y^{1/2+i\nu} + b_0 y^{1/2-i\nu}. \tag{3.5}$$

This proves the theorem. $\qquad\square$

Corollary 3.1. *If $f(z)$ is a Maass form for Γ, then its Fourier expansion is*

$$f(z) = a_0 y^{1/2+i\nu} + b_0 y^{1/2-i\nu}$$
$$+ \sum_{n\neq0} a_n \sqrt{y} K_{i\nu}(2\pi|n|y) e(nx). \tag{3.6}$$

For $\lambda \geq 1/4$, a square-integrable Maass form $f(z)$ must be a cusp form, and its Fourier expansion takes the form

$$f(z) = \sum_{n\neq0} a_n \sqrt{y} K_{i\nu}(2\pi|n|y) e(nx). \tag{3.7}$$

Proof. Comparing the asymptotic formulae

$$I_{i\nu}(y) = \frac{e^y}{\sqrt{2\pi y}} \left(1 + O\left(\frac{1+|\nu|^2}{y}\right)\right), \tag{3.8}$$

$$K_{i\nu}(y) = e^{-y} \sqrt{\frac{\pi}{2y}} \left(1 + O\left(\frac{1+|\nu|^2}{y}\right)\right), \tag{3.9}$$

for $y \to \infty$ with Condition (iii) of Definition 2.1, we see that b_n must be zero, and (3.6) follows from (3.1).

If $\lambda \geq 1/4$, then ν is real and

$$\int_{\Gamma\backslash\mathbb{H}} |c_0(y)|^2 dz \geq \int_{-1/2}^{1/2} dx \int_1^\infty \frac{1}{y} |a_0 y^{i\nu} + b_0 y^{-i\nu}|^2 dy$$
$$\gg \int_1^\infty \frac{dy}{y} = \infty$$

if $a_0^2 + b_0^2 > 0$. Hence for a square-integrable Maass form f, we must have $c_0(y) = 0$, and its Fourier expansion takes the form of (3.2). This proves the Corollary.   □

Let $\iota : \mathbb{H} \to \mathbb{H}$ be the antiholomorphic involution

$$\iota(x + iy) = -x + iy.$$

As in the case of (2.2), we make the convention to define the action of ι on f by $f \circ \iota$:

$$(\iota f)(x + iy) = (f \circ \iota)(x + iy) = f(-x + iy).$$

We note the following facts: *If f is an eigenfunction of Δ, then $f \circ \iota$ is an eigenfunction with the same eigenvalue. Its eigenvalues are ± 1.* Indeed,

$$\Delta(f \circ \iota)(z) = \Delta(f(\iota(z))) = \Delta f(-x + iy) = \lambda f(-x + iy) = \lambda(f \circ \iota)(z),$$

whence the first assertion follows. The second assertion follows on noting that $\iota^2 = 1$, and

$$f = \iota^2 f = \alpha(\iota f) = \alpha^2 f,$$

whence $\alpha = \pm 1$.

We may therefore diagonalize the Maass cusp forms with respect to ι. If $f \circ \iota = f$, we call f *even*. In this case $a_n = a_{-n}$. If $f \circ \iota = -f$, then we call f *odd*. Then $a_n = -a_{-n}$.

3.2. *Analytic continuation of Eisenstein series*

We will establish the analytic continuation of $E^*(z, s)$ via its Fourier expansion.

Theorem 3.2.

(i) The Eisenstein series $E^*(z, s)$, originally defined for $\sigma > 1$, has the meromorphic continuation to the whole s-plane $\mathbb{C}$ in view of its Fourier expansion

$$E^*(z, s) = \xi(2s)y^s + \xi(2 - 2s)y^{1-s}$$
$$+ 2 \sum_{r \neq 0} |r|^{s-\frac{1}{2}} \sigma_{1-2s}(|r|) \sqrt{y} K_{s-\frac{1}{2}}(2\pi|r|y)e(rx), \qquad (3.10)$$

where

$$\sigma_a(n) = \sum_{d|n} d^a \qquad (3.11)$$

is the sum-of-divisors function. It is analytic except at $s = 1$ and $s = 0$, where it has simple poles. The residues at $s = 1$ and $s = 0$ are the constant functions $z = 1/2$ and $-1/2$ respectively. This is also known as the Selberg-Chowla integral formula.

(ii) The Eisenstein series satisfies the functional equation

$$E^*(z, s) = E^*(z, 1 - s). \tag{3.12}$$

(iii) We have

$$E^*(x + iy, s) \ll y^{\max(\sigma, 1-\sigma)}, \quad y \to \infty. \tag{3.13}$$

It follows that the Eisenstein series $E^*(z, s)$ is a Maass form with Laplace eigenvalue $s(1 - s)$.

Proof. (i) By Theorem 2.1, we know that $E^*(z, s)$ has the Fourier expansion (3.1) with

$$i\nu = \sqrt{-s(1 - s) + \frac{1}{4}} = s - \frac{1}{2}.$$

We compute the Fourier coefficient $c_r = c_r(y)$ anew:

$$E^*(z, s) = \sum_{r \in \mathbb{Z}} c_r(y, s) e(rx), \tag{3.14}$$

where

$$c_r(y, s) = \int_0^1 E^*(x + iy, s) e(rx) dx.$$

Recalling (2.2), we see that the contribution from the terms with $m = 0$, being independent of x, is only to a_0, and it is

$$\pi^{-s} \Gamma(s) y^s \sum_{n=1}^{\infty} n^{-2s} = \pi^{-s} \Gamma(s) \zeta(2s) y^s = \xi(2s) y^s, \tag{3.15}$$

which corresponds to $a_0 y^s$ in (3.1) and is part of c_0, as we shall see below.

Next, we consider the contribution from the terms with $m \neq 0$. Since $(-m, -n)$ and (m, n) contribute equally, we may simply sum the terms with $m > 0$. The contribution to a_r equals

$$C_r(y, s) = \pi^{-s} \Gamma(s) y^s \sum_{m=1}^{\infty} \sum_{n=-\infty}^{\infty} \int_0^1 \frac{e(rx)}{\{(mx + n)^2 + m^2 y^2\}^s} dx, \tag{3.16}$$

which becomes, on classifying the inner sum modulo m,

$$\pi^{-s} \Gamma(s) y^s \sum_{m=1}^{\infty} \sum_{n \bmod m} \int_{-\infty}^{\infty} \frac{e(rx)}{\{(mx + n)^2 + m^2 y^2\}^s} dx. \tag{3.17}$$

By making the substitution $x \mapsto x - n/m$, this becomes

$$\pi^{-s}\Gamma(s)y^s \sum_{m=1}^{\infty} m^{-2s} \sum_{n \bmod m} e\left(\frac{rn}{m}\right) \int_{-\infty}^{\infty} \frac{e(rx)}{(x^2+y^2)^s} dx$$

whose double sum is

$$\sum_{m|r} m^{1-2s} = \sigma_{1-2s}(r)$$

for $r \neq 0$, and is $\zeta(1-2s)$ for $r = 0$. Hence

$$C_r(y,s) = \begin{cases} \left(\dfrac{y}{\pi}\right)^s \Gamma(s)\sigma_{1-2s}(r) \displaystyle\int_{-\infty}^{\infty} \dfrac{e(rx)}{(x^2+y^2)^s} dx, & \text{if } r \neq 0, \\[3mm] \left(\dfrac{y}{\pi}\right)^s \Gamma(s)\zeta(2s-1) \displaystyle\int_{-\infty}^{\infty} \dfrac{1}{(x^2+y^2)^s} dx, & \text{if } r = 0. \end{cases}$$

Now we claim that

$$\left(\frac{y}{\pi}\right)^s \Gamma(s) \int_{-\infty}^{\infty} \frac{e(rx)}{(x^2+y^2)^s} dx$$

$$= \begin{cases} 2|r|^{s-1/2}\sqrt{y}K_{s-1/2}(2\pi|r|y), & \text{if } r \neq 0, \\[2mm] \pi^{-s+1/2}\Gamma(s-1/2)y^{1-s}, & \text{if } r = 0. \end{cases} \tag{3.18}$$

Using this, we see that the Fourier coefficient $c_r(y,s) = C_r(y,s)$ with $r \neq 0$ is given by

$$c_r(y,s) = 2|r|^{s-1/2}\sigma_{1-2s}(|r|)\sqrt{y}K_{s-\frac{1}{2}}(2\pi|r|y). \tag{3.19}$$

On the other hand, for $r = 0$, by (3.18), the contribution to c_0 is

$$\pi^{\frac{1}{2}-s}\Gamma\left(s-\frac{1}{2}\right)\zeta(2s-1)y^{1-s} = \xi(2s-1)y^{1-s}, \tag{3.20}$$

which corresponds to the last term $b_0(s)y^{1-s}$ in (3.6) and will turn into it in view of the functional equation $\xi(s) = \xi(1-s)$. Equations (3.15) and (3.20) give the Fourier coefficient

$$c_0(y,s) = \xi(2s)y^s + \xi(2-2s)y^{1-s}. \tag{3.21}$$

Now (3.19) and (3.21) establish (3.10) for $\sigma > 1$. Therefore the theorem now follows from an examination of the expansion (3.10). Each individual term of the series has analytic continuation to all s, except that c_0 has simple poles at $s = 0$ and $s = 1$. Each of the two terms in (3.21) has a pole at $s = 1/2$, but these cancel. The convergence of the infinite series follows from the rapid decay of the Bessel function. Thus we obtain the continuation.

It is easy to check that the residues of c_0 at the poles at $s = 1$ and $s = 0$ are the constant functions $z = 1/2$ and $z = -1/2$, respectively.

(ii) To get the functional equation (3.12), we observe that

$$c_r(y, s) = c_r(y, 1 - s).$$

This is clearly true if $r = 0$; if $r \neq 0$, this follows from (3.19), the relation $K_\nu(z) = K_{-\nu}(z)$ and the identity

$$r^s \sigma_{-2s}(r) = \prod_{d_1 d_2 = r} d_1^s d_2^{-s} = r^{-s} \sigma_{2s}(r).$$

Hence we have (3.12).

(iii) As regards (3.13), it follows from (3.19) that the non-constant term (3.19) decay rapidly as $y \to 0$, and so asymptotically the behavior of $E^*(z, s)$ is the same as its constant term (3.21).

Thus, we conclude that $E^*(z, s)$ is a Maass form.

It remains to establish the claim (3.18). Expressing $\Gamma(s) \left(\frac{y}{\pi(x^2 + y^2)} \right)^s$ as an integral, we write the left-hand side of (3.18) as

$$\int_0^\infty \int_{-\infty}^\infty \exp\left(-\pi t \frac{x^2 + y^2}{y} \right) t^s e(rx) dx \frac{dt}{t},$$

where we have changed the order of integration, which is justified by absolute convergence. Now we have

$$\int_{-\infty}^\infty e^{-t\pi x^2/y} e(rx) dx = \begin{cases} \sqrt{y/t} \, \exp(-y\pi r^2/t), & \text{if } r \neq 0, \\ \sqrt{y/t}, & \text{if } r = 0, \end{cases}$$

where the $r \neq 0$ case is a well-known formula in the theory of Fourier transforms, and the $r = 0$ case is the probability integral. We substitute this into the previous integral. If $r = 0$, then we get (3.18); if $r \neq 0$, the substitution $t \mapsto |r|t$ also gives (3.18). $\square$

We note that (3.18) in the case $r \neq 0$ is the following well-known formula of Basset:

$$K_\nu(z) = \frac{1}{\sqrt{\pi}} (2z)^s \Gamma\left(\nu + \frac{1}{2} \right) \int_0^\infty \frac{\cos t}{(t^2 + z^2)^{2\nu+1}} dt, \qquad (3.22)$$

for $\Re \nu > -1/2$ and $|\arg z| < \pi/2$.

We restate (3.10) in the form of a Fourier series for a non-analytic Eisenstein series in the following

Corollary 3.2. *The non-analytic Eisenstein series $E(z,s)$ has the Fourier expansion*

$$E(z,s) = \sum_{r \in \mathbb{Z}} a_r(y,s)e(rx)$$

with

$$a_0(y,s) = y^s + \varphi(s)y^{1-s},$$

where

$$\varphi(s) = \frac{\pi^{2s-1}\Gamma(1-s)\zeta(2-2s)}{\Gamma(s)\zeta(2s)} = \frac{\xi(2-2s)}{\xi(2s)},$$

with $\xi(s)$ defined immediately after (2.5), and, for $r \neq 0$,

$$a_r(y,s) = \frac{2}{\xi(2s)}|r|^{s-1/2}\sigma_{1-2s}(|r|)\sqrt{y}K_{s-1/2}(2\pi|r|y).$$

4. Spectral decomposition of non-Euclidean Laplacian Δ

A function $f : \mathbb{H} \to \mathbb{C}$ is said to be strictly automorphic with respect to Γ (cf. Definition 2.1(i)) if

$$f(gz) = f(z), \quad \text{for all } g \in \Gamma.$$

Therefore, this f lives on $\Gamma\backslash\mathbb{H}$. We denote the space of such functions by $\mathcal{A}(\Gamma\backslash\mathbb{H})$. Our objective is to extend automorphic functions to automorphic forms subject to a suitable growth condition. The main results hold in the Hilbert space

$$\mathcal{L}(\Gamma\backslash\mathbb{H}) = \{f \in \mathcal{A}(\Gamma\backslash\mathbb{H}) : \|f\| < \infty\},$$

where the norm $\|f\| = \langle f, f \rangle$ is induced from the inner product

$$\langle f, g \rangle = \int_{\Gamma\backslash\mathbb{H}} f(z)\bar{g}(z)\frac{dxdy}{y^2}.$$

Note that the measure $d\mu(z) = \frac{dxdy}{y^2}$ is Γ-invariant.

Theorem 4.1. *On $\mathcal{L}(\Gamma\backslash\mathbb{H})$, the non-Euclidean Laplace operator Δ is positive semi-definite and a fortiori has its self-adjoint extension denoted by the same symbol and referred to as the automorphic Laplacian.*

Outline of Proof. Define

$$\mathcal{D}(\Gamma\backslash\mathbb{H}) = \{f \in \mathcal{A}(\Gamma\backslash\mathbb{H}) : f, \Delta f \text{ smooth and bounded}\}.$$

Then one can show that $\mathcal{D}(\Gamma\backslash\mathbb{H})$ is dense in $\mathcal{L}(\Gamma\backslash\mathbb{H})$.

One may also check that Δ is positive semi-definite and symmetric on $\mathcal{D}(\Gamma\backslash\mathbb{H})$. To this end, one checks that, for $f, g \in \mathcal{D}(\Gamma\backslash\mathbb{H})$,

$$\langle \Delta f, g \rangle = \int_{\Gamma\backslash\mathbb{H}} \nabla f \overline{\nabla g} dxdy, \qquad (4.1)$$

where $\nabla f = (f_x, f_y)$ is the gradient. Consequently

$$\langle \Delta f, g \rangle = \langle f, \Delta g \rangle,$$

and

$$\langle \Delta f, f \rangle = \int_{\Gamma\backslash\mathbb{H}} |\nabla f|^2 dxdy \geq 0,$$

from which follows the symmetry and positive semi-definiteness of Δ. We apply Green's formula to establish (4.1). Suppose $P = P(x, y)$ and $Q = Q(x, y)$ and smooth in $D \subset \mathbb{R}^2$. Then

$$\iint_D (Q_x - P_y)dxdy = \int_{\partial D} Pdx - Qdy. \qquad (4.2)$$

Choosing $P = -\bar{g}f_y$ and $Q = \bar{g}f_x$ gives

$$Q_x - P_y = \Delta^e f \bar{g} + \nabla f \cdot \overline{\nabla g}.$$

Hence (4.2) reads

$$\iint_D (\Delta^e f \bar{g} + \nabla f \cdot \overline{\nabla g})dxdy = \int_{\partial D} \bar{g}(-f_y dx + f_x dy).$$

Now below $-f_y dx + f_x dy$ is shown to be invariant under Γ. The boundary ∂D contains of two half-lines reversely oriented, and the left and right arcs reversely oriented. Hence the last integral is 0, so that

$$-\iint_D \Delta^e f \bar{g} dxdy = \iint_D \nabla f \cdot \overline{\nabla g} dxdy,$$

which is (4.1).

When $w = u + iv$ and $z = x + iy$ are related by $g(z) = w$, $g \in \Gamma$, we write $f(x, y) = F(u, v)$. Then by the chain rule and the Cauchy-Riemann equation

$$\begin{aligned}
&- f_y dx + f_x dy \\
&= -(F_u u_y + F_v v_y)dx + (F_u u_x + F_v v_x)dy \\
&= (F_u v_x - F_v u_x)dx + (F_u v_y + F_v u_y)dy \\
&= -F_v(u_x dx + u_y dy) + F_y(v_x dy + v_y dy) \\
&= -F_v du + F_u dv,
\end{aligned}$$

which proves the invariance of the differential form in question. Note that this amounts to the invariance of the non-Euclidean outer-normal derivative.

Finally, by Friedrichs' theorem in functional analysis, Δ has a unique self-adjoint extension to $\mathcal{L}(\Gamma\backslash\mathbb{H})$. The theorem is proved. $\square$

It follows from the argument in the above proof that the eigenvalue $\lambda = s(1-s)$ of an eigenfunction $f \in \mathcal{D}(\Gamma\backslash\mathbb{H})$ is real and non-negative. Therefore, either $s = 1/2 + it$ with $t \in \mathbb{R}$, or $0 < s < 1$.

Let $\psi(y)$ be a smooth function with compact support on $(0, +\infty)$. In analogy to (2.6), an incomplete Eisenstein series is defined by

$$E(z|\psi) = \sum_{g \in \Gamma_\infty \backslash \Gamma} \psi(\Im gz),$$

where Γ_∞ is defined after (2.6). Clearly, $E(z|\psi)$ is invariant under Γ. Since ψ has compact support, $E(z|\psi) \in \mathcal{L}(\Gamma\backslash\mathbb{H})$. It is not an automorphic form, since it fails to be an eigenfunction of Δ. However, by Mellin's inversion, one can represent the incomplete Eisenstein series as a contour integral involving the Eisenstein series

$$E(z|\psi) = \frac{1}{2\pi i} \int_{(\sigma)} E(z,s)\hat{\psi}(s)ds, \qquad (4.3)$$

where $\sigma > 1$ and

$$\hat{\psi}(s) = \int_0^\infty \psi(y)y^{-s}\frac{dy}{y}$$

is the Mellin transform of ψ.

To pursue the analysis, we select two linear subspaces of $\mathcal{L}(\Gamma\backslash\mathbb{H})$:

- $\mathcal{B}(\Gamma\backslash\mathbb{H})$, the space of smooth and bounded automorphic functions,
- $\mathcal{E}(\Gamma\backslash\mathbb{H})$, the space of incomplete Eisenstein series.

We have the inclusions

$$\mathcal{E}(\Gamma\backslash\mathbb{H}) \subset \mathcal{B}(\Gamma\backslash\mathbb{H}) \subset \mathcal{L}(\Gamma\backslash\mathbb{H}) \subset \mathcal{A}(\Gamma\backslash\mathbb{H}). \qquad (4.4)$$

The space $\mathcal{B}(\Gamma\backslash\mathbb{H})$ is dense in $\mathcal{L}(\Gamma\backslash\mathbb{H})$, but $\mathcal{E}(\Gamma\backslash\mathbb{H})$ need not be.

Let us examine the orthogonal complement of $\mathcal{E}(\Gamma\backslash\mathbb{H})$ in $\mathcal{B}(\Gamma\backslash\mathbb{H})$.

Take $f \in \mathcal{B}(\Gamma \backslash \mathbb{H})$ and $E(\cdot | \psi) \in \mathcal{E}(\Gamma \backslash \mathbb{H})$, and then apply the unfolding process. Since f and $d\mu$ are invariant under Γ, it follows that

$$
\begin{aligned}
\langle E(\cdot | \psi), f \rangle &= \int_{\Gamma \backslash \mathbb{H}} E(z|\psi) \bar{f}(z) \frac{dxdy}{y^2} \\
&= \sum_{g \in \Gamma_\infty \backslash \Gamma} \int_{\Gamma_\infty \backslash \mathbb{H}} \psi(\Im gz) \bar{f}(gz) d\mu(gz) \\
&= \int_{\Gamma_\infty \backslash \mathbb{H}} \psi(y) \bar{f}(z) d\mu(z).
\end{aligned}
$$

Since

$$
\Gamma_\infty \backslash \mathbb{H} = \{x + iy : 0 < x \le 1, y > 0\},
$$

we have

$$
\langle E(\cdot | \psi), f \rangle = \int_0^\infty \psi(y) \left\{ \int_0^1 \bar{f}(z) dx \right\} \frac{dy}{y^2}.
$$

Therefore, $f \in \mathcal{B}(\Gamma \backslash \mathbb{H})$ is orthogonal to $\mathcal{E}(\Gamma \backslash \mathbb{H})$ if and only if

$$
\int_0^1 f(z) dx = 0 \quad \text{for all } y > 0,
$$

i.e. its Fourier expansion has the form

$$
f(z) = \sum_{n \ne 0} c_n(y) e(nx) \tag{4.5}
$$

with $c_0(y) = 0$. Such a function is sometimes referred to as a parabolic forms or a cusp function. Denote by $\mathcal{C}(\Gamma \backslash \mathbb{H})$ the subspace of these cusp functions f. Then we have the orthogonal decomposition

$$
\mathcal{B}(\Gamma \backslash \mathbb{H}) = \mathcal{C}(\Gamma \backslash \mathbb{H}) \oplus \mathcal{E}(\Gamma \backslash \mathbb{H}).
$$

Noting that $\mathcal{B}(\Gamma \backslash \mathbb{H})$ is dense in $\mathcal{L}(\Gamma \backslash \mathbb{H})$, we have

Theorem 4.2. *We have the orthogonal decomposition*

$$
\mathcal{L}(\Gamma \backslash \mathbb{H}) = \widetilde{\mathcal{C}}(\Gamma \backslash \mathbb{H}) \oplus \widetilde{\mathcal{E}}(\Gamma \backslash \mathbb{H}),
$$

where the tilde stands for the closure in the Hilbert space $\mathcal{L}(\Gamma \backslash \mathbb{H})$ with respect to the norm topology.

Clearly,

$$
\Delta : \mathcal{C}(\Gamma \backslash \mathbb{H}) \to \mathcal{C}(\Gamma \backslash \mathbb{H}), \quad \Delta : \mathcal{E}(\Gamma \backslash \mathbb{H}) \to \mathcal{E}(\Gamma \backslash \mathbb{H}).
$$

We describe the spectral decomposition on $\mathcal{C}(\Gamma\backslash\mathbb{H})$ and $\mathcal{E}(\Gamma\backslash\mathbb{H})$ in the following two theorems.

Theorem 4.3. *The automorphic Laplacian Δ has a purely point spectrum on $\mathcal{C}(\Gamma\backslash\mathbb{H})$, i.e. the space $\mathcal{C}(\Gamma\backslash\mathbb{H})$ is spanned by cusp forms. The eigenvalues are*

$$0 = \lambda_0 < \lambda_1 \leq \lambda_2 \leq \ldots \to \infty$$

and the eigenspaces have finite dimension. For any complete orthonormal system of cusp forms $\{u_j\}$, every $f \in \mathcal{C}(\Gamma\backslash\mathbb{H})$ has the expansion

$$f(z) = \sum_{j=0}^{\infty} \langle f, u_j \rangle u_j(z),$$

converging in the norm topology. If $f \in \mathcal{C}(\Gamma\backslash\mathbb{H}) \cap \mathcal{D}(\Gamma\backslash\mathbb{H})$, then the series converges absolutely and uniformly on compacta.

Indeed, applying Δ to the Fourier expansion (4.3) of f, we see that the Fourier expansion of Δf will also have zero constant term, so that $\Delta\mathcal{C}(\Gamma\backslash\mathbb{H}) \subset \mathcal{C}(\Gamma\backslash\mathbb{H})$. Since Δ is self-adjoint, $\mathcal{C}(\Gamma\backslash\mathbb{H})$ is spanned by the eigenfunctions of Δ, i.e. spanned by the Maass cusp forms.

On the other hand, in the space $\mathcal{E}(\Gamma\backslash\mathbb{H})$ the spectrum turns out to be continuous. Here the analytic continuation of the Eisenstein series is the key issue. After this is established, the spectral resolution of Δ in $\mathcal{E}(\Gamma\backslash\mathbb{H})$ will evoke from (4.1) by contour integration. The eigenpacket of the continuous spectrum consists of the Eisenstein series $E(z, s)$ on the line $\sigma = 1/2$ (analytically continued).

Theorem 4.4. *The spectrum of Δ on $\mathcal{E}(\Gamma\backslash\mathbb{H})$ is absolutely continuous; it covers the segment $[1/4, +\infty)$ uniformly with multiplicity 1. Every $f \in \mathcal{E}(\Gamma\backslash\mathbb{H})$ has the expansion*

$$f(z) = \frac{1}{4\pi} \int_{-\infty}^{\infty} \langle f, E(\cdot, 1/2 + it) \rangle E(z, 1/2 + it) dt, \qquad (4.6)$$

which converges in the norm topology. If $f \in \mathcal{E}(\Gamma\backslash\mathbb{H}) \cap \mathcal{D}(\Gamma\backslash\mathbb{H})$, then the series converges pointwise absolutely and uniformly on compacta.

Combining Theorems 4.2-4.4, one gets the spectral decomposition of the whole space $\mathcal{L}(\Gamma\backslash\mathbb{H})$,

$$f(z) = \sum_{j=0}^{\infty} \langle f, u_j \rangle u_j(z) + \frac{1}{4\pi} \int_{-\infty}^{\infty} \langle f, E(\cdot, 1/2 + it) \rangle E(z, 1/2 + it) dt. \quad (4.7)$$

Theorem 4.5. *The Laplace eigenvalue of any Maass cusp form for Γ in the discrete spectrum is $> 3\pi^2/2$.*

Theorem 4.6. *(Weyl's law) Let $N_\Gamma(x)$ denote the number of Laplace eigenvalues up to x. Then*

$$N_\Gamma(x) = \frac{x}{12} + O(x^{1/2}\log x).$$

The proof of Theorem 4.6 depends on Selberg's trace formula which will not be presented in this notes. It follows from Theorem 4.6 that Maass cusp forms exist.

5. Hecke's theory for Maass forms

For $n \geq 1$, define the set

$$\Gamma_n = \left\{ \begin{pmatrix} a & b \\ c & d \end{pmatrix} : \ a, b, c, d \in \mathbb{Z}, \ ad - bc = n \right\}. \tag{5.1}$$

In particular, $\Gamma_1 = \Gamma$ is the modular group. The group Γ_1 naturally acts on Γ_n. For $n \geq 1$, the Hecke operator

$$T_n : \mathcal{L}(\Gamma_1 \backslash \mathbb{H}) \to \mathcal{L}(\Gamma_1 \backslash \mathbb{H})$$

is defined by

$$(T_n f)(z) = \frac{1}{\sqrt{n}} \sum_{g \in \Gamma_1 \backslash \Gamma_n} f(gz). \tag{5.2}$$

Picking up specific representations of $\Gamma_1 \backslash \Gamma_n$, we can also write

$$(T_n f)(z) = \frac{1}{\sqrt{n}} \sum_{ad=n} \sum_{b (\bmod d)} f\left(\frac{az+b}{d}\right). \tag{5.3}$$

Clearly, this is a finite sum, the number of terms being

$$[\Gamma_n : \Gamma_1] = \sum_{d|n} d = \sigma(n) = \sigma_1(n),$$

and therefore T_n is bounded on $\mathcal{L}(\Gamma_1 \backslash \mathbb{H})$ by $\sigma(n)n^{-1/2}$.

We first examine the action of T_n on a Maass cusp form f. To this end, we rewrite the Fourier expansion for f in (3.7) as

$$f(z) = \sum_{m \neq 0} a(m)\sqrt{y}K_{i\nu}(2\pi|m|y)e(mx), \tag{5.4}$$

which emphasizes that the coefficient $a(m)$ is a function of m.

Theorem 5.1. *Let f be a Maass cusp form with eigenvalue $1/4 + \nu^2$ with Fourier expansion (5.4). Then*

$$(T_n f)(z) = \sum_{m \neq 0} t_n(m) \sqrt{y} K_{i\nu}(2\pi|m|y) e(mx), \qquad (5.5)$$

with

$$t_n(m) = \sum_{d|(m,n)} a\left(\frac{mn}{d^2}\right). \qquad (5.6)$$

Proof. Since

$$(T_n f)(z) \qquad (5.7)$$

$$= \frac{1}{\sqrt{n}} \sum_{0 < d|n} \sum_{m \neq 0} a(m) \sqrt{\frac{ny}{d^2}} K_{i\nu}\left(\frac{2\pi|m|ny}{d^2}\right) e\left(\frac{mn}{d^2}x\right) \sum_{b \bmod d} e\left(\frac{m}{d}b\right)$$

and the innermost sum on the right is d for $d|m$ and 0 for $d \nmid m$, we have

$$(T_n f)(z) = \sqrt{y} \sum_{d|n} \sum_{m \neq 0} \sum_{d|m} a(m) K_{i\nu}\left(2\pi \frac{|m|n}{d^2} y\right) e\left(\frac{mn}{d^2}x\right).$$

Writing $m = m'd$, $m' \neq 0$ and then $m'\frac{n}{d} = m$, we get (5.5) and (5.6).   □

Now we are ready to establish some important facts about the Hecke operators T_n.

Theorem 5.2. *Let T_n be defined as above. Then*

(i) We have

$$T_m T_n = \sum_{d|(m,n)} T_{mn/d^2},$$

so that in particular T_m and T_n commute.
(ii) The Hecke operators commute with the Laplace operator Δ.
(iii) T_n is also self-adjoint in $\mathcal{L}(\Gamma_1 \backslash \mathbb{H})$, i.e.

$$\langle T_n f, g \rangle = \langle f, T_n g \rangle.$$

Therefore, in the space $\mathcal{C}(\Gamma_1 \backslash \mathbb{H})$ of cusp functions, an orthonormal basis $\{u_j(z)\}$ can be chosen which consists of simultaneous eigenfunctions for all T_n, i.e.

$$T_n u_j(z) = \lambda_j(n) u_j(z), \quad j \geq 1, \ n \geq 1, \qquad (5.8)$$

where $\lambda_j(n)$ is the eigenvalue of T_n for $u_j(z)$.

(iv) Up to a constant, $\lambda_j(n)$ and the Fourier coefficient $a_j(n)$ are equal. More precisely,

$$\lambda_j(n)a_j(1) = a_j(n), \quad n \geq 1, \ j \geq 1. \tag{5.9}$$

(v) The Eisenstein series $E(z, 1/2 + it)$ is an eigenfunction of all the Hecke operators T_n with eigenvalue $\eta_t(n)$, i.e.

$$T_n E(z, 1/2 + it) = \eta_t(n)E(z, 1/2 + it); \quad n \geq 1, \ t \in \mathbb{R}, \tag{5.10}$$

where

$$\eta_t(n) = \sum_{ad=n} \left(\frac{a}{d}\right)^{it}. \tag{5.11}$$

Proof. We prove only (5.9). Erase the j for simplicity, and let $u(z)$ denote $u_j(z)$. Suppose $u(z)$ is given by its Fourier series (5.4). Then by (5.8),

$$(T_n u)(z) = \sum_{m \neq 0} \lambda(n)a(m)\sqrt{y}K_{i\nu}(2\pi|m|y)e(mx).$$

Comparing this with (5.5), we obtain $\lambda(n)a(m) = t_n(m)$. Hence by (5.6),

$$\lambda(n)a(m) = \sum_{d|(m,n)} a\left(\frac{mn}{d^2}\right), \tag{5.12}$$

and the desired result (5.9) follows on letting $m = 1$. □

6. The Kuznetsov trace formula

We state without proof the following Kuznetsov trace formula [12]. Let $\{u_j\}$ be an orthonormal basis of the space of Maass cusp forms for Γ. Denote by $1/4 + \nu_j^2$ the Laplace eigenvalue for u_j. Let

$$u_j(z) = (y \cosh \pi\nu_j)^{1/2} \sum_{n \neq 0} \lambda_j(n)K_{i\nu_j}(2\pi|n|y)e(nx)$$

be the Fourier expansion of u_j.

Theorem 6.1. *(Kuznetsov) Let $h(r)$ be an even function of complex variable, which is analytic in $-\Delta \leq \Im r \leq \Delta$ for some $\Delta \geq 1/4$. Assume in this region that $h(r) \ll r^{-2-\delta}$ for some $\delta > 0$ as $r \to \infty$. Then for any n,*

$m \geq 1$,

$$\sum_{u_j} h(\nu_j)\lambda_j(m)\overline{\lambda_j(n)} + \frac{1}{\pi}\int_{\mathbb{R}} \tau_{ir}(n)\tau_{ir}(m)h(r)\frac{1}{|\zeta(1+2ir)|^2}dr$$

$$= \frac{\delta_{n,m}}{\pi^2}\int_{\mathbb{R}} \tanh(\pi r)h(r)dr$$

$$+ \frac{2i}{\pi}\sum_{c\geq 1}\frac{S(n,m;c)}{c}\int_{\mathbb{R}} J_{2ir}\left(\frac{4\pi\sqrt{mn}}{c}\right)\frac{h(r)r}{\cosh(\pi r)}dr. \qquad (6.1)$$

Here

$$\tau_v(n) = \sum_{ab=|n|}\left(\frac{a}{b}\right)^v = |n|^{-v}\sigma_{2v}(|n|)$$

and $S(m,n;c)$ is the classical Kloosterman sum defined as

$$S(m,n;c) = \sum_{a \bmod c}^{*} e\left(\frac{am + \bar{a}n}{c}\right).$$

The left-hand side of (6.1) is the spectral side, and the right the geometric side of the trace formula. The integral on the spectral side represents the continuous spectrum of the Laplace operator; we recall that the divisor function $\tau_{ir}(n)$ is the Fourier coefficients of Eisenstein series.

The integral on the left-hand side converges absolutely; this can be seen from

$$|\zeta(1+2ir)| \gg \frac{1}{\log(2+|r|)},$$

a bound of de la Vallée Poussin. The absolute convergence follows then from the fact that $\tau_{ir}(n) \ll_n 1$ and $h(r) \ll r^{-2-\delta}$. The same bound for $h(r)$ also give us convergence of the first integral on the right-hand side of (6.1).

The sum of Fourier coefficients on the left side of (6.1) is an infinite sum. The normalization for u_j is crucial here. Without suitable normalization of the Maass forms u_j, there would be no reason to have this sum convergent. In fact, the Kuznetsov trace formula (6.1) is only valid for Maass forms u_j which form an orthonormal basis of the space of the Maass cusp forms.

For a proof of the Kuznetsov trace formula, as well as a discussion of convergence, see Kuznetsov [12] or Iwaniec [5].

7. Automorphic *L*-functions

7.1. *Automorphic L-functions attached to Maass forms*

Let f be a Maass cusp form with Laplace eigenvalue $1/4 + \nu_f^2$ and Fourier expansion

$$f(z) = \sum_{n \neq 0} \lambda_f(n) \sqrt{y} K_{i\nu}(2\pi |n| y) e(nx). \tag{7.1}$$

Here we have normalized f so that $\lambda_f(1) = 1$. Then, for $\sigma > 1$, the *L*-function attached to f is defined as

$$L(s, f) = \sum_{n \geq 1} \frac{\lambda_f(n)}{n^s} = \prod_p \left(1 - \frac{\lambda_f(p)}{p^s} + \frac{1}{p^{2s}} \right)^{-1}. \tag{7.2}$$

From the normalization $\lambda_f(1) = 1$ we see that the *L*-function has the first term equal to 1. By the trivial bound $\lambda_f(n) \ll n^{1/2} \tau(n)$, the series and the product in (7.2) converge absolutely for $\sigma > 3/2$. Using the Rankin-Selberg method, one can further prove that

$$\sum_{n \leq x} |\lambda_f(n)|^2 \ll_f x,$$

and hence (7.2) indeed converges absolutely for $\sigma > 1$.

Theorem 7.1. *The series in (7.2) converges absolutely for $\sigma > 1$.*

We define $\alpha_f(p)$ and $\beta_f(p)$ by

$$\alpha_f(p) + \beta_f(p) = \lambda_f(p), \qquad \alpha_f(p)\beta_f(p) = 1.$$

Then the *L*-function can be written as, for $\sigma > 1$,

$$L(s, f) = \prod_p \left(1 - \frac{\alpha_f(p)}{p^s} \right)^{-1} \left(1 - \frac{\beta_f(p)}{p^s} \right)^{-1}.$$

The Generalized Ramanujan Conjecture actually predicts that

$$|\alpha_f(p)| = |\beta_f(p)| = 1.$$

By (5.11), the conjecture is obviously true in the space of continuous spectrum. In the cuspidal space, the best known bound towards the conjecture is

$$|\alpha_f(p)| \leq p^\theta, \qquad |\beta_f(p)| \leq p^\theta, \tag{7.3}$$

with $\theta = 7/64$; this is due to Kim-Sarnak [9]. The trivial bound is

$$\theta = 1/2. \tag{7.4}$$

Theorem 7.2. *Let f be a Maass form with eigenvalue $1/4 + \nu_f^2$. Let $\epsilon = 0$ or 1 according as f is even or odd. Let*

$$\Phi(s, f) = \pi^{-s} \Gamma\left(\frac{s + \epsilon + i\nu_f}{2}\right) \Gamma\left(\frac{s + \epsilon - i\nu_f}{2}\right) L(s, f).$$

Then $\Phi(s, f)$ has an analytic continuation to all $s \in \mathbb{C}$ and satisfies the functional equation

$$\Phi(s, f) = (-1)^\epsilon \Phi(1 - s, f). \tag{7.5}$$

It is indeed an entire function.

Proof. This is one of the many cases where the function equation follows from the theta transformation formula. First consider the case where f is even, and start from the integral

$$\int_0^\infty f(iy) y^{s - 1/2} \frac{dy}{y}. \tag{7.6}$$

As $y \to \infty$, the integrand is vanishingly small because of the rapid decay of the Bessel function; and as $y \to 0$, one uses the theta transformation

$$f(iy) = f(i/y) \tag{7.7}$$

to see that the integrand is small. Hence (7.6) is convergent for all $s \in \mathbb{C}$. If σ is large, one substitutes the Fourier expansion of f and uses

$$\int_0^\infty K_{i\nu_f}(y) y^s \frac{dy}{y} = 2^{s-2} \Gamma\left(\frac{s + i\nu_f}{2}\right) \Gamma\left(\frac{s - i\nu_f}{2}\right),$$

to see that (7.6) equals $\frac{1}{2}\Phi(s, f)$. The functional equation now follows from (7.7).

If f is odd, then we consider instead the integral

$$\int_0^\infty g(iy) y^{s + 1/2} \frac{dy}{y},$$

where

$$g(z) = \frac{1}{4\pi i} \frac{\partial f}{\partial x}(z) = \sum_{n \geq 1} \lambda_f(n) n \sqrt{y} K_{i\nu_f}(2\pi i |n| y) \cos(2\pi n x).$$

In this case, we have the theta transformation formula

$$g(iy) = -\frac{1}{y^2} g\left(\frac{i}{y}\right),$$

and the functional equation (7.5) follows. $\qquad\square$

It follows that the function $L(s, f)$ has a functional equation and analytic continuation. We also know that it is nonzero on $\sigma = 1$. The Generalized Riemann Hypothesis (GRH in brief) predicts that all the nontrivial zeros of $L(s, f)$ are on the critical line $\sigma = 1/2$.

7.2. *Rankin-Selberg automorphic L-functions*

The next L-function we will consider is the Rankin-Selberg L-function. Let f and g be two Maass cusp forms with Laplacian eigenvalue $1/4 + \nu_f^2$ and $1/4 + \nu_g^2$, respectively, and have Fourier expansion as in (7.1). Their Rankin-Selberg L-function is defined by, for $\sigma > 1$,

$$
L(s, f \times g) = \prod_p \left(1 - \frac{\alpha_f(p)\alpha_g(p)}{p^s}\right)^{-1} \left(1 - \frac{\alpha_f(p)\beta_g(p)}{p^s}\right)^{-1}
$$
$$
\times \left(1 - \frac{\beta_f(p)\alpha_g(p)}{p^s}\right)^{-1} \left(1 - \frac{\beta_f(p)\beta_g(p)}{p^s}\right)^{-1}
$$
$$
= \zeta(2s) \sum_{n \geq 1} \frac{\lambda_f(n)\lambda_g(n)}{n^s}, \tag{7.8}
$$

where $\zeta(s)$ is the Riemann zeta-function.

Theorem 7.3. *Let f and g be two Maass cusp forms, and have Fourier expansion as in* (7.1). *Then*

(i) *The product and the series in* (7.8) *are absolutely convergent for $\sigma > 1$.*
(ii) *The function $L(s, f \times g)$ has a functional equation and analytic continuation. When $f = g$, it has a simple pole at $s = 1$.*

Maass forms for Hecke congruence subgroups, as well as their automorphic L-functions, will be introduced in the next section.

8. Maass forms for $\Gamma_0(N)$ and their L-functions

The theory developed for $\Gamma = SL_2(\mathbb{Z})$ so far can be generalized to the group

$$
\Gamma_0(N) = \left\{ \begin{pmatrix} a & b \\ c & d \end{pmatrix} \in SL_2(\mathbb{Z}) : N | c \right\},
$$

where N is a positive integer. This $\Gamma_0(N)$ is called the Hecke congruence subgroup of level N. In this convention, $\Gamma_0(1) = SL_2(\mathbb{Z})$, and the index of $\Gamma_0(N)$ in the modular group is

$$
\nu(N) = [\Gamma_0(1) : \Gamma_0(N)] = N \prod_{p | N} \left(1 + \frac{1}{p}\right).
$$

In this section, we will briefly review the basic theory of Maass forms for $\Gamma_0(N)$ and their automorphic L-functions. Materials presented here can be found in books on Maass forms, for example Iwaniec [5].

8.1. *Maass forms for $\Gamma_0(N)$*

The group $\Gamma_0(N)$ acts on the upper half-plane $\mathbb{H}$ by fractional linear transformation. A Maass cusp form f is an eigenfunction of non-Euclidean Laplacian Δ with eigenvalue

$$\lambda = s(1-s) = \frac{1}{4} + \nu_f^2, \qquad s = \frac{1}{2} + i\nu_f.$$

Since $\lambda \geq 0$, we have either $\Re s = 1/2$ or $1/2 \leq s < 1$. Thus, the ν_f above is not necessarily real. In general, the Fourier expansion of f takes the form (7.1). The space of cusp forms for $\Gamma_0(N)$ is denoted by $\mathcal{C}(\Gamma_0(N)\backslash\mathbb{H})$; it is a Hilbert space with respect to the inner product

$$\langle f, g\rangle = \int_{\Gamma_0(N)\backslash\mathbb{H}} f(z)\bar{g}(z)\frac{dxdy}{y^2}.$$

For $n \geq 1$, define the Hecke operators T_n as in (5.3); nevertheless, only those T_n with $(n, N) = 1$ are interesting. As before, T_m and T_n commute. Moreover, the Hecke operators commute with the non-Euclidean Laplace operator Δ. For every $n \geq 1$, T_n is also self-adjoint on the space $\mathcal{C}(\Gamma_0(N)\backslash\mathbb{H})$ of cusp forms, i.e.

$$\langle T_n f, g\rangle = \langle f, T_n g\rangle, \quad (n, N) = 1.$$

Therefore, in the space $\mathcal{C}(\Gamma_0(N)\backslash\mathbb{H})$, an orthonormal basis $\{u_j\}_{j=1}^\infty$ can be chosen which consists of simultaneous eigenfunctions for all T_n, i.e.

$$T_n u_j = \lambda_j(n)u_j, \quad \text{for all } j \geq 1, \ (n, N) = 1, \tag{8.1}$$

where $\lambda_j(n)$ is the eigenvalue of T_n for u_j. Up to a constant, $\lambda_j(n)$ and the Fourier coefficient $a_j(n)$ of u_j are equal; see (7.1) for the Fourier expansion for u_j. More precisely,

$$\lambda_j(n)a_j(1) = a_j(n), \quad \text{for all } (n, N) = 1, \ j \geq 1. \tag{8.2}$$

Note that if $a_j(1) = 0$, then all $a_j(n) = 0$ for $(n, N) = 1$.

Fix any j, we let $\{\lambda_j(n)\}_{n=1}^\infty$ be the sequence of eigenvalues for all T_n as in (8.1). Then the Hecke eigenvalues $\{\lambda_j(n)\}_{n=1}^\infty$ are real; the Hecke eigenvalues are multiplicative in the following sense:

$$\lambda_j(m)\lambda_j(n) = \sum_{d|(m,n)} \lambda_j\left(\frac{mn}{d^2}\right), \quad (n, N) = 1,$$

and

$$\lambda_j(m)\lambda_j(p) = \lambda_j(mp), \quad p|N. \tag{8.3}$$

It follows that

$$\lambda_j(p)^2 = \lambda_j(p^2) + 1, \quad (p, N) = 1. \tag{8.4}$$

We cannot deduce from (8.2) that $a_j(n) \neq 0$. Thus, we need to work with the new forms.

If u_j is a new form with eigenvalue $s_j(1 - s_j)$, then (8.1) holds for all $n \geq 1$. The first coefficient in the Fourier expansion of u_j does not vanish, so one can normalize u_j by setting $a_j(1) = 1$. In this case, $a_j(n) = \lambda_j(n)$ for all $n \geq 1$, and hence

$$u_j(z) = \sum_{n \neq 0} \lambda_j(n)\sqrt{y}K_{i\nu_j}(2\pi|n|y)e(nx). \tag{8.5}$$

This will be important in §§9-10.

Let λ_* be the smallest eigenvalue of Δ acting on $\mathcal{C}(\Gamma_0(N)\backslash\mathbb{H})$. Corresponding to Theorem 4.5, Selberg [23] proved that $\lambda_* \geq 3/16$, and conjectured that $\lambda_* \geq 1/4$, for all $N \geq 1$. The latter is called Selberg's Eigenvalue Conjecture; the best known bound

$$\lambda_* \geq \frac{1}{4} - \left(\frac{7}{64}\right)^2$$

towards the conjecture follows from Kim-Sarnak [9]. A result like Theorem 4.6 can also be established using Selberg's trace formula, and consequently there are many cusp forms.

8.2. *Automorphic L-functions for Maass forms for* $\Gamma_0(N)$

In this subsection, we review some properties of the automorphic L-function $L(s, f)$ attached to a Maass new form f. Take f be one of the u_j in the above section. Then (8.5) becomes

$$f(z) = \sum_{n \neq 0} \lambda_f(n)\sqrt{y}K_{i\nu_f}(2\pi|n|y)e(nx). \tag{8.6}$$

Also, for this f, (8.3) and (8.4) become, respectively,

$$\lambda_f(m)\lambda_f(p) = \lambda_f(mp), \quad p|N, \tag{8.7}$$

and

$$\lambda_f(p)^2 = \lambda_f(p^2) + 1, \quad (p, N) = 1. \tag{8.8}$$

The Hecke relations (8.7) and (8.8) say that $\lambda_f(p)$ and $\lambda_f(p^2)$ cannot be small simultaneously, and this fact will be used essentially in §10.

To a Maass new form f of level N on $\Gamma_0(N)$ as in (8.6), we may attach, for $\sigma > 1$, an automorphic L-function

$$L(s, f) = \sum_{n=1}^{\infty} \frac{\lambda_f(n)}{n^s}. \tag{8.9}$$

The complete L-function is defined as

$$\Phi(s, f) = \pi^{-s} \Gamma\left(\frac{s + \epsilon + \nu_f}{2}\right)$$
$$\times \Gamma\left(\frac{s + \epsilon - \nu_f}{2}\right) L(s, f), \tag{8.10}$$

where ϵ is the eigenvalue of ι introduced in §3. The complete L-function satisfies the functional equation

$$\Phi(s, f) = \varepsilon_f N^{1/2 - s} \Phi(1 - s, f), \tag{8.11}$$

where ε_f is a complex number of modulus 1. For $\sigma > 1$, the function $L(s, f)$ admits the Euler product

$$L(s, f) = \prod_p \left(1 - \frac{\lambda_f(p)}{p^s} + \frac{\chi_N^0(p)}{p^{2s}}\right)^{-1}, \tag{8.12}$$

where χ_N^0 is the principal character modulo N.

Recall that all eigenvalues $\lambda_f(n)$ of a new form f are real. This explains why on the right-hand side of (8.11) we have $\Phi(1 - s, f)$ instead of $\Phi(1 - s, \bar{f})$. We may also factor the Hecke polynomial in (8.12) and formulate the Generalized Ramanujan Conjecture in this case. The strongest bound towards this Generalized Ramanujan Conjecture is still that of Kim-Sarnak [9].

The following subconvexity is a new result of Michel-Venkatesh [19].

Theorem 8.1. *(Michel-Venkatesh). Let f be a Maass new form on $\Gamma_0(N)$ with Laplace eigenvalue $1/4 + \nu_f^2$. Then*

$$L(1/2 + it, f) \ll \{(3 + |t + \nu_f|^2)N\}^{1/4 - \delta}, \tag{8.13}$$

where δ is a positive absolute constant.

Subconvexity bounds for any one of the three aspects ν_f, N, or t have been studied extensively in the literature, but uniform subconvexity bound is only known of the shape (8.13), where $\delta > 0$ is not specified. See Michel [18] for a survey in this direction, and Michel-Venkatesh [19] for recent developments.

9. Linnik-type problems

9.1. *Linnik's original problem*

In view of Dirichlet's theorem that there are infinitely many primes in the arithmetic progression $m \equiv l \,(\mathrm{mod}\, q)$ with $(q, l) = 1$, it is a natural question how big the least prime, denoted by $P(q, l)$, is in this arithmetic progression. Linnik [15] [16] proved that there is an absolute constant $L > 0$ such that

$$P(q, l) \ll q^L,$$

and this constant L was named after him. Since then, a number of authors have established acceptable numerical values for Linnik's constant L. The first numerical value for L was obtained by Pan [21], while the best result known is $L = 5.5$ by Heath-Brown [4]. We remark that these results depend on, among other things, numerical estimates concerning zero-free regions and the Deuring-Heilbronn phenomenon of Dirichlet L-functions.

Linnik's philosophy can be summarized, very roughly, as follows. Suppose we have an infinite sequence $\{a_n\}_{n=1}^{\infty}$ contained in a set $\mathcal{P}$. Let $Q(\mathcal{P})$ be the conductor of $\mathcal{P}$, meaning that $Q(\mathcal{P})$ contains all the necessary parameters of $\mathcal{P}$. Then, can we bound a_1 from above in terms of the conductor $Q(\mathcal{P})$? In the case of Linnik's original problem, $\mathcal{P}$ is the set of m in the arithmetic progression $m \equiv l \,(\mathrm{mod}\, q)$, and $\{a_n\}_{n=1}^{\infty}$ is the set of infinitely many primes lying in $\mathcal{P}$, guaranteed by the Dirichlet theorem. Here $Q(\mathcal{P}) = q$, where we understand that q is the main parameter for $\mathcal{P}$.

Linnik's problem is a rich resource for further mathematical thoughts, and there are a number of problems that can be formulated in the direction of Linnik's philosophy, which are called Linnik-type problems in the following.

9.2. *A Linnik-type problem for Maass forms*

Let f be a Maass eigenform that is a new form of level N on $\Gamma_0(N)$. Thus, f has a Fourier expansion of the form (8.6). In (8.6), we may normalize so that

$$\lambda_f(1) = 1. \tag{9.1}$$

Since N is the exact level of f, all its Fourier coefficients $\{\lambda_f(n)\}_{n=1}^{\infty}$ are real. It is pointed out at the end of Knopp-Kohnen-Pribitkin [10] that, by a classical method of Landau [13], the sequence $\{\lambda_f(n)\}_{n=1}^{\infty}$ has infinitely many sign changes, i.e. there are infinitely many n such that $\lambda_f(n) > 0$, and infinitely many n such that $\lambda_f(n) < 0$.

Then a natural question arises:

Question Q. *When will the first negative $\lambda_f(n)$ occur?*

The size of the first n such that $\lambda_f(n) < 0$ will be measured by Q_f, the conductor of f, which is defined as

$$Q_f = (3 + |\nu_f|^2)N. \tag{9.2}$$

We understand that this Q_f captures all important parameters of the form f, and the definition (9.2) indeed follows the philosophy described in Iwaniec-Sarnak [8]. The following theorem of Qu [22] answers Question Q.

Theorem 9.1. *Let f be a Maass eigenform that is a new form of level N on $\Gamma_0(N)$, with Laplace eigenvalue $1/4 + \nu_f^2$. Then there is a positive integer n satisfying*

$$n \ll Q_f^{1/2 - \delta}, \quad (n, N) = 1, \tag{9.3}$$

such that $\lambda_f(n) < 0$, where δ is a positive absolute constant, and the implied constant is absolute.

Question Q for holomorphic Hecke eigenform g will be explained in §9.3. At the end of §9.3, we will comment about the connection and differences between the proof of Theorem 9.1 and that for holomorphic Hecke eigenform g. Corresponding results for automorphic representations for higher rank groups will be mentioned in §9.4. Theorem 9.1 will finally be proved in §10.

9.3. *A Linnik-type problem for holomorphic forms*

Let g be a Hecke eigenform that is a new form of level N of even integral weight k on $\Gamma_0(N)$, with Fourier expansion

$$g(z) = \sum_{n \geq 1} \lambda_g(n) n^{(k-1)/2} e(nz), \tag{9.4}$$

where we also normalize so that $\lambda_g(1) = 1$. In this case, N is the exact level of g, and all its Fourier coefficients $\{\lambda_g(n)\}_{n=1}^{\infty}$ are real. Applying the afore-mentioned classical theorem of Landau, one shows that the sequence $\{\lambda_g(n)\}_{n=1}^{\infty}$ must have infinitely many sign changes. Question Q for this sequence has been considered by a number of authors, and the reader is referred to Kohnen [11] for a survey. In particular, Iwaniec-Kohnen-Sengupta [6] proved that there is some n with

$$n \ll (k^2 N)^{29/60}, \quad (n, N) = 1, \tag{9.5}$$

such that $\lambda_g(n) < 0$. To prove this, the following three ingredients are used:

(i) arithmetic properties of $\lambda_g(n)$, in particular, Hecke's relation that

$$\lambda_g(p)^2 = \lambda_g(p^2) + 1, \quad (p, N) = 1; \tag{9.6}$$

(ii) the Ramanujan conjecture for $\lambda_g(n)$ proved by Deligne [3];
(iii) sieve methods.

It should be noted that the conductor Q_g of g can also be defined in the spirit of Iwaniec-Sarnak [8], and computed as $Q_g \asymp k^2 N$. Thus, the right-hand side of (9.5) takes the form $Q_g^{29/60}$.

The more precise question, that how long is the sequence of Hecke eigenvalues that keep the same sign, has been studied by several authors, and best known result is obtained in Lau-Wu [14].

A new difficulty one meets in proving Theorem 9.1 is that, for the Maass cusp form f, the generalized Ramanujan conjecture has not been proved yet. This difficulty makes it impossible here to apply the sieve methods as did in [6]. To get around of this difficulty, one applies the new subconvexity bound in Theorem 8.1.

9.4. *Linnik-type problems for higher rank groups*

We end this section by mentioning some corresponding results for automorphic representations for higher rank groups. Let $m \geq 2$ be an integer, and π an irreducible unitary cuspidal representation for $GL_m(\mathbb{A}_\mathbb{Q})$, whose attached automorphic L-function is denoted by $L(s, \pi)$. Let $\{\lambda_\pi(n)\}_{n=1}^\infty$ be the sequence of coefficients in the Dirichlet series expression

$$L(s, \pi) = \sum_{n=1}^\infty \frac{\lambda_\pi(n)}{n^s}, \quad \sigma > 1. \tag{9.7}$$

It is also proved in Qu [22] that, if π is such that the sequence $\{\lambda_\pi(n)\}_{n=1}^\infty$ is real, then there are infinitely many sign changes in the sequence $\{\lambda_\pi(n)\}_{n=1}^\infty$, and the first sign change occurs at some $n \ll Q_\pi^{m/2+\varepsilon}$, where Q_π is the conductor of π, and the implied constant depends only on m and ε. This generalizes the previous results for GL_2. A result of the same quality is also established for $\{\Lambda(n)a_\pi(n)\}_{n=1}^\infty$, the sequence of coefficients in the Dirichlet series expression

$$-\frac{L'}{L}(s, \pi) = \sum_{n=1}^\infty \frac{\Lambda(n)a_\pi(n)}{n^s}, \quad \sigma > 1, \tag{9.8}$$

where $\Lambda(n)$ is the von Mangoldt function. Proof of these two results are different from that of Theorem 9.1; the reader is referred to [22] for details.

10. The first negative Fourier coefficient of Maass forms

Now we give a proof of Theorem 9.1, applying the properties of f and of $L(s, f)$, described in §8.

Proof of Theorem 9.1. The idea is to consider the sum

$$S(x) := \sum_{\substack{n \leq x \\ (n,N)=1}} \lambda_f(n) \log \frac{x}{n},$$

assuming that

$$\lambda_f(n) \geq 0 \quad \text{for } n \leq x \text{ and } (n, N) = 1. \tag{10.1}$$

The desired result will follow from upper and lower bound estimates for $S(x)$.

To get an upper bound for $S(x)$, we apply Perron's formula to the Dirichlet series (8.9), getting

$$\sum_{n \leq x} \lambda_f(n) \log \frac{x}{n} = \frac{1}{2\pi i} \int_{2-i\infty}^{2+i\infty} L(s, f) \frac{x^s}{s^2} ds.$$

Moving the contour to the vertical line $\sigma = 1/2$, where we apply the Michel-Venkatesh bound (8.13) for $L(s, f)$, we obtain

$$\sum_{n \leq x} \lambda_f(n) \log \frac{x}{n} = \frac{1}{2\pi i} \int_{1/2-i\infty}^{1/2+i\infty} L(s, f) \frac{x^s}{s^2} ds$$

$$\ll \int_{-\infty}^{\infty} \{(3 + |t + \nu_f|^2)N\}^{1/4-\delta} \frac{x^{1/2}}{|t|^2 + 1} dt$$

$$\ll Q_f^{1/4-\delta} x^{1/2}.$$

To recover an estimate for $S(x)$ from the above result, we introduce the condition $(n, N) = 1$ by means of the Möbius inversion formula, which gives

$$S(x) = \sum_{d|N} \mu(d) \sum_{dm \leq x} \lambda_f(dm) \log \frac{x}{dm}.$$

Since $d|N$, we may apply the mutiplicativity property (8.7), which states that $\lambda_f(dm) = \lambda_f(d)\lambda_f(m)$ in the current situation. It follows that

$$S(x) = \sum_{d|N} \mu(d)\lambda_f(d) \sum_{dm \le x} \lambda_f(m) \log \frac{x}{dm}$$

$$\ll \sum_{d|N} |\mu(d)\lambda_f(d)| \left| \sum_{m \le x/d} \lambda_f(m) \log \frac{x/d}{m} \right|$$

$$\ll Q_f^{1/4-\delta} x^{1/2} \sum_{d|N} \frac{|\lambda_f(d)|}{d^{1/2}}.$$

The Kim-Sarnak bound states that $\lambda_f(d) \ll d^{7/64+\varepsilon}$, and therefore,

$$\sum_{d|N} \frac{|\lambda_f(d)|}{d^{1/2}} \ll \tau(N) \ll N^\varepsilon,$$

where $\tau(N)$ is the divisor function. We note that the trivial bound $\lambda_f(d) \ll d^{1/2+\varepsilon}$ works equally well in the above argument. Consequently, we conclude that

$$S(x) \ll Q_f^{1/4-\delta} x^{1/2+\varepsilon}. \tag{10.2}$$

To get a lower bound for $S(x)$ under the assumption (10.1), we first get rid of the weight $\log(x/n)$ in a simple way:

$$S(x) \gg \sum_{\substack{n \le x/2 \\ (n,N)=1}} \lambda_f(n).$$

We now restrict the summation to $n = pq$, where p and q are primes satisfying

$$p \le \sqrt{x/2}, \quad q \le \sqrt{x/2}, \quad (p,N) = 1, \quad (q,N) = 1,$$

and use the formulae

$$\begin{cases} \lambda_f(pq) = \lambda_f(p)\lambda_f(q) & \text{if } p \ne q, \ (p,N)=1, \ (q,N)=1, \\ \lambda_f(p^2) = \lambda_f(p)^2 - 1 & \text{if } p = q, \ (p,N)=1. \end{cases}$$

We get

$$S(x) \gg \sum_{\substack{p \le \sqrt{x/2} \\ (p,N)=1}} \sum_{\substack{q \le \sqrt{x/2} \\ (q,N)=1}} \lambda_f(pq)$$

$$= \left\{ \sum_{\substack{p \le \sqrt{x/2} \\ (p,N)=1}} \lambda_f(p) \right\}^2 - \sum_{\substack{p \le \sqrt{x/2} \\ (p,N)=1}} 1.$$

Recalling the assumption (10.1), we have $\lambda_f(p^2) \geq 0$ for $p \leq \sqrt{x/2}$ and $(p, N) = 1$, and therefore (8.8) gives

$$\lambda_f(p)^2 = \lambda_f(p^2) + 1 \geq 1,$$

that is $\lambda_f(p) \geq 1$. It follows from this and the prime number theorem that

$$S(x) \geq \left\{ \sum_{\substack{p \leq \sqrt{x/2} \\ (p,N)=1}} 1 \right\}^2 - \left\{ \sum_{p \leq \sqrt{x/2}} 1 \right\}$$

$$\geq \left\{ \sum_{p \leq \sqrt{x/2}} 1 - \sum_{p \mid N} 1 \right\}^2 - \sqrt{x}$$

$$\gg \left(\frac{\sqrt{x}}{\log x} - \tau(N) \right)^2 - \sqrt{x}. \tag{10.3}$$

If

$$x \gg N^\varepsilon \tag{10.4}$$

where $\varepsilon > 0$ may be arbitrarily small, then (10.3) gives

$$S(x) \gg \frac{x}{\log^2 x}. \tag{10.5}$$

We remark that (10.4) is a mild requirement, and is of course satisfied.

Comparing (10.5) with (10.2), we get

$$\frac{x}{\log^2 x} \ll S(x) \ll Q_f^{1/4-\delta} x^{1/2+\varepsilon},$$

that is $x \ll Q_f^{1/2-\delta}$. This proves the theorem. $\qquad\square$

Acknowledgements. The first half of the notes is an outcome from the first half of my lecture series given at Postech, Korea, in February 25-27, 2007, whose aim was to explain the basic concepts and ideas of Maass forms. I would like to thank YoungJu Choie for her invitation and hospitality. Shigeru Kanemitsu seriously and constantly encouraged me to publish my Postech lecture notes; he read through the notes carefully, and gave many detailed valuable suggestions. To him, I would also like to express my heartfelt thanks. The second half of my Postech lecture series dealt with subconvexity of Rankin-Selberg L-functions in the eigenvalue aspect, which is not included here. Instead, in the latter half of the notes, I present a different subject, i.e. sign changes of Fourier coefficients of Maass eigenforms. The author is supported in part by the 973 Program, and by NSFC Grant # 10531060.

References

1. A. Borel, *Automorphic forms on $SL_2(\mathbb{R})$*, Cambridge University Press, 1999.
2. D. Bump, *Automorphic forms and representations*, Cambridge University Press, 1996.
3. P. Deligne, La conjecture de Weil, I, *Inst. Hautes Études Sci. Publ. Math.* **43** (1974), 273-307.
4. D. R. Heath-Brown, Zero-free regions for Dirichlet L-functions, and the least prime in an arithmetic progression, *Proc. London Math. Soc.* (3) **64** (1992), 265-338.
5. H. Iwaniec, *Introduction to the Spectral Theory of Automorphic Forms*, Biblioteca de la Revista Matemática Iberoamericana, Madrid, 1995. Second edition: *Spectral Methods of Automorphic Forms*, Amer. Math. Soc. and Revista Matemática Iberoamericana, Providence, 2002.
6. H. Iwaniec, W. Kohnen, and J. Sengupta, The first negative Hecke eigenvalue, *Intern. J. Number Theory* **3** (2007), 355-363.
7. H. Iwaniec and E. Kowalski, *Analytic Number Theory*, Amer. Math. Soc. Colloquium Publ. 53, Amer. Math. Soc., Providence, 2004.
8. H. Iwaniec and P. Sarnak, Perspectives in the analytic theory of L-functions, *Geom. Funct. Anal. special issue* (2000), 705-741.
9. H. Kim and P. Sarnak, Appendix 2: Refined estimates towards the Ramanujan and Selberg conjectures, *J. Amer. Math. Soc.* **16** (2003), 175-181.
10. M. Knopp, W. Kohnen, and W. Pribitkin, On the signs of Fourier coefficients of cusp forms, *Ramanujan J.* **7** (2003), 269-277.
11. W. Kohnen, Sign changes of Fourier coefficients and eigenvalues of cusp forms, in: *Number Theory: Sailing on the sea of number theory, Proceedings of the 4th China-Japan Seminar*, eds. S. Kanemitsu and J. Y. Liu, World Sci. Publ., Hackensack, NJ, 2007, 97-107.
12. N. V. Kuznetsov, Petersson's conjecture for cusp forms of weight zero and Linnik's conjecture. Sums of Kloosterman sums, *Mat. Sb. (N.S.)* **111 (153)** (1980), 334-383, 479 (in Russian). English translation: Math. USSR Sbornik, **39** (1981), 299-342.
13. E. Landau, Über die Anazahl der Gitterpunkte in gewissen Bereichen (Part II), *Gött. Nachr.*, (1915), 209-243.
14. Yuk-Kam Lau and Jie Wu, The number of Hecke eigenvalues of same signs, to appear in *Math. Z.*
15. Yu. V. Linnik, On the least prime in an arithmetic progression, I: the basic theorem, *Math. Sbornik (N.S.)* **15** (1944), 139-178.
16. Yu. V. Linnik, On the least prime in an arithmetic progression, II: the Deuring-Heilbronn phenomenon, *Math. Sbornik (N.S.)* **15** (1944), 347-368.
17. H. Maass, Über eine neue Art von nichtanalytischen automorphen Funktionen und die Bestimmung Dirichletscher Reihen durch Funktionalgleichungen, *Math. Ann.* **121** (1949), 141-183.
18. P. Michel, Analytic number theory and families of automorphic L-functions, in: *Automorphic Forms and Applications*, P. Sarnak and F. Shahidi eds., IAS/Park City Math. Series, vol. 12, Amer. Math. Soc. and Institute for Advanced Study, 2007, 181-295.

19. P. Michel and A. Venkatesh, The subconvexity problem of GL_2, preprint.
20. Ram Murty, Oscillations of Fourier coefficients of modular forms, *Math. Ann.* **262** (1983), 431-446.
21. C. D. Pan, On the least prime in an arithmetic progression, *Sci. Record (N.S.)* **1** (1957), 311-313.
22. Yan Qu, *Linnik-type problems for automorphic L-functions*, Ph.D. Thesis, Université de Nancy 1 & Shandong University, 2008.
23. A. Selberg, On the estimation of Fourier coefficients of modular forms, in: *Proc. Symp. Pure Math. 8, Amer. Math. Soc., Providence (1965)*, 1-15.
24. C. L. Siegel, *Lectures on advanced analytic number theory*, Tata Inst. Bombay.
25. Yangbo Ye, *Automorphic forms and trace formulae* (in Chinese), Peking University Press, 2001.

THE NUMBER OF NON-ZERO COEFFICIENTS
OF A POLYNOMIAL-SOLVED
AND UNSOLVED PROBLEMS

ANDRZEJ SCHINZEL

Institute of Mathematics, The Polish Academy of Sciences,
P.O. Box 137, ul. Sniadeckich 8, 00-950 Warszawa 10, Poland
E-mail: a.schinzel@impan.gov.pl

Let $N(f)$ denote the number of non-zero coefficients of a polynomial f. We start with a simple theorem (for $K = \mathbb{Q}$, see [5]).

Theorem 1. *Let K be any field. For every $a \in K^*$, the binomial $x^n - a$ has a factor f irreducible over K such that $N(f) \leq 3$.*

Proof. Assume first that either char $K = 2$ or K contains a primitive fourth root of unity ζ_4. Then we shall prove by induction on n the existence of f with $N(f) = 2$.

For $n = 1$ the equality is true.

Assume that it is true for all exponents less than n. If $x^n - a$ is irreducible, we take $f = x^n - a$; if $x^n - a$ is reducible, then by Capelli's theorem (see [3, Theorem 428] $a = b^p$, where $b \in K$ and p is a prime dividing n. Then $x^{n/p} - b \mid x^n - a$ and by the inductive assumption there exists f irreducible over K such that $f \mid x^{n/p} - b$ and $N(f) = 2$. This f satisfies our condition. Now, if char $K \neq 2$ and ζ_4 is not in K, then we apply our result to the field $K(\zeta_4)$ and infer that $x^n - a$ has a factor f_0 irreducible over $K(\zeta_4)$ with $N(f_0) = 2$. Let $f_0 = X^m - b$. If $b \in K$, then we take $f = f_0$. Otherwise, $x^m - b' \mid x^n - a$, where b' is the conjugate to b over K. Since $b \neq b'$ we have $(x^m - b, x^m - b') = 1$ and $(x^m - b)(x^m - b') \mid x^n - a$. It suffices to take $f = (x^m - b)(x^m - b')$. $\square$

Theorem 1 suggests the following Problem 1 (1963).

Problem 1. *Does there exist a constant $C(\mathbb{Q})$ such that every trinomial over $\mathbb{Q}$ has an irreducible factor f with $N(f) \leq C(\mathbb{Q})$?*

Example 1 (John Abbott).

$$4x^{20} + 7x^{18} + 64 = (2x^{10} + 5x^9 + 8x^8 + 8x^7 + 4x^6 + 8x^3 + 16x^2 + 16x + 8)$$
$$\times (2x^{10} - 5x^9 + 8x^8 - 8x^7 + 4x^6 - 8x^3 + 16x^2 - 16x + 8).$$

Corollary 1. $C(\mathbb{Q}) \geq 9$.

Problem 2. *Let $C(K)$ have the meaning of Problem 1 with $\mathbb{Q}$ replaced by K. Are the constants $C(K)$, if they exist, bounded?*

In 1947 Rényi [4] constructed a polynomial f of degree 28 such that $N(f) = 29$ and $N(f^2) = 28$. In 1949 Erdös [2] proved the existence of a constant $c_2 < 1$ such that for an infinite sequence of polynomials $f_n \in \mathbb{R}[x]$:

$$\lim N(f_n) = \infty$$

and

$$N(f_n^2) << N(f_n)^{c_2}.$$

The same year Verdenius [11] proved that $c_2 = \log 8 / \log 13$ has the indicated property. He has also proved an analogue for cubes, where c_2 is replaced by $\log 270 / \log 271$. These results have been generalized by Davenport and Coppersmith [1], who proved that for any polynomial $F \in \mathbb{C}[x]$ there exist a constant $c_F < 1$ and an infinite sequence of polynomials $f_n \varepsilon \mathbb{C}[x]$ such that $N(f_n) = n$ and $N(F(f_n)) << n^{c_F}$. They have also improved on the work of Rényi by the following Example 2.

Example 2 (D. Coppersmith and J. Davenport). $N(f^2) < N(f)$ for

$$f(x) = (1 + 2x - 2x^2 + 4x^3 - 10x^4 + 50x^5 + 125x^6)(1 - 110x^6).$$

In 1987 I have proved [6] the conjecture of Erdös and Rényi that $\lim N(f_n)$
$= \infty \Rightarrow \lim N(f_n^2) = \infty$ in the following more general form. If f is in $K[x]$, char $K = 0$ or char $K > l$ deg f and $N(f) > 1$, then $N(f^l) \geq l + 1 + 1/\log 2 \cdot \log(1 + \log(N(f) - 1)/(l \log 4l - l))$. Recently, Zannier and I, in a paper not yet published [10] improved this as follows.

Theorem 2. *Under the above assumptions,*

$$N(f^l) \geq 2 + (\log(N(f) - 1)) / \log 4l.$$

There is a big gap between this result and the result of Verdenius, which suggests

Problem 3. *Does there exist a constant $c(l) > 0$ such that if $f \in K[x]$, char $K = 0$, then $N(f^l) >> N(f)^{c_l}$?*

The gap is still greater in the case of a general polynomial F. Namely, Zannier [12] has proved that there exists a computable function $B(n)$, such that if F is in $K[x] \setminus K$, char $K = 0$ and $N(F(f)) < n$, then $N(f) < B(n)$.

In 1976 P. Weinberger proposed the following

Problem 4. *Does there exist a function $A(r, s)$ such that, if f, g are in $\mathbb{Q}[x]$, $N(f) = r$, $N(g) = s$, then $N((f, g)) \leq A(r, s)$?*

The same problem can be asked with $\mathbb{Q}$ replaced by any field K, only then the function $A(r, s)$ should be replaced by $A(K, r, s) = sup N((f, g))$, where the supremum is taken over all pairs of polynomials f, g in $K[x]$ such that $N(f) = r$, $N(g) = s$. I have proved [8]

Theorem 3. $A(K, 2, 2) = 2$, $A(K, 2, 3) = 3$ *and if* $r > 1$, $s > 1$, $\langle r, s \rangle \neq \langle 2, 2 \rangle$, $\langle 2, 3 \rangle$, $\langle 3, 2 \rangle$,
$\langle$char K, r, $s \rangle \neq \langle 0, 3, 3 \rangle$, *then* $A(K, r, s) = \infty$.

Thus the only unsolved case is $\langle$char$K, r, s \rangle = \langle 0, 3, 3 \rangle$. The best known example is

Example 3 (S. Chaładus). *For*
$$f(x) = x^9 + 9x^2 + 27, \; g(x) = x^{15} - 27x^9 + 729$$
is
$$(f, g) = x^5 + 3x^4 + 6x^3 + 9x^2 + 9x + 9.$$

Corollary 2. $A(\mathbb{Q}, 3, 3) \geq 6$.

Problem 4 leads in a natural way to the question of reducibility of trinomials. Since trinomials $x^n + Ax^m + B$ and $x^n + A/Bx^{n-m} + 1/B$ are simultaneously reducible we may assume $n \geq 2m$. We have [7]

Theorem 4. *Let* n, m *be integers,* $n \geq 2m > 0$, K *a field such that char* K *does not divide* $nm (n - m)$. *If* A, B *are in* $K(y)$, $A^{-n}B^{n-m}$ *is not in* K, *then* $x^n + Ax^m + B$ *is reducible over* $K(y)$, *if and only if either*

(i) $x^{n_1} + Ax^{m_1} + B$ *has a proper divisor of degree* ≤ 2, $n_1 = n/(n, m)$, $m_1 = m/(n, m)$,

(ii) there exists an integer l such that $\langle n/l, m/l \rangle = \langle s, r \rangle \in S_0 =$

$$\bigcup_{p \text{ prime}} \langle 2p, p \rangle \cup \{\langle 6, 1 \rangle, \langle 6, 2 \rangle, \langle 7, 1 \rangle, \langle 8, 2 \rangle, \langle 8, 4 \rangle, \langle 9, 3 \rangle,$$

$$\langle 10, 2 \rangle, \langle 10, 4 \rangle, \langle 12, 2 \rangle, \langle 12, 3 \rangle, \langle 12, 4 \rangle, \langle 15, 5 \rangle\}$$

and $A = u^{s-r} A_{s,r}(v), B = u^s B_{s,r}(v)$, where u, v are in $K(y)$ and polynomials $A_{s,r}$, $B_{s,r}$ are explicitly given.

This result is similar to Capelli's theorem, but compared with it has one great disadvantage, namely the condition $A^{-n} B^{n-m} \notin K$. This condition is avoided in the following Theorem 5.

Theorem 5. *Let n, m be integers, $n \geq 2m > 0$, K a number field. If A, B are in K^*, then $x^n + Ax^m + B$ is reducible over K, if and only if either (i) or (ii) holds with $K(y)$ replaced by K or*

(iii) *there exists a number l such that $\langle n/l, m/l \rangle = \langle s, r \rangle$ is in $S_1 = \{\langle 7, 2 \rangle,$*
   $\langle 7, 3 \rangle, \langle 8, 1 \rangle, \langle 9, 1 \rangle,$
   $\langle 14, 2 \rangle, \langle 21, 7 \rangle\}$ and $A = u^{s-r} A_{s,r}(v, w),$
   $B = u^s B_{s,r}(v, w)$, where $u \in K$, $\langle v, w \rangle \in E_{s,r}(K)$ and polynomials $A_{s,r}, B_{s,r}$ and the elliptic curve $E_{s,r}$ are explicitly given,
(iv) *there exists an integer l such that $\langle n/l, m/l \rangle = \langle s, r \rangle$ and $A = u^{s-r} A_0,$*
   $B = u^s B_0$, where $u \in K$ and $\langle A_0, B_0 \rangle$ is in a certain finite, possibly empty, set $F_{s,r}(K)$.

Conjecture. *Sets $F_{s,r}(K)$ can be chosen so that*

$$\bigcup_{s,r} \bigcup_{\langle A_0, B_0 \rangle \in F_{s,r}(K)} \{x^s + A_0 x^r + B_0\}$$

is finite.

Consequence 1. For every number field K there exists a constant $C_1(K)$ such that, if $n_1 > C_1(K)$, then $x^n + Ax^m + B$ is reducible over K, if and only if $x^{n_1} + Ax^{m_1} + B$ has a proper divisor of degree ≤ 2.

Example 4 (J. Browkin). $x^{52} + 2^{34} \times 3 \times 53x + 2^{35} \times 103$ *is divisible by* $x^3 + 2x^2 + 4x + 4$.

Corollary 3. $C_1(Q) \geq 52$.

Consequence 2. The answer to Problem 1 is YES.

Consequence 3. There exists a constant $C_2(\mathbb{Q})$ such that if $b \in \mathbb{Z}$, $|b| > C_2(\mathbb{Q})$, $n > m > 0$, $n \neq 2m$, then $x^n + bx^m + 1$ is irreducible over Q.

Example 5 (A. Bremner). $x^{33} + 67x^{11} + 1 = (x^3 + x + 1)(x^{30} - \cdots + 1)$.

Corollary 4. $C_2(Q) \geq 67$.

Concerning Problem 4 I have the following partial result [9].

Theorem 6. *For every number field K and for all positive integers $n > m$ there exist finite sets $E_{n,m}(K)$ such that if $T_i = x^{n_i} + A_i x^{m_i} + B_i$ and $A^{-n_i} B^{n_i - m_i}$ is not in $E_{n_i, m_i}(K)$, then $N((T_1, T_2)) \leq 10$.*

References

1. D. Coppersmith and J. Davenport, Polynomials whose powers are sparse, *Acta. Arith.* **58** (1991), 79–87.
2. P. Erdös, On the number of terms of a square of a polynomial, *Nieuw Arch. Wiskund.* (2) **23** (1949), 63–65.
3. L. Rédei, *Algebra*, Erster Teil, Akademische Verlaggesellschaft, Leipzig, 1959.
4. A. Rényi, On the minimal number of terms in the square of a polynomial, *Selected papers*, **1**, Budapest, (1976), 92–97.
5. A. Schinzel, Some unsolved problems on polynomials, Selecta **1**, 703–708, Zürich. 2007.
6. A. Schinzel, On the number of terms of a power of a polynomial, ibid. 450–465.
7. A. Schinzel, On reducible trinomials, ibid. 466–548.
8. A. Schinzel, On the greatst common divisor of two univariate polynomials, ibid, 632–645.
9. A. Schinzel, On the greatest common divisor of two univariate polynomials, ibid. 646–657.
10. A. Schinzel and U. Zannier, The number of terms of a power of a polynomial, to appear.
11. W. Verdenius, On the number of terms of the square and the cube of polynomials, *Indag. Math.* **11** (1949), 459–465.
12. U. Zannier, On composite polynomials and the proof of a conjecture of Schinzel, to appear in *Invent Math.*

OPEN PROBLEMS ON EXPONENTIAL AND CHARACTER SUMS

IGOR E. SHPARLINSKI

Department of Computing, Macquarie University
Sydney, NSW 2109, Australia
E-mail: igor@ics.mq.edu.au

1. Introduction

This is a collection of open questions most of them being unrelated to each other, at various levels of difficulty, however, all related to exponential and multiplicative character sums. One may certainly notice a large proportion of self-references in the bibliography. By no means should this be considered as an indication of anything else than the fact that the choice of problems reflects the taste and interests of the author. Thus it is not surprising that many (but not all) of the problems are related to some previous results of the author.

One can find the necessary background and a wealth of information about exponential and character sums in [63] (mainly in the context of finite fields) and in [56] (in a more general contexts), see also [60,62,79,91].

Some of the problems below are likely to be very difficult, some may constitute a feasible long term project (such as doctoral thesis), some are not so hard and may be used by the beginners in this area as an entry point to gain some immediate "hands-on" experience with exponential and character sums. In particular, part of the motivation for writing this list of problems has been the frequent requests to the author to suggest a research problem on exponential sums.

One can also notice that the problems fall into two categories. In one category we talk about obtaining new bounds of exponential and character sums, while in the other category we seek new applications, problems from the former are motivated by those from the latter.

As a general rule, the well known problems, which are described in the literature, are not presented. We have also avoided open ended problems of the form "*Improve the result of ...*" unless we are able to give some ideas about how it may be possible to achieve this.

2. Notation

For an integer $m \geq 1$, we denote

$$\mathbf{e}_m(z) = \exp(2\pi i z/m).$$

We always follow the convention that arithmetic operations in the arguments of $\mathbf{e}_m$ are performed modulo m.

We use (n/m) to denote the Jacobi symbol modulo an odd integer $m \geq 3$ (that is the Legendre symbol in the case that $m = p$ is prime).

The letter p (possibly subscripted) always denotes a prime; k, m and n always denote integers (and so do K, M and N).

As usual, we say that $n \geq 2$ is y-smooth if all prime divisors p of n satisfy $p \leq y$.

We use ε to denote a small positive parameter on which implied constants may depend. Calligraphic letters, for example, $\mathcal{A} = (a_n)$, usually denote sets or sequences of integers.

For a prime power q, we use $\mathbb{F}_q$ to denote the finite field of q elements.

For an integer m, we use $\mathbb{Z}/m\mathbb{Z}$ to denote the residue ring modulo m.

We use some standard notations for most common arithmetic functions. For $m \geq 2$ be an integer, we denote by:

- $P(m)$ the largest prime divisor of m,
- $\varphi(m)$ the Euler (totient) function of m,
- $\omega(m)$ the number of distinct prime divisors of m,
- $\tau(m)$ the number of positive integer divisors of m.

We also define $P(1) = \omega(1) = 0$ and $\tau(1) = \varphi(1) = 1$.

Letting $x \geq 0$ be a real number, we denote by:

- $\pi(x)$ the number of primes $p \leq x$,
- $\pi(x; q, a)$ the number of primes $p \leq x$ such that $p \equiv a \pmod{q}$.

Finally, $\log x$ denotes the natural logarithm; we always assume the argument is large enough for the whole expression to make sense (as well as in the case of iterated logarithms).

3. Problems

3.1. *Exponential functions*

Problem 3.1. *Obtain analogues of the results of J. Bourgain, A. A. Glibichuk and S. V. Konyagin [24] for multiplicative character sums*

$$\sum_{x_1,\cdots,x_k\in\mathcal{X}}\chi(x_1\cdots x_k + a) \qquad and \qquad \sum_{x=1}^{N}\chi(g^x + a)$$

with values of N very small relative to p, where $\mathcal{X}\subseteq\mathbb{Z}/p\mathbb{Z}$, $\gcd(g,p)=1$ and χ is a nonprincipal multiplicative character modulo p, see also [16–18, 20,22].

Problem 3.2. *Obtain explicit forms, with all constants explicitly evaluated, of the results of J. Bourgain [16–18,20], J. Bourgain and M.-C. Chang [22], J. Bourgain, A. A. Glibichuk and S. V. Konyagin [24] and several other similar results on short exponential sums with exponential functions*

$$\sum_{x=1}^{N}\mathbf{e}_m(a_1g_1^x + \cdots + a_kg_k^x), \qquad and \qquad \sum_{x=1}^{M}\sum_{y=1}^{N}\mathbf{e}_m(ag^x + bg^y + cg^{xy}),$$

where $a_1,\ldots,a_k, a, b, c\in\mathbb{Z}$, $\gcd(gq_1\cdots g_k, m) = 1$, $M, N\in\mathbb{Z}$, and with the products

$$\sum_{x_1\in\mathcal{X}_1}\cdots\sum_{x_k\in\mathcal{X}_k}\mathbf{e}_m(ax_1\cdots x_k), \qquad a\in\mathbb{Z},$$

where $\mathcal{X}_1,\ldots,\mathcal{X}_k\subseteq\mathbb{Z}_m^*$.

Comments: J. Bourgain and M. Z. Garaev [23] and M.-C. Chang and C. Z. Yao [32] have recently obtained a series of very interesting results in this direction.

Problem 3.3. *Obtain stronger bounds for the sums of Problem 3.2 on average over prime moduli m = p.*

Comments: The ideas used in the proof of [60, Theorem 5.5] and of [3, Lemma 2.10] can possibly be of help.

Problem 3.4. *Use the method of [60, Chapter 5] to estimate exponential sums*

$$\sum_{x=1}^{N}\mathbf{e}_m(ag^x), \qquad a\in\mathbb{Z},$$

where $\gcd(g, m) = 1$, *for values of N very small relative to m, for "almost all" m.*

Comments: There are amazingly strong and general results of J. Bourgain and M.-C. Chang, see [18,20,22], which apply to very short and general sums and thus have significantly reduced the interest to this question. On the other hand, the method of [60, Chapter 5] may lead to more explicit and stronger bounds (but which hold only for "almost all" m rather than all m).

Problem 3.5. *Estimate exponential sums*

$$\sum_{x=1}^{T} \mathbf{e}_p \left(ag^{x^2} + bg^x \right), \qquad a, b \in \mathbb{Z},$$

where T is a multiplicative order of g modulo p with $\gcd(g, p) = 1$.

Comments: J. Bourgain [20] has obtained a nontrivial estimate for a large class of primes p, but in the general case there is no nontrivial bound (even if g is a primitive root modulo p). In the case of $b = 0$ these sums are estimated in [40], provided $T \geq p^{1/2+\varepsilon}$, see also [44,64]. Bounds of similar sums with a composite denominator m can be found in [43].

Problem 3.6. *Estimate exponential sums*

$$\sum_{x=1}^{N} \mathbf{e}_m \left(ag^{\lfloor f(x) \rfloor} \right), \qquad a \in \mathbb{Z},$$

where $f(X) \in (R)[X]$ *is a polynomial with real coefficients or some sufficiently smooth function.*

Comments: In the case of polynomials $f(X) \in \mathbb{Z}[X]$ with integers coefficients and prime $m = p$ such that $ell^2 \mid p - 1$ for another prime $ell \geq p^\varepsilon$ (with arbitrary $\varepsilon > 0$) a nontrivial bound on the above sums are given by M.-C. Chang [29]. Also for $f(X) \in \mathbb{Z}[X]$ and prime $m = p$, in [83], a nontrivial bound is given "on average" over all primitive roots g modulo p. In fact in both [29] and [83] more general sums over arbitrary finite fields are estimated. The general case of polynomials, even for polynomials $f(X) \in \mathbb{Z}[X]$, probably requires some new ideas. On the other hand, the case of smooth and slowly growing functions $f(X)$ can probably be studied by the method of [13]. Results of these type have a natural interpretation of studying polynoially growinng sequences on orbits of the dynamical system generated by the map $u \to gu$ in the ring $\mathbb{Z}_m$, see [14,46,47] and references therein for various results of such type in the settings of ergodic theory.

Problem 3.7. *Estimate incomplete exponential sums*

$$\sum_{\substack{x=1 \\ \gcd(x,T)=1}}^{N} \mathbf{e}_m\left(ag^{1/x}\right), \qquad a \in \mathbb{Z},$$

where T is a multiplicative order of g modulo m with $\gcd(g, m) = 1$.

Comments: For a prime $m = p$ and $T \geq p^\varepsilon$ this has been done by J. Bourgain and I. E. Shparlinski [25], see also [86] for a more explicit estimate in the case of $T \geq p^{1/2+\varepsilon}$.

Problem 3.8. *Estimate complete exponential sums*

$$\sum_{\substack{x=1 \\ \gcd(x,T)=1}}^{T} \mathbf{e}_m\left(ag^x + bg^{1/x}\right), \qquad a, b \in \mathbb{Z},$$

where T is a multiplicative order of g modulo m with $\gcd(g, m) = 1$.

Comments: Even the case of prime $m = p$ is of interest. Also, for a prime $m = p$, some bounds on average over $a, b \in \mathbb{F}_p$ are given in [80].

3.2. Short character sums

Problem 3.9. *Extend the Burgess bound, see [56, Theorems 12.6], in full generality, to the settings of arbirary finite fields.*

Comments: Very promising results in this direction have recently been obtained by M.-C. Chang [30,31].

Problem 3.10. *Prove that for arbitrary integers $m \geq 2$, $M \geq 0$ and $N \geq 1$, the bound*

$$\sum_{\chi \neq \chi_0} \left| \sum_{z=M+1}^{M+N} \chi(z) \right|^4 \leq m^{1+o(1)} N^2$$

holds, where the outter sum is taken over all nonprincipal multiplicative characters χ modulo m.

Comments: For $M = 0$ this bound is established by J. B. Friedlander and H. Iwaniec [41]. Furthermore, if $m = p$ then it has also been obtained for any $M \geq 0$ by A. Ayyad, T. Cochrane and Z. Zheng [1]. In full generality Problem 3.10 is still open.

Problem 3.11. *Prove that*

$$\max_{0 \leq a < b \leq p-1} \left| \sum_{x=1}^{h} \left(\frac{(x+a)(x+b)}{p} \right) \right| = o(h)$$

for any $h \geq p^\alpha$ with some fixed $\alpha < 1/2$.

Comments: V. Shoup [76] gives a factorization algorithm for polynomials over $\mathbb{F}_p$ of complexity $O(n^{2+o(1)}p^{1/2+o(1)})$ (where n is the degree of the polynomial $f(X) \in \mathbb{F}_p[X]$ to be factored), which is based on the Weil bound

$$\max_{0 \leq a < b \leq p-1} \left| \sum_{x=1}^{h} \left(\frac{(x+a)(x+b)}{p} \right) \right| = O\left(p^{1/2} \log p \right).$$

Any improvement of this bound immediately leads to a better complexity estimate. Even obtaining such a bound only for "almost all" p is already of great interest.

Problem 3.12. *Extend the result and method of A. A. Karatsuba [58] to double character sums modulo a composite.*

Problem 3.13. *Estimate sums of Jacobi symbols over the solutions to the congruence $xy \equiv a \pmod m$ in a given box, that is,*

$$\sum_{\substack{xy \equiv a \pmod{m} \\ 1 \leq x \leq X, 1 \leq y \leq Y \\ y \equiv 1 \pmod 2}} \left(\frac{x}{y} \right), \qquad a \in \mathbb{Z},$$

for X and Y reasonably small compared to m.

Comments: For very small m some results of this kind can be extracted from the work of G. Yu [93], see also [92] for applications of results of this kind to studying the distribution of Selmer groups of some families of elliptic curves.

Problem 3.14. *Assuming the Generalised Riemann Hypothesis estimate the sums*

$$S_m(\chi, N) = \sum_{n=1}^{N} \chi(n)$$

with a nontrivial multiplicative character χ modulo m.

Comments: The bound $S_m(\chi, N) = N^{1/2} m^{o(1)}$ has been a part of folk-lore and in particular is quoted in [66, Bound (13.2)] as a consequence of the weaker *Generalised Lindelöf Hypotheis*. However it seems that a full proof has never been published. Furthermore, it is interesting to get an explicit expression for the term $q^{o(1)}$.

3.3. *Smooth numbers, S-units and primes*

Problem 3.15. *Estimate exponential and multiplicative character sums*

$$\sum_{\substack{n \leq x \\ n \text{ is } y\text{-smooth}}} \mathbf{e}_m\left(f(n)\right) \qquad and \qquad \sum_{\substack{n \leq x \\ n \text{ is } y\text{-smooth}}} \chi(f(n))$$

with a polynomial $f(T) \in \mathbb{Z}[T]$, where $a \in \mathbb{Z}$ and χ is a multiplicative character modulo m.

Comments: Several bounds for exponential sums over smooth numbers are given by Fouvry & Tenenbaum [45] and, more recently, by de la Bretèche & Tenenbaum [26]. Character sums in the case of a linear polynomial $f(T) = T - a$ are estimated [82]. The approach of [82] has also been used by Gong [52] for more general sums. The arguments of [26,45] and [52,82] are quite different and it is quite possible that combining them, or using them in different scenarios, one can obtain many new interesting results about exponential and character sums over y-smooth integers in wide ranges of the parameters m, x and y.

Problem 3.16. *For a set $(S) = \{p_1, \ldots, p_s\}$ of s primes and an integer N we denote by $(U)((S), N)$ the set of integral S-units up to N, that is the set of integers $u \leq N$ composed of primes from the set (S). Estimate character sums*

$$U_k((S), N) = \sum_{u_1, \ldots, u_k \in (U)((S), N)} \chi\left(a_1 u_1 + \cdots + a_k u_k\right), \qquad a_1, \cdots, a_k \in \mathbb{Z},$$

and

$$W_k((S), N) = \sum_{u_1, \ldots, u_k \in (U)((S), N)} \chi\left(a_1 u_1 + \cdots + a_k u_k + 1\right), \qquad a_1, \ldots, a_k \in \mathbb{Z},$$

with a multiplicative character χ modulo m.

Comments: In the case $s = k = 1$ the sums $W_k((S), N)$ are estimated by E. Dobrowolski and K. S. Williams [39] and H. B. Yu [94] (in fact in this case the only generator p_1 need not be prime. Note that we ask for estimates which make use of larger values of s or k in a substantial way rather than merely reduce the question to the case $s = k = 1$. There should be a strong connection between estimates on $U_{k+1}((S), N)$ and $W_k((S), N)$.

Problem 3.17. *In the case of a positive solution to Problem 3.16 use its result together with the square sieve of R. Heath-Brown [53] to estimate the number of perfect squares of the form*

$$u_1 + \cdots + u_k \qquad and \qquad \frac{w_1 - 1}{w_2 - 1}$$

where $u_1, \ldots, u_k, w_1, w_2 \in (U)((S), N)$ and similar.

Comments: The question is related to some finiteness conjectures about the number of perfect squares and higher powers of the above and similar types made by P. Corvaja [38]. As for Problem 3.35, only sums with the Jacobi symbol modulo products of two distinct primes $m = p_1 p_2$ are relevant.

Problem 3.18. *Make use of the Burgess bound, see [56, Theorems 12.6], in the argument of Z. Kh. Rakhmonov [75] in order to improve the bound [75] on the sums*

$$\sum_{p \leq N} \chi(p + a), \qquad a \in \mathbb{Z},$$

with multiplicative characters modulo m, where $\gcd(a, m) = 1$. Obtain a result which is nontrivial for $N \geq m^{1-\alpha+\varepsilon}$ with some fixed $\alpha > 0$.

Comments: Z. Kh. Rakhmonov [75] uses the Polya-Vinogradov bound; accordingly, his estimate is nontrivial only for $N \geq m^{1+\varepsilon}$, see also a result of A. A. Karatsuba [57] for the case of prime modulus $m = p$ which is nontrivial starting with $N \geq p^{1/2+\varepsilon}$.

3.4. *Combinatorial sequences*

Problem 3.19. *Let $\mathcal{P}_L$ be that set of all palindromes of length L, that is, the set of n for which the base $g \geq 2$ representation*

$$n = \sum_{k=0}^{L-1} a_k(n)g^k,$$

where

$$a_k(n) \in \{0, 1, \ldots, g-1\}, \quad k = 0, 1, \ldots, L-1, \qquad and \qquad a_{L-1}(n) \neq 0,$$

satisfies the symmetry condition:

$$a_k(n) = a_{L-1-k}(n), \qquad k = 0, 1, \ldots, L-1.$$

Estimate exponential and character sums

$$\sum_{n \in \mathcal{P}_L} \mathbf{e}_m(f(n)) \qquad and \qquad \sum_{n \in \mathcal{P}_L} \chi(n),$$

where $f(X) \in \mathbb{Z}[X]$ and χ is a multiplicative character modulo m.

Comments: W. Banks, D. Hart and M. Sakata [6] have used bounds of twisted Kloosterman sums to estimate exponential sums with palindromes in the case $f(X) = X$, see also [8]. However, the method of [6] does not seem to apply to either exponential sums with more general polynomials or character sums. Even the case of prime $m = p$ is still open.

Problem 3.20. *Obtain nontrivial upper bounds on exponential and character sums with factorials modulo a prime p, such as*

$$\sum_{n=1}^{N} \mathbf{e}_p(an!) \qquad and \qquad \sum_{n=1}^{N} \chi(n! + a), \qquad a \in \mathbb{Z},$$

where χ is a multiplicative character modulo p.

Comments: Double exponential sums with $m!n!$ and single multiplicative character sums as above with $a = 0$ are estimated in [48] and [49], respectively.

Problem 3.21. *Estimate double exponential sums with binomial coefficients*

$$\sum_{n=0}^{N} \sum_{m=0}^{n} \mathbf{e}_p \left(a \binom{n}{m} \right), \qquad a \in \mathbb{Z},$$

for $0 \leq N < p$.

Comments: Several estimates of single and double exponential and character sums with binomial coefficients and other combinatorial numbers are given by M. Z. Garaev, F. Luca and I. E. Shparlinski in [48–51].

Problem 3.22. *Estimate exponential sums*

$$\sum_{n=1}^{N} \sum_{k=0}^{n} \mathbf{e}_m \left(as(n, k) \right) \qquad and \qquad \sum_{n=1}^{N} \sum_{k=0}^{n} \mathbf{e}_m \left(aS(n, k) \right), \qquad a \in \mathbb{Z},$$

with Stirling numbers of the first and second kind, respectively.

Problem 3.23. *Estimate exponential sums*

$$\sum_{2\leq n\leq x} \mathbf{e}_n\left(aw_n\right) \qquad and \qquad \sum_{2\leq n\leq x} \mathbf{e}_{P(n)}\left(aw_n\right), \qquad a \in \mathbb{Z},$$

where $w_1 = 1$ and

$$w_n = \binom{2n-1}{n-1} = \frac{1}{2}\binom{2n}{n}, \qquad n \geq 2.$$

Comments: This question has also been posed in [28].

Problem 3.24. *Obtain nontrivial upper bounds on exponential and character sums with n^n modulo an integer m, such as*

$$\sum_{n=1}^{N} \mathbf{e}_m\left(an^n\right) \qquad and \qquad \sum_{n=1}^{N} \chi\left(n^n + a\right), \qquad a \in \mathbb{Z},$$

where χ is a multiplicative character modulo m.

Comments: Probably the case of prime $m = p$ and the range $N = p-1$ is of most interest. In this case the bound $O\left(p^{35/18+o(1)}\right)$ of [27] on the number of solutions to the congruence $k^k \equiv n^n \pmod{p}$, $1 \leq k, n < p$, is equivalent to the estimate of the average values

$$\frac{1}{p}\sum_{a=1}^{p}\left|\sum_{n=1}^{p-1}\mathbf{e}_m\left(an^n\right)\right|^2 \leq p^{35/18+o(1)}$$

and

$$\frac{1}{p}\sum_{a=1}^{p}\left|\sum_{n=1}^{p-1}\chi\left(n^n + a\right)\right|^2 \leq p^{35/18+o(1)}.$$

3.5. *Polynomial discriminants*

Problem 3.25. *Estimate exponential and character sums*

$$\sum_{\substack{\deg f=n,\ H(f)\leq H \\ f\in\mathbb{Z}[X]\ \text{monic}}} \mathbf{e}_m\left(aD(f)\right) \qquad and \qquad \sum_{\substack{\deg f=n,\ H(f)\leq H \\ f\in\mathbb{Z}[X]\ \text{monic}}} \chi\left(D(f)\right),$$

with $a \in \mathbb{Z}$ and nontrivial multiplicative characters χ modulo m over discriminants $D(f)$ of monic polynomials

$$f(X) = X^n + a_{n-1}X^{n-1} + \cdots + a_0 \in \mathbb{Z}[X]$$

of degree n and height $H(f) = \max\{|a_0|, \ldots, |a_{n-1}|\} \leq H$.

Comments: Certainly the determinant is a polynomials of degree n in the coefficients $a_0, \ldots, a_{n-1}$. So many general estimates for exponential and character sums with polynomials can be applied. For example, for a prime $m = p$, an immediate application of the Weil bound implies that for $H \leq p$ each of the above sums is $O(H^{n-1} p^{1/2} \log p)$. Certainly only estimates which are better than such bounds and which make use of some special properties of $D(f)$ are of interest. We also recall that if $f \in \mathbb{F}_p[X]$ is squarefree then by the Stickelberger theorem, see [15, Theorem 6.68],

$$\left(\frac{D(f)}{p} \right) = (-1)^{n-s},$$

where $n = \deg f$ and s is the number of distinct irreducible factors of f. Therefore if χ is the Legendre symbol then the corresponding sum is a polynomial analogue of the sum of Möbius function.

Problem 3.26. *Estimate double character sums*

$$\sum_{n=1}^{N} \sum_{k=1}^{n-1} \chi \left(\Delta(n, k) \right),$$

for a nontrivial multiplicative character modulo m, where

$$\Delta(n, k) = (-1)^{n(n-1)/2} \left(n^n - (-1)^n k^k (n - k)^{n-k} \right)$$

is the discriminant of the trinomial $X^n + X^k + 1$.

Comments: The question is of interest even for a prime $m = p$, and especially for $m = p_1 p_2$ which is a product of two primes since in this case one can use the square sieve of R. Heath-Brown [53] to study the square-free part of $\Delta(n, k)$, see [67,85] for some related problems about the discriminants of trinomials. If such a bound is obtained then again the Stickelberger theorem, see the comments to Problem 3.25, can be used to extract some nontrivial information about factorisations of trinomials.

3.6. *Arithmetic functions*

Problem 3.27. *Estimate exponential sums with the divisor function*

$$\sum_{n=1}^{N} \mathbf{e}_m(a\tau(n)), \qquad a \in \mathbb{Z},$$

for an odd $m \geq 3$.

Comments: One can use the fact that "typically" $\tau(n) = 2^k a$ for a small a and rather large k. Then one can try to use bounds of exponential sums with exponential function from [16–19,22,24,60]. In fact the case of $m = p^k$ where p is fixed can be a good starting point where one can use the result of N. M. Korobov [61] for the relevant exponential sums.

Problem 3.28. *Estimate exponential and character sums with the largest squarefree divisor of n and with the squarefree part of n, that is,*

$$\sum_{n=1}^{N} \mathbf{e}_m(aQ(n)) \qquad and \qquad \sum_{n=1}^{N} \mathbf{e}_m(aS(n)), \qquad a \in \mathbb{Z},$$

where

$$Q(n) = \prod_{p \mid n} p \qquad and \qquad S(n) = \min\{s \ : \ \sqrt{n/s} \in \mathbb{Z}\}.$$

Comments: Bounds of character sums with these functions are given by H. Liu and W. Zhang [65], but the method of [65] does not apply to exponential sums. Exponential sums with $P(n)$ have been estimated in [5]; exponential and character sums with the Euler function are estimated in [2, 7,11].

Problem 3.29. *For an integer $g > 1$ estimate exponential sums*

$$\sum_{\substack{n=1 \\ n \text{ composite}}}^{N} \mathbf{e}_n\left(a(g^{n-1} - 1)\right), \qquad a \in \mathbb{Z}.$$

Comments: This question has also been posed in [3] where one can also find estimates of some other related sums with fractions $(g^{n-1} - 1)/n$ and alike.

3.7. *Beatty sequences*

Problem 3.30. *Extend the bounds of character sums with Beatty sequences $\lfloor \alpha n + \beta \rfloor$ of W. D. Banks and I. E. Shparlinski [9,10] to composite moduli.*

Problem 3.31. *Obtain explicit versions of the bounds of character sums with Beatty sequences $\lfloor \alpha n + \beta \rfloor$ of W. D. Banks and I. E. Shparlinski [12] in the case when α is of approximation type 1.*

Problem 3.32. *Estimate character sums with Beatty sequences $\lfloor \alpha p + \beta \rfloor$ over primes $p \leq x$.*

Problem 3.33. *Estimate character sums with Beatty sequences* $\lfloor \alpha s + \beta \rfloor$ *over smooth y-smooth integers* $s \leq x$.

Problem 3.34. *Estimate double character sums with Beatty sequences*

$$\sum_{k \in \mathcal{K}} \sum_{m \in \mathcal{M}} \chi \left(\lfloor \alpha(k + m) + \beta \rfloor \right)$$

on sumsets, where χ *is a multiplicative character modulo* $m \geq 2$ *and* $\mathcal{K}$ *and* $\mathcal{M}$ *are sufficiently dense sets of integers of the interval* $[1, N]$. *Even the case of* $m = p$ *and very large* N *is still open.*

Problem 3.35. *In the case of positive solutions to Problems 3.30 and 3.31 use their results together with the square sieve of R. Heath-Brown [53] in order to get a lower bound on the number of distinct quadratic fields of the form* $\mathbb{Q}\left(\sqrt{\lfloor \alpha n + \beta \rfloor} \right)$, $n = 1, \ldots, N$.

Comments: Only sums with the Jacobi symbol modulo products of two distinct primes $m = p_1 p_2$ are relevant.

Problem 3.36. *Extend the bounds of character sums with integer parts* $\lfloor f(n) \rfloor$ *of "smooth" functions of W. D. Banks and I. E. Shparlinski [13] to composite moduli.*

Problem 3.37. *In the case of a positive solution to Problem 3.36 use its result together with the square sieve of R. Heath-Brown [53] in order to get a lower bound on the number of distinct quadratic fields of the form* $\mathbb{Q}\left(\sqrt{\lfloor f(n) \rfloor} \right)$, $n = 1, \ldots, N$.

Comments: As for Problem 3.35, only sums with the Jacobi symbol modulo products of two distinct primes $m = p_1 p_2$ are relevant.

3.8. *Sparse polynomials*

Problem 3.38. *Obtain analogues of the result of T. Cochrane, C. Pinner and J. Rosenhouse [34] on exponential sums*

$$\sum_{x=1}^{m} \mathbf{e}_m \left(a_1 x^{k_1} + \cdots + a_s x^{k_s} \right), \qquad a_1, \ldots, a_s \in \mathbb{Z},$$

with sparse polynomials of large degree, containing only s *monomials, modulo a composite* $m \geq 2$.

Comments: The result of T. Cochrane, C. Pinner and J. Rosenhouse [34] is a generalisation of the bound of S. V. Konyagin [59] which corresponds to the case $s = 1$. Extending the bound of [59] to composite moduli is already an interesting problem. The bound of [34] has been applied to various problems in number theory [8], computer science [87] and cryptography [88]. Several other bounds (stronger, but more restrictive on the class of polynomials to which they apply) on exponential sums with sparse polynomials can be found in [17,21,35,78].

Problem 3.39. *Obtain bounds on exponential sums*

$$\sum_{\substack{x=1 \\ \gcd(g(x),m)=1}}^{m} \mathbf{e}_m\left(f(x)/g(x)\right),$$

with sparse rational functions $f(X)/g(X)$, where

$$f(X) = a_1 X^{h_1} + \cdots + a_r X^{h_r} \quad \text{and} \quad g(X) = b_1 X^{k_1} + \cdots + b_s X^{k_s},$$

which are better than general estimates of T. Cochrane and Z. Y. Zheng [36] depending only on the degrees of f and g, see also [37].

Problem 3.40. *Let us define the constants*

$$A(n) = \sup_{m \geq 1} \max_{\gcd(a,m)=1} \left|\sum_{x=1}^{q} \mathbf{e}_m\left(ax^n\right)\right| m^{-1+1/n}.$$

Compute

$$A = \max_{n \geq 2} A(n).$$

Comments: S. B. Stechkin [89] has given the bound

$$A(n) = \exp\left(O(\ln \ln n)^2\right)$$

and conjectured that $A(n) = O(1)$. This conjecture is proved in [77] in the following stronger form $A(n) = 1 + O(n^{-1/4+o(1)})$, which is improved to

$$A(n) = 1 + O(n^{-1}\tau(n)\log n)$$

in [60, Theorem 6.7]. Thus the constant A is correctly defined, and since the methods of [60,77] are effective, computing A is a feasible task (but may require significant efforts in making the estimates of [60,77] fully explicit). In fact it is quite possible that $A = A(2) = \sqrt{2}$. Problem 3.40 is also posed as [60, Question 6.9]. Finally, we recall that by [60, Theorem 6.8]

$$A(n) \geq 1 + n^{-1}\exp\left(0.43\frac{\log n}{\log \log n}\right).$$

for infinitely many n.

Problem 3.41. *Use the bound of W. D. Banks, A. Harcharras and I. E. Shparlinski [4] on analogues of short Kloosterman sums with polynomials over finite fields to sharpen the Brun-Titchmarsh theorem over $\mathbb{F}_q[X]$, see [55].*

Comments: One may try to imitate the approach of J. B. Friedlander and H. Iwaniec [42].

Problem 3.42. *Estimate multiplicative character sums*

$$\sum_{n \leq N} \chi\left(\frac{n^{p-1} - 1}{p}\right) \quad and \quad \sum_{n \leq N} \chi\left(\frac{n^p - n}{p}\right)$$

where χ is a multiplicative character modulo p.

Similar exponential sums are estimated by D. R. Heath-Brown and S. V. Konyagin [54].

3.9. *Nonlinear recurrence sequences*

Problem 3.43. *For a given polynomial $f \in \mathbb{F}_q[X]$ and $u_0 \in \mathbb{F}_q$ we define the following sequence of elements of $\mathbb{F}_q$*

$$u_n = f(u_{n-1}), \qquad n = 1, 2, \ldots.$$

Clearly this sequence eventually becomes periodic, that is $u_{n+T} = u_n$ for some $T \geq 1$ whenever $n \geq N_0$. In fact, without loss of generality, one can assume that it is purely periodic, that is, $N_0 = 0$. Improve the bound of H. Niederreiter and A. Winterhof [72] on character sums

$$\sum_{n=1}^{N} \psi(au_n), \qquad a \in \mathbb{F}_q^*,$$

with a nontrivial additive character ψ of $\mathbb{F}_q$ and $N \leq T$.

Comments: The bound of [72] develops some ideas suggested in [90] and also improves the previous result of [68]; however it is still nontrivial only if T is very close to its largest possible value p. Even constructing some special (but general enough) families of polynomials for which such improvement is possible is of interest. In the multidimensional case (that is, for iterations of multivariate polynomials) such families are constructed in [74]. Furthermore, A. Ostafe [73] has modified the construction of [74] in a way that the method of [70] applies to this constructions and has led to much better results "on average" over all initial values. Unfortunately the constructions of [73,74] do not work in the univariate case. This problem has close links with pseudorandom number generation, see [69,71,90] for these links and further references.

Problem 3.44. *For a prime p, a primitive root g modulo p and an integer v_0, the following sequence of integers*

$$v_n \equiv g^{v_{n-1}} \pmod{p}, \ 0 \leq v_n \leq p-1, \qquad n = 1, 2, \ldots.$$

As before we note that this sequence eventually becomes periodic with some period T and we also assume that it is purely periodic. Estimate exponential sums

$$\sum_{n=1}^{N} \psi(a v_n), \qquad a \in \mathbb{Z},$$

for $N \leq T$.

3.10. *Evaluation of Kloosterman sums*

Problem 3.45. *Find an algorithm to evaluate or approximate Kloosterman sums*

$$\sum_{x \in \mathbb{F}_q^*} \psi(x + a x^{-1}), \qquad a \in \mathbb{F}_q^*,$$

with an additive character ψ of $\mathbb{F}_q$ more efficiently than directly from the definition.

Comments: This question has some links with and motivation coming from quantum computing, see [33]. Results of [81,84] imply lower bounds on the complexity of computation of Kloosterman sum in some standard computational models.

Acknowledgements

The author is very grateful to his colleagues and especially to his co-authors, for joint projects or short discussions on exponential and characters sums, number theory and mathematics in general. This list of problems is a tribute to the sacrificed beer and chalk.

Special thanks also go to Bill Banks for the careful reading of the preliminary version and many valuable comments.

References

1. A. Ayyad, T. Cochrane and Z. Zheng, The congruence $x_1x_2 \equiv x_3x_4 \pmod{p}$, the equation $x_1x_2 = x_3x_4$ and the mean value of character sums, *J. Number Theory* **59** (1996), 398–413.

2. S. Balasuriya, I. E. Shparlinski and D. Sutantyo, Multiplicative character sums with the Euler function, *Studia Sci. Math. Hungarica* (to appear).

3. W. D. Banks, M. Z. Garaev, F. Luca and I. E. Shparlinski, Uniform distribution of fractional parts related to pseudoprimes, *Canad. J. Math.* **61** (2009), 481-502.

4. W. D. Banks, A. Harcharras and I. E. Shparlinski, Short Kloosterman sums for polynomials over finite fields, *Canad. J. Math.* **55** (2003), 225–246.

5. W. D. Banks, G. Harman and I. E. Shparlinski, Distributional properties of the largest prime factor, *Michigan Math. J.* **53** (2005), 665–681.

6. W. D. Banks, D. Hart and M. Sakata, Almost all palindromes are composite, *Math. Res. Lett.* **11** (2004), 853–868.

7. W. Banks and I. E. Shparlinski, Congruences and exponential sums with the Euler function, *High Primes and Misdemeanours: Lectures in Honour of the 60th Birthday of Hugh Cowie Williams* Amer. Math. Soc., 2004, 49–60.

8. W. D. Banks and I. Shparlinski, Prime divisors of palindromes, *Period. Math. Hungar.* **51** (2005), 1–10.

9. W. D. Banks and I. E. Shparlinski, Non-residues and primitive roots in Beatty sequences, *Bull. Austral. Math. Soc.* **73** (2006), 433–443.

10. W. D. Banks and I. E. Shparlinski, Short character sums with Beatty sequences, *Math. Res. Lett.* **13** (2006), 539–547.

11. W. Banks and I. E. Shparlinski, Congruences and rational exponential sums with the Euler function, *Rocky Mountain J. Math.* **36** (2006), 1415–1426.

12. W. D. Banks and I. E. Shparlinski, Character sums with Beatty sequences on Burgess-type intervals, *Analytic Number Theory – Essays in Honour of Klaus Roth* Cambridge Univ. Press, Cambridge, 2009, 15-21.

13. W. D. Banks and I. E. Shparlinski, Multiplicative character sums with twice-differentiable functions, *Quart. J. Math.* (to appear).

14. V. Bergelson, A. Leibman and E. Lesigne, Intersective polynomials and the polynomial Szemeredi theorem, *Advances in Math.* **219** (2008), 369–388.

15. E. R. Berlekamp, *Algebraic coding theory* McGraw-Hill, NY, 1968.

16. J. Bourgain, Estimates on exponential sums related to Diffie-Hellman distributions, *Geom. and Func. Anal.* **15** (2005), 1–34.

17. J. Bourgain, Mordell's exponential sum estimate revisited, *J. Amer. Math. Soc.* **18** (2005), 477–499.

18. J. Bourgain, Exponential sum estimates on subgroups of $\mathbb{Z}_q$, q arbitrary, *J. Anal. Math.* **97** (2005), 317–355.

19. J. Bourgain, Exponential sum estimates in finite commutative rings and applications, *J. Anal. Math.* **101** (2007), 325–355.

20. J. Bourgain, On an exponential sum related to the Diffie-Hellman cryptosystem, *Intern. Math. Research Notices* **2008** (2008), Article ID 61271, 1–15.

21. J. Bourgain, Estimates of polynomial exponential sums, *Israel J. Math.* (to appear).

22. J. Bourgain and M.-C. Chang, Exponential sum estimates over subgroups and almost subgroups of $\mathbb{Z}_q$, where q is composite with few prime factors, *Geom. and Func. Anal.* **16** (2006), 327–366.

23. J. Bourgain and M. Z. Garaev, On a variant of sum-product estimates and explicit exponential sum bounds in prime fields, *Math. Proc. Cambridge Phil. Soc.* **146** (2009), 1–21.

24. J. Bourgain, A. A. Glibichuk and S. V. Konyagin, Estimates for the number of sums and products and for exponential sums in fields of prime order, *J. Lond. Math. Soc.* **73** (2006), 380–398.

25. J. Bourgain and I. E. Shparlinski, Distribution of consecutive modular roots of an integer, *Acta Arith.* **134** (2008), 83–91.

26. R. de la Bretèche and G. Tenenbaum, Sommes d'exponentielles friables d'arguments rationnels, *Funct. Approx. Comment. Math.* **37** (2007), 31–38.

27. K. A. Broughan and I. E. Shparlinski, On the number of solutions of exponential congruences, *Preprint* 2009.

28. K. A. Broughan, F. Luca and I. E. Shparlinski, Some divisibility properties of binomial coefficients and Wolstenholme's conjecture, *Preprint* 2009.

29. M.-C. Chang,On a problem of Arnold on uniform distribution, *J. Funcional Analysis* **242** (2007), 272–280.

30. M.-C. Chang, On a question of Davenport and Lewis on character sums and primitive roots in finite fields, *Duke Math. J.* **145** (2008), 409–442.

31. M.-C. Chang, Burgess inequality in $\mathbb{F}_{p^2}$, *Geom. Funct. Anal.* (to appear).

32. M.-C. Chang and C. Z. Yao, An explicit bound on double exponential sums related to DiffieHellman distributions, *SIAM J. Discr. Math.* **22** (2008), 348–359.

33. A. M. Childs, L. J. Schulman and U. V. Vazirani, Quantum algorithms for hidden nonlinear structures, *Proc. 48th IEEE Symp. on Found. Comp. Sci.* IEEE, 2007, 395–404.

34. T. Cochrane, C. Pinner and J. Rosenhouse, Bounds on exponential sums and the polynomial Waring problem mod p, *J. London Math. Soc.* **67** (2003), 319–336.

35. T. Cochrane, C. Pinner and J. Rosenhouse, Sparse polynomial exponential sums, *Acta Arith.* **108** (2003), 37–52.

36. T. Cochrane and Z. Y. Zheng, Exponential sums with rational function entries, *Acta Arith.* **95** (2000), 67–95.

37. T. Cochrane and Z. Y. Zheng, A survey on pure and mixed exponential sums modulo prime powers, *Proc. Illinois Millennial Conf. on Number Theory, Vol.1* A.K. Peters, Natick, MA, 2002, 271–300.

38. P. Corvaja, Problems and results on integral points on ratonal surfaces, *Diophantine Geometry* CRM Series, Vol. 4, Scuola Normale Superiore, Pisa, 2007, 123–141.

39. E. Dobrowolski and K. S Williams, An upper bound for the sum $\sum_{n=a+1}^{a+H} f(n)$ for a certain class of functions f, *Proc. Amer. Math. Soc.* **114** (1992), 29–35.

40. J. B. Friedlander, J. Hansen and I. E. Shparlinski, On character sums with exponential functions, *Mathematika* **47** (2000), 75–85.

41. J. B. Friedlander and H. Iwaniec, The divisor problem for arithmetic progressions, *Acta Arith.* **45** (1985), 273–277.
42. J. Friedlander and H. Iwaniec, The Brun–Titchmarsh theorem, *Analytic Number Theory* Lond. Math. Soc. Lecture Note Series **247**, 1997, 363–372.
43. J. B. Friedlander, S. V. Konyagin and I. E. Shparlinski, Some doubly exponential sums over $\mathbb{Z}_m$, *Acta Arith.* **105** (2002), 349–370.
44. J. B. Friedlander, D. Lieman and I. E. Shparlinski, On the distribution of the RSA generator, *Proc. Intern. Conf. on Sequences and Their Applications (SETA'98), Singapore* Springer-Verlag, London, 1999, 205-212.
45. É. Fouvry and G. Tenenbaum, Entiers sans grand facteur premier en progressions arithmétiques, *Proc. London Math. Soc.* **63** (1991), 449–494.
46. N. Frantzikinakis, Multiple recurrence and convergence for Hardy sequences of polynomial growth, *Preprint* 2009, (available from http://arxiv.org/abs/0903.0042).
47. H. Furstenberg, From the Erdős–Turán conjecture to ergodic theory — The contribution of combinatorial number theory to dynamics, *Paul Erdős and His Mathematics* Springer-Verlag, Berlin, 2002, 261–277.
48. M. Z. Garaev, F. Luca and I. E. Shparlinski, Character sums and congruences with $n!$, *Trans. Amer. Math. Soc.* **356** (2004), 5089–5102.
49. M. Z. Garaev, F. Luca and I. E. Shparlinski, Exponential sums and congruences with factorials, *J. Reine Angew. Math.* **584** (2005), 29–44.
50. M. Z. Garaev, F. Luca and I. E. Shparlinski, Catalan and Apéry numbers in residue classes, *J. Combin. Theory, Ser. A* **113** (2006), 851–865.
51. M. Z. Garaev, F. Luca and I. E. Shparlinski, Exponential sums with Catalan numbers' *Indag. Math.* **18** (2007), 23–37.
52. K. Gong, On certain character sums over smooth numbers, *Glasnik Math.* (to appear).
53. D. R. Heath-Brown, The square sieve and consecutive squarefree numbers, *Math. Ann.* **266** (1984), 251–259.
54. D. R. Heath-Brown and S. V. Konyagin, New bounds for Gauss sums derived from kth powers, and for Heilbronn's exponential sum, *Quart. J. Math. Oxford* **51** (2000), 221–235.
55. C.-N. Hsu, The Brun-Titchmarsh theorem in function fields, *J. Number Theory* **79** (1999), 67–82.
56. H. Iwaniec and E. Kowalski, *Analytic number theory* Amer. Math. Soc., Providence RI, 2004.
57. A. A. Karatsuba, Sums of characters with prime numbers, *Izv. Akad. Nauk Ser. Mat.* **34** (1970) 299–321.
58. A. A. Karatsuba, The distribution of values of Dirichlet characters on additive sequences, *Doklady Acad. Sci. USSR* **319** (1991), 543–545 (in Russian).
59. S. V. Konyagin, On estimates of Gaussian sums and the Waring problem modulo a prime, *Trudy Matem. Inst. Acad. Nauk USSR* Moscow, **198** (1992), 111–124 (in Russian).
60. S. V. Konyagin and I. E. Shparlinski, *Character sums with exponential functions and their applications* Cambridge Univ. Press, Cambridge, 1999.
61. N. M. Korobov, On the distribution of digits in periodic fractions, *Matem.*

Sbornik **89** (1972), 654–670 (in Russian).

62. N. M. Korobov, *Exponential sums and their applications* Kluwer Acad. Publ., Dordrecht, 1992.

63. R. Lidl and H. Niederreiter, *Finite fields* Cambridge Univ. Press, Cambridge, 1997.

64. D. Lieman and I. E. Shparlinski On a new exponential sum, *Canad. Math. Bull.* **44** (2001), 87–92.

65. H. Liu and W. Zhang, On the squarefree and squarefull numbers, *J. Math. Kyoto Univ.* **45** (2005), 247–255.

66. H. L. Montgomery, *Topics in multiplicative number theory Lect. Notes in Math.* Springer-Verlag, Berlin, **227** (1971).

67. A. Mukhopadhyay, M. R. Murty and K. Srinivas, Counting squarefree discriminants of trinomials under *abc, Proc. Amer. Math. Soc.* **137** (2009), 3219–3226.

68. H. Niederreiter and I. E. Shparlinski, On the distribution and lattice structure of nonlinear congruential pseudorandom numbers, *Finite Fields and Their Appl.* **5** (1999), 246–253.

69. H. Niederreiter and I. E. Shparlinski, Recent advances in the theory of nonlinear pseudorandom number generators, *Proc. Conf. on Monte Carlo and Quasi-Monte Carlo Methods, 2000* Springer-Verlag, Berlin., 2002, 86–102.

70. H. Niederreiter and I. E. Shparlinski, On the average distribution of inversive pseudorandom numbers, *Finite Fields and Their Appl.* **8** (2002), 491–503.

71. H. Niederreiter and I. E. Shparlinski, Dynamical systems generated by rational functions, *Lect. Notes in Comp. Sci.* Springer-Verlag, Berlin, **2643** (2003), 6–17.

72. H. Niederreiter and A. Winterhof, Exponential sums for nonlinear recurring sequences, *Finite Fields Appl.* **14** (2008), 59–64.

73. A. Ostafe, Multivariate permutation polynomial systems and nonlinear pseudorandom number generators, *Preprint* (to appear).

74. A. Ostafe and I. E. Shparlinski, On the degree growth in some polynomial dynamical systems and nonlinear pseudorandom number generators, *Math. Comp.* (to appear).

75. Z. Kh. Rakhmonov, On the distribution of values of Dirichlet characters and their applications, *Proc. Steklov Inst. Math.* **207** (1995), 263–272.

76. V. Shoup, On the determenistic complexity of factoring polynomials over finite fields, *Inform. Proc. Letters* **33** (1990), 261–267.

77. I. E. Shparlinski, On bounds of Gaussian sums, *Matem. Zametki* **50** (1991), 122–130 (in Russian).

78. I. E. Shparlinski, On exponential sums with sparse polynomials and rational functions, *J. Number Theory* **60** (1996), 233–244.

79. I. E. Shparlinski, *Finite fields: Theory and computation* Kluwer Acad. Publ., Dordrecht, 1999.

80. I. E. Shparlinski, Exponential function analogue of Kloosterman sums, *Rocky Mountain J. Math.* **34** (2004), 1497–1502.

81. I. E. Shparlinski, On the nonlinearity of the sequence of signs of Kloosterman sums, *Bull. Aust. Math. Soc.* **71** (2005), 405–409.

82. I. E. Shparlinski, Character sums over shifted smooth numbers, *Proc. Amer. Math. Soc.* **135** (2007), 2699–2705.

83. I. E. Shparlinski, On some dynamical systems in finite fields and residue rings, *Discr. and Cont. Dynam. Syst., Ser.A* **17** (2007), 901–917.

84. I. E. Shparlinski, On the distribution of Kloosterman sums, *Proc. Amer. Math. Soc.* **136** (2008), 403–407.

85. I. E. Shparlinski, On quadratic fields generated by discriminants of irreducible trinomials, *Proc. Amer. Math. Soc.* (to appear).

86. I. E. Shparlinski, Exponential sums with consecutive modular roots of an integer, *Quart. J. Math.* (to appear).

87. I. E. Shparlinski and A. Winterhof, Noisy interpolation of sparse polynomials in finite fields, *Appl. Algebra in Engin., Commun. and Computing* **16** (2005), 307–317.

88. I. E. Shparlinski and A. Winterhof, A hidden number problem in small subgroups, *Math. Comp.* **74** (2005), 2073–2080.

89. S. B. Stechkin, An estimate for Gaussian sums, *Matem. Zametki* **17** (1975), 342–349 (in Russian).

90. A. Topuzoğlu and A. Winterhof, Pseudorandom sequences, *Topics in Geometry, Coding Theory and Cryptography* Springer-Verlag, 2006, 135–166.

91. R. C. Vaughan, *The Hardy-Littlewood method* Cambridge Univ. Press, Cambridge, 1981.

92. M. Xiong and A. Zaharescu, Distribution of Selmer groups of quadratic twists of a family of elliptic curves, *Advances Math.* **219** (2008), 523–553.

93. G. Yu, Rank 0 quadratic twists of a family of elliptic curves, *Compos. Math.* **135** (2003), 331–356.

94. H. B. Yu, Estimates of character sums with exponential function, *Acta Arith.* **97** (2001), 211–218.

Errata to
"A GENERAL MODULAR RELATION
IN ANALYTIC NUMBER THEORY"
(Number Theory: Sailing on the Sea of Number Theory,
World Scientific, 2007, 214–236)

HARUO TSUKADA

Department of Information and Computer Sciences,
School of Humanity-Oriented Science and Engineering,
University of Kinki, Iizuka, Fukuoka 820-8555, Japan
E-mail: tsukada@fuk.kindai.ac.jp

page 223, line 18, after the paragraph ending with "after analytic continuation.", insert the following paragraph:

Let

$$\Delta = \begin{pmatrix} \{(1-a_j, A_j)\}_{j=1}^{n}; & \{(a_j, A_j)\}_{j=n+1}^{p} \\ \{(b_j, B_j)\}_{j=1}^{m}; & \{(1-b_j, B_j)\}_{j=m+1}^{q} \end{pmatrix} \in \Omega$$

so that

$$\Delta(s) = \frac{\displaystyle\prod_{j=1}^{m} \Gamma(b_j + B_j s) \ \prod_{j=1}^{n} \Gamma(a_j - A_j s)}{\displaystyle\prod_{j=n+1}^{p} \Gamma(a_j + A_j s) \ \prod_{j=m+1}^{q} \Gamma(b_j - B_j s)}.$$

page 227, line 15, "the disciplines" should read "other disciplines".
page 228, (4.27), "$= \sqrt{z}$" should read "$= \sqrt{\pi}$".

BIBLIOGRAPHY ON DETERMINANTAL EXPRESSIONS OF RELATIVE CLASS NUMBERS OF IMAGINARY ABELIAN NUMBER FIELDS

KEN YAMAMURA

Department of Mathematics, National Defense Academy,
Hashirimizu Yokosuka 239-8686, Japan
E-mail: yamamura@cc.nda.ac.jp

L. Carlitz and F. R. Olson [1] (1955) were the first who obtained the closed formula for Maillet's determinant:

$$\det(R(rs'))_{1 \leq r,s \leq (p-1)/2} = (-1)^{\frac{p-3}{2}} p^{\frac{p-3}{2}} h_p^*,$$

where p is an odd prime and h_p^* is the realtive class number of the p th cyclotomic field. Here, for an integer r with $(r,p) = 1$, r' is the least nonnegative integer with $rr' \equiv 1 \bmod p$ and $R(r)$ denotes the least nonnegative integer congruent to r modulo p. Thenceforth several authors have studied relationships between deteminants with rational (integer) entries and realtive class numbers h_K^* of imaginary abelian number fields K, most of which are of the form

$$\det R_K = c_K h_K^*,$$

where R_K and c_K denote a square matrix with rational entries and a rational constant, possibly zero, respectively, both explicitly determined by K.

We collect here all the relevant papers which are more than sixty in number and present them with review and digitization information for researcher's convenience. As an appendix, we also present papers on determinantal expressions of relative class numbers of imaginary function fields.

References

1. L. Carlitz and F. R. Olson, *Maillet's determinant,* Proc. Amer. Math. Soc. **6** (1955), no. 2, 265–269; doi:10.2307/2032352 [PDF], JSTOR; MR0069207 (**16**, **999d**); Zbl 0065.02703.

2. K. Inkeri, *Über die Klassenzahl des Kreiskörpers des l^{ten} Einheitswurzeln*, Ann. Acad. Sci. Fenn. Ser. A I No. 199 (1955), 12 pp.; reprinted in Collected Papers of Kustaa Inkeri (T. Metsänkylä and P. Ribenboim, eds., with a preface by Ribenboim), Queen's Papers in Pure and Applied Mathematics, Volume 91, Queen's Univ., Kingston, ON, Canada, 1992, 169, pp. 243–254; MR0079042 (**18, 20d**); Zbl 0065.26903.

3. L. Carlitz, *A genralization of Maillet's determinant and a bound for the first factor of the class number*, Proc. Amer. Math. Soc. **12** (1961), no. 2, 256–261; doi:10.2307/2034318 [PDF], JSTOR; MR0121354 (**22** #12093); Zbl 0131.03602.

4. Z. I. Borevich and I. R. Shafarevich, *[Number Theory]* (Russian) Izdat. "Nauka", Moskov, 1964; 3rd ed., 1985; German translation by H. Koch : S. I. Borewicz und I. R. Šafarevič : *Zahlentheorie*, Lehrbücher und Monogaraphien aus dem Gebiete der Exakten Wissenschften, Mathematische Reihe, Band 32. Birkhäuser Verlag, Basel und Stuttgart, 1966; English translation by N. Greenleaf : Z. I. Borevich and I. R. Shafarevich : *Number theory*, Pure and Applied Math., vol. 20, Academic Press, New York and London, 1966, Chap. 5, Sec. 6, Problem 2; doi:10.1016/S0079-8169(08)61696-7 [PDF]; Traduction faite d'après l'édition originale russe par Myrian et J.-L. Verley : Z. I. Borevich et I. R. Chafarevitch : *Théorie des nombres*, Monographies Internationales de Mathématiques Modernes, No. 8. Gauthier-Villars, Paris, 1967;reprint by Éditions Jacques Gabey, Sceaux; Japanese translation by Y. Sasaki, Yoshiokashoten, Kyoto, 1971-1972.

5. T. Lepistö, *On the first factor of the class number of the cyclotomic number field and Dirichlet's L-functions*, Ann. Acad. Sci. Fenn. Ser. A I No. 387 (1966), 53 pp.; MR0195843 (**33** #4041); Zbl 0137.02302.

6. T. Metsänkylä, *Bemerkungen über den ersten Faktor Klassenzahl des Kreiskörpers*, Ann. Univ. Turku. Ser. A I No. 105 (1967), 15 pp.; MR0213325 (**35** #4189); Zbl 0171.01202.

7. S. Hyyrö, *Über eine Determinantenidentität und den ersten Factor der Klassenzahl des Kreiskörpers*, Ann. Acad. Sci. Fenn. Ser. A I No. 398 (1967), 7 pp.; MR0219511 (**36** #2592); Zbl 0314.12007.

8. T. Metsänkylä, *Über den ersten Faktor Klassenzahl des Kreiskörpers*, Ann. Acad. Sci. Fenn. Ser. A I No. 416 (1967), 48 pp.; MR0228466 (**37** #4046); Zbl 0153.37804.

9. T. Metsänkylä, *Congruences modulo 2 for class number factors in cyclotomic fields*, Ann. Acad. Sci. Fenn. Ser. A I No. 453 (1969), 12 pp.; MR0251007 (**40** #4238); Zbl 0176.33303.

10. M. Newman, *A table of the fisrt factor for prime cyclotomic fields*, Math. Comp. **24** (1970), no. 109, 215–219; doi:10.2307/2004891 [PDF], JSTOR; MR0257029 (**41** #1684); Zbl 0198.36902.

11. V. M. Galkin, *[The first factor of the ideal class number of a cyclotomic field]* (Rusiian), Uspekhi Mat. Nauk XXVII (1972), no. 6 (168), 233–234; [PDF]; MR0392914 (**52** #13727); Zbl 0262.12006.

12. J. Kühnova, *Maillet's determinant $D_{p^{n+1}}$*, Arch. Math. (Brno) XV (1979), no. 4, 209–212; [PDF], DML-CZ; [PDF], GDZ; [PDF], DigiZeitschriften;

MR0563149 (82e:12003); Zbl 0456.10029.

13. L. Skula, *Another proof of Iwasawa's class number formula*, Acta Arith. XXXIX (1981), no. 1, 1–6; [PDF], icm; MR0638737 (83c:12010); Zbl 0372.12012.

14. K. Tateyama, *Maillet's determinant*, Sci. Papers College Gen. Ed. Tokyo **32** (1982), no. 2, 97–100; [PDF], URI: http://hdl.handle.net/2261/21161; MR0694835 (85c:11095); Zbl 0508.12006.

15. S. Okada, *Generalized Maillet determinant*, Proc. 29th Symposium on Algebra (Iwate, 1983), 1983, pp. 271–279.

16. S. Okada, *Generalized Maillet determinant*, Nagoya Math. J. **94** (1984), 165–170; euclid.nmj/1118787495 [PDF]; MR0748097 (85j:11147); Zbl 0535.12005.

17. T. Metsänkylä, *Maillet's matrix and irregular primes*, Ann. Univ. Turku. Ser. A I No. 186 (1984), 72–79; MR0748520 (85j:11145); Zbl 0531.12003.

18. K. Wang, *On Maillet's determinant*, J. Number Theory **18** (1984), no. 3, 306–312; doi:10.1016/0022-314X(84)90064-7 [PDF]; MR0746866 (85m:11072) (= Zbl 0535.12004); Zbl 0535.12004.

19. A. Endô, *The relative class numbers of certain imaginary abelian fields*, Abh. Math. Sem. Univ. Hamburg **58** (1988), 237–243; DOI: 10.1007/BF02941380 [PDF]; MR1027444 (90m:11169); Zbl 0699.12015.

20. A. Endô, *The relative class numbers of certain imaginary abelian number fields and determinants*, J. Number Theory **34** (1990), no. 1, 13–20; doi:10.1016/0022-314X(90)90048-V [PDF]; MR1039763 (91b:11121); Zbl 0695.12004.

21. T. Funakura, *On Kronecker's limit formula for Dirichlet series with periodic coefficients*, Acta Arith. LV (1990), no. 1, 59–73; MR1056115 (91j:11064); Zbl 0654.10039.

22. F. Hazama, *Demjanenko matrix, class number, and Hodge group*, J. Number Theory **34** (1990), no. 2, 174–177; doi:10.1016/0022-314X(90)90147-J [PDF]; MR1042490 (90m:11090); Zbl 0697.12003.

23. A. Endô, *On an index formula for the relative class numbers of a cyclotomic field*, J. Number Theory **36** (1990), no. 3, 332–338; doi:10.1016/0022-314X(90)90094-8 [PDF]; MR1077712 (91m:11092); Zbl 0715.11062.

24. A. Endô, *On the Stickelberger ideal of* $(2, \cdots, 2)$*-extensions of a cyclotomic number field*, Manuscripta Math. **69** (1990), fasc. 2, 107–132; DOI: 10.1007/BF02567915 [PDF]; [PDF], GDZ; [PDF], DigiZeitschriften; MR1072984 (91i:11144); Zbl 0715.11061.

25. G. Fujisaki, *A genralization of Carlitz's determinants*, Sci. Papers College Arts Sci. Univ. Tokyo **40** (1990), no. 2, 63–68; [PDF], URI: http://hdl.handle.net/2261/21201; MR1093185 (91m:11012); Zbl 0722.11014.

26. K. Girstmair, *A recursion formula for the relative class number of the* p^n*-th cyclotomic field*, Abh. Math. Sem. Univ. Hamburg **61** (1991), 131–138; DOI: 10.1007/BF02950757 [PDF]; MR1138278 (92j:11119); Zbl 0747.11046.

27. K. Girstmair, *On the cosets of the 2q-power group in the unit group modulo p*, Abh. Math. Sem. Univ. Hamburg **62** (1992), 217–232; DOI: 10.1007/BF02941628 [PDF]; MR1182851 (94a:11165); Zbl 0779.11052.

28. K. Girstmair, *The relative class numbers of imaginary cyclic fields of degrees* 4, 6, 8 *and* 10, Math. Comp. **61** (1993), no. 204, 881–887, S25–S27; doi:10.2307/2153259 [PDF], JSTOR; MR1195428 (94a:11170); Zbl 0787.11046.

29. W. Schwarz, *Demjanenko matrix and 2-divisibility of class numbers*, Arch. Math. (Basel) **60** (1993), No. 2, 154–156; DOI: 10.1007/BF01199101 [PDF]; MR1199672 (94g:11095); Zbl 0807.11053.

30. K. Dohmae, *Demjanenko matrix for imaginary abelian fields of odd conductors*, Proc. Japan Acad. Ser. A Math. Sci. **70** (1994), No. 9, 292–294; euclid.pja/1195510898 [PDF]; MR1313181 (96e:11134); Zbl 0829.11055.

31. J. W. Sands and W. Schwarz, *A Demjanenko matrix for abelian fields of prime power conductors*, J. Number Theory **52** (1995), no. 1, 85–97; doi:10.1006/jnth.1995.1057 [PDF]; MR1331767 (96b:11141); Zbl 0829.11054.

32. L. Skula, *The Stickelberger ideal and the Demjanenko matrix*, Algebraic cycles and related topics (Kitasakado, 1994) (F. Hazama, ed.), World Sci. Publishing, River Edge NJ, 1995, pp. 69–76; MR1414437 (97e:11136); Zbl 0861.11062.

33. T. Agoh and L. Skula, *Kummer type congruences and Stickelberger subideals*, Acta Arith. LXXV (1996), no. 3, 235–250; [PDF], icm; MR1387862 (97j:11052); Zbl 0841.11012.

34. H. Tsumura, *On a direct sum decomposition of the Dem'yanenko matrix*, Acta Arith. LXXVII (1996), no. 1, 71–76; [PDF], icm; MR1404977 (97e:11138); Zbl 0848.11051.

35. H. Tsumura, *On Demjanenko's matrix and Maillet's determinant for imaginary abelian number fields*, J. Number Theory **60** (1996), no. 1, 70–79; doi:10.1006/jnth.1996.0113 [PDF]; MR1405726 (98g:11122a); Zbl 0866.11063.

36. M.-H. Le, *A determinant concerning the relative class number of the cyclotomic field* $\mathbb{Q}(\zeta_{p^n})$, Discuss. Math. Algebra Stochastic Methods **16** (1996), no. 1, 61–65; MR1429798 (97k:11146); Zbl 0867.11078.

37. C. Hélou, *Proof of a conjecture of Terjanian for regular primes*, C. R. Math. Rep. Acad. Sci. Canada **43** (1996), no. 5, 193–198; MR1425291 (98a:11142); Zbl 0882.11059.

38. A. Endô, *The relative class numbers of certain imaginary abelian number field of odd conductors*, Proc. Japan Acad. Ser. A Math. Sci. **72** (1996), No. 3, 64–68; euclid.pja/1195510421 [PDF (504 KB)]; MR1391900 (97e:11137); Zbl 0862.11062.

39. T. Metsänkylä, *Letter to the editor. Comment on: "On Demyanenko's matrix and Maillet's determinant for imaginary abelian number fields" [J. Number Theory 60 (1996), no. 1, 70–79] by Tsumura*, J. Number Theory **64** (1997), no. 1, 162–163; doi:10.1006/jnth.1997.2139 [PDF (199 K)]; MR1450490 (98g:11122b); Zbl 0881.11076.

40. F. Tsumura, [*On Demjanenko matrix for general imaginary abelian fields*] (Japanese), Algebraic number theory and Fermat's problem (Kyoto, 1995), Sūrikaisekikenkyūsho Kōkyūroku 971, 1996, 125–133; [PDF]; [PDF], URI: http://hdl.handle.net/2433/60693; MR1481237 (98j:11088); Zbl 1043.11551.

41. M. Hirabayashi, *A relative class number formula of an imaginary abelian field by means of Demyanenko matrix*, Proceedings of the conference on analytic and elementary number theory: a satellite conference of the European Congress on Mathematics '96 (Vienna, 1996) (W. G. Nowak and J. Schoißengeier, eds.), Univ. Wien, 1996, Wien, pp. 81–91; Zbl 0882.11062.

42. F. Hazama, *Hodge cycles on the jacobian variety of the Catalan curve*, Compositio Math. **107** (1997), no. 3, 339–353; doi:10.1023/A:1000106427229 [PDF, Springer; PDF, CJO]; MR1458755 (98d:11065); Zbl 1044.11587.

43. P. Fuchs, *Maillet's determinant and a certain basis of Stickelberger ideal*, Number theory (Liptovský, Ján, 1995), Tatra Mt. Math. Publ. **11** (1997), 121–128; [PostScript]; MR1475509 (99g:11131); Zbl 0978.11059.

44. S. Huang, [*On the representation of the first factor of the class number of the cyclotomic field* $\mathbf{Q}(\zeta_p)$ *by determinants*] (Chinese), J. Zhanjiang Normal College Nat. Sci. Ed. **18** (1997), no. 1, 23–24; DOI: cnki:ISSN:1006-4702.0.1997-01-006 [CAJ, PDF].

45. M. Hirabayashi, *A genralization of Maillet and Demyanenko determinants*, Acta Arith. LXXXIII (1998), no. 4, 391–397; [PDF], icm; MR1610490 (99b:11122); Zbl 0895.11045.

46. S. Kanemitsu and T. Kuzumaki, *On a genralization of the Maillet determinant*, Number theory (Eger, 1996) (K. Győry, A. Pethő and V. T. Sós, eds.), de Gruyter, Berlin, 1998, pp. 271–287; MR1628848 (99h:11122); Zbl 0920.11071.

47. R. Kučera, *Formulae for the relative class numbers of an imaginary abelian number field in the form of a determinant*, Nagoya Math. J. **163** (2001), 167–191; euclid.nmj/1114631625 [PDF]; MR1855194 (2002j:11129); Zbl 1002.11079.

48. S. Kanemitsu and T. Kuzumaki, *On a genralization of the Maillet determinant II*, Acta Arith. XCIX (2001), no. 4, 343–361; [PDF]; MR1845690 (2002h:11115); Zbl 0984.11056.

49. M. Hirabayashi, *A genralization of Maillet and Demjanenko determinants for the cyclotomic* $\mathbb{Z}_p$-*extension*, Abh. Math. Sem. Univ. Hamburg **71** (2001), 15–28; DOI: 10.1007/BF02941457 [PDF]; MR1872709 (2002k:11194); Zbl 1030.11059.

50. M. Hirabayashi, *Inkeri's determinant for an imaginary abelian number field*, Arch. Math. (Basel) **79** (2002), No. 3, 175–181; DOI: 10.1007/s00013-002-8302-5 [PDF]; MR1933375 (2003j:11133); Zbl 1006.11062.

51. M. Hirabayashi, *A relative class number formula for an imaginary abelian number field by means of Dedekind sum*, Manuscripta Math. **109** (2002), fasc. 2, 223–227; DOI: 10.1007/s00229-002-0304-x [PDF]; MR1935030 (2003j:11135); Zbl 1045.11078.

52. R. Kučera, *Formulae for the relative class number of an imaginary abelian number field in the form of a product of determinants*, Acta Math. Inform. Univ. Ostraviensis **10** (2002), no. 1, 79–83; [PDF], DML-CZ; MR1943026 (2004c:11203); Zbl 1056.11067.

53. A. Endô, *A genralization of the Maillet determinant and the Demyanenko determinant*, Abh. Math. Sem. Univ. Hamburg **73** (2003), 181–193; DOI:

10.1007/BF02941275 [PDF]; MR2028513 (2005a:11168); Zbl 1050.11092.

53. M.-H. Le, [*A class of determinants and its applications*] (Chinese), J. Wuhan Univ. Sci Engrg. **16** (2003), no. 1, 89–91; DOI: cnki:ISSN:1005-4790.0.2003-01-021 [CAJ, PDF].

54. M. Hirabayashi, *A formula of Girstmair and a remark on relative class number formula*, Proceedings of the 2003 Nagoya Conference "Yokoi-Chowla Conjecture and Related Problems" (Nagoya, 2003) (S.-I. Katayama, C. Levesque and T. Nakahara, eds.), Saga Univ., Saga, 2004, pp. 11–24; MR2109018 (2005i:11159).

55. J. Ahn and H. Jung, *Determinant formulas and class numbers in cyclotomic fields*, Proceedings of the Japan-Korea joint seminar on number theory (Kuju, 2004) (H. K. Kim, ed.), Kyushu Univ., 2004, pp. 1–15; Zbl 1099.11063.

56. M. Hirabayashi, *Genralizations of Girstmair's formulas*, Abh. Math. Sem. Univ. Hamburg **75** (2005), 83–95; DOI: 10.1007/BF02942037 [PDF]; MR2187580 (2006g:11223); Zbl pre02245670.

57. M. Hirabayashi and H. Tsumura, *Multiple Dedekind sums and relative class number formulae*, Math. Nachr. **278** (2005), Issue 14, 1673–1680; DOI: 10.1002/mana.200310334 [PDF]; MR2176271 (2006h:11130); Zbl 1084.11061.

58. M. Hirabayashi, [*On relative class number formulas for imaginary abelian number fields*] (Japanese), Proceedings of the Workshop on Number Theory 2005 (Waseda Univ., 2005), 2005, pp. 75–88.

59. T. Agoh and T. Taniguchi, *A study of Inkeri's class number formula*, Exposition. Math. **24** (2006), Issue 1, 53–79; doi:10.1016/j.exmath.2005.06.002 [HTML, PDF]; MR2195183 (2007b:11173); Zbl 1122.11072.

60. H.-Y, Jung and J. Ahn, *Group determinant formulas and class numbers of cyclotomic fields*, J. Korean Math. Soc. **44** (2007), No. 3, 499–509; [PDF]; MR2314824 (2008c:11145); Zbl pre05202700.

61. M. Hirabayashi, *A determinant formula for the quotient of the relative class numbers of imaginary abelian number fields of relative degree 2*, Proceedings of the Symposium on Algebraic Number theory and Related Topics (K. Hashimoto, Y. Nakajima and H. Tsunogai, eds.) RIMS Kôkyûroku Bessatsu B4, Research Institute for Mathematical Sciences, Kyoto Univ., Kyoto, 2007, pp. 21–33; Zbl pre05258105.

62. M. Hirabayashi, *A genralization of Newmann's formula*, Arch. Math. (Basel) **90** (2008), No. 3, 223–229; DOI: 10.1007/s00013-007-2445-3 [PDF (130.8 KB)]; MR2391356; Zbl pre05263985.

Bibliography on determinantal expressions of
relative class numbers of imaginary function fields
(last update: May 2008)

References

1. M. Rosen, A note on the relative class number in function fields, *Proc. Amer. Math. Soc.* (5) **125** (1997), 1299–1303; DOI: 10.1090/S0002-9939-97-03748-9 [PDF]; [PDF], JSTOR; MR1371139 (97g:11133); Zbl 0872.11045.

2. S. Bae and P.-L. Kang, Class numbers of cyclotomic function fields, *Acta Arith.* (3) **102** (2002), 251–259; [PDF]; MR1884718 (2002m:11098); Zbl 0989.11064.

3. H. Jung and J. Ahn, Demyanenko matrix and recursion formula for the relative class number over function fields, *J. Number Theory* (1) **98** (2003), 55–66; doi:10.1016/S0022-314X(02)00023-9 [PDF]; MR1950438 (2003i:11170); Zbl 1057.11057.

4. H. Jung and J. Ahn, On the relative class number of cyclotomic function fields, *Acta Arith.* (1) **107** (2003), 91–101; [PDF]; MR1956987 (2003j:11143); Zbl 1057.11058.

5. S. Bae, H. Jung, and J. Ahn, Class numbers of some abelian extensions of rational function fields, *Math. Comp. (245)* **73** *(2004), 377–386; DOI: 10.1090/S0025-5718-03-01528-X [PDF]; MR2034128 (2004j:11144);* Zbl 1057.11056.

6. *H. Jung, S. Bae, and J. Ahn, Determinant formulas for class numbers in function fields, Math. Comp.* (250) **74** (2005), 953–965; DOI: 10.1090/S0025-5718-04-01671-0 [PDF]; MR2114658 (2006a:11151); Zbl 1082.11075.

7. J. Ahn, S. Choi, and H. Jung, Class number formulae in the form of a product of determinants in function fields, *J. Austral. Math. Soc. Ser. A* **78** (2005), Part 2, 227–238; [PDF]; MR2141879 (2005m:11217); Zbl 1092.11044.

8. H.-Y, Jung and J. Ahn, Group determinant formulas and class numbers of cyclotomic fields, *J. Korean Math. Soc.* (3) **44** (2007), 499–509; [PDF]; MR2314824 (2008c:11145); Zbl pre05202700.

INDEX

2012 年 8 月 30 日,时年 43 岁的日本数学家,京都大学教授望月新一在数学系主页上发布了 4 篇论文,他宣称自己解决了数学史上最富传奇色彩的数论未解猜想:ABC 猜想,于是数论和日本数学家又开始充斥于各级媒体.

ABC 在我们日常语言中常用,有入门的含义,但这里我们所说的 ABC 却是个超级大猜想.

2016 年,"阿尔法狗横扫李世石"成为当年的十大科学传播事件之一后,有人提出:ABC 正在成为我们时代的主题:A 是 AI,人工智能;B 是 Big Data,大数据;C 是 Cloud,云计算.

但我们数学圈内流传的 ABC 猜想是在 27 年前由 Masser 和 Oesterle 分别独立提出的,俄罗斯甚至还以此猜想为背景命制了一道竞赛试题:

正整数 n 的所有质因数的乘积(每一个质因数只出现一次)称为基根基,记为 $\mathrm{rad}(n)$,例如 $\mathrm{rad}(120) = 2 \times 3 \times 5 = 30$. 是否存在三个两两互质的正整数 a,b,c,使得 $a + b = c$,且 $c > 1\ 000 \cdot \mathrm{rad}(abc)$?

有选手找到了形如 $a = 10^n - 1, b = 1, c = 10^n$ 的例子.

现在取 k,使得 $3^k > 1\ 000$,而正整数 n,由 Euler 定理:由于 $\varphi(3^{k+1}) = 2 \cdot 3^k$,所以 $10^{2 \cdot 3^k} - 1$ 是 3^{k+1} 的倍数,那么就可取 $n = 2 \cdot 3^k$,于是

$$\mathrm{rad}(abc) = \mathrm{rad}((10^n - 1)10^n) = 10\mathrm{rad}(10^n - 1) =$$

$$3 \cdot 10\mathrm{rad}\left(\frac{10^n - 1}{3^{k+1}}\right) < 10\frac{10^n - 1}{3^k} <$$

$$\frac{10^{n+1}}{10\ 000} = \frac{c}{1\ 000}.$$

著名的 ABC 猜想则是断言:对于任何 $\varepsilon > 0$,都存在这样的常数 k,使得任何满足关系式 $A + B = C$ 的两两互质的正整数 A,B,C 都满足不等式 $C < k \cdot \mathrm{rad}(ABC)^{1+\varepsilon}$.

由这个猜想的正确性,可以推出数论中的一系列著名论断. 例如,只要 ABC 猜想成立,那么就不难推出,在 $n > 2$ 时,Fermat 方程 $x^n + y^n = z^n$ 就只有有限个解. 而上面所引的这道俄罗斯奥数试题所给出的信息就是:不能将 ABC 猜想中的 $1 + \varepsilon$ 换成 1.

ABC 猜想在我国关注度不高,但还是有. 超椭圆曲线存在一个由其到 p' 的二重覆盖映射的代数曲线,具有一些特殊的算术和几何性质,其中某些性质和椭圆曲线的类似,但是由于亏格的原因不再具有 Mordell-Weil 群的结构,也有很多性质和椭圆曲线差别较大,椭圆曲线上的 Szpiro 猜想是丢番图逼近理论中很重要的一个猜想,该猜想和著名的 ABC 猜想等价. 清华大学的一位硕士在邱林生教授的指导下,在承认 ABC 猜想的情况下,在一定条件下将在超椭圆曲线上的 Szpiro 猜想推广到了更一般的情况,而且由于他们采用了显式的 ABC 猜想的表达形式,所以还可以看出 ABC 猜想中的控制常数是怎样影响结论的.

数域上的 ABC 猜想是这样的:k 是一个数域. $O_{k,s}$ 是 k 的 s – 整数环,设 $a,b,c \in O_{k,s}$ 满足 $a + b + c = 0$,那么对于 $\forall \varepsilon > 0$

$$H_k(a,b,c) \ll \mathrm{rad}_{s,k}(abc)^{1+\varepsilon}$$

这里的隐性常数只依赖于 k,s 和 ε.

当然还有一般代数数域上的 ABC 猜想,不过是引进了数域 k 的共轭差积 (discrimnant).

日本从来就不乏传奇的数论学家,以在攻克 Fermat 大定理的征途中出现的日本数学家为例就有:日本东京大学的志村五郎 (Goro Shimura) 和谷山 (Yutaka Taniyama).

谷山在 1955 年 9 月召开的东京日光会议上,与志村联手研究了椭圆曲线的参数化问题,提出了谷山 — 志村猜想,即:任一椭圆曲线都是模曲线. 1986 年贝特由塞尔猜想证明了谷山 — 志村猜想,这样要证明费马大定理,只需要证对半稳定椭圆曲线谷山 — 志村猜想成立,而怀尔斯恰好就是这么做的.

整个20世纪70年代谷山—志村猜想都在数学家中引起了惊惶,因为它的蔓延之势不可阻挡,怀尔斯后来回忆说:"我们构造了越来越多的猜想,它们不断地向前方延伸,但如果谷山—志村猜想不是真的,那么它们全都会显得滑稽可笑.因此我们必须证明谷山—志村猜想,才能证明我们满怀希望地勾勒出来的对未来的整个设计是正确的.

费马定理像一块试金石,它检验着世界各国的数学水平,日本还曾出现了一位对此颇有贡献的数学家,他就是日本数学界的骄傲——宫冈洋一先生.宫冈先生是东京都立大学数学教授,曾在德国波恩访问进修.1988年整个数学界被闹得沸沸扬扬,有关宫冈证明了费马大定理的新闻传遍了世界各个角落,那么宫冈洋一真的成功了吗?现在我们已经从1988年4月8日 *The Independent* 发表的一篇评论中知道:"不幸,宫冈博士试图在一个相关的领域——代数数论中,得到一种基变换,但这一点似乎是行不通的."

日本数论专家浪川幸彦以《波恩来信》的形式讲述了这一事件的经过.他的讲述既通俗又有趣,他写道:

"收到贺年信一直想要回信,转眼之间过了一个月,而且到了月底.不过托您的福我可以报告一个本世纪的大新闻.

历史上最古老而著名的问题之一费马猜想很可能已被在德国波恩逗留的宫冈洋一(从理论上)证明了.目前正处在细节的完成阶段,还要花些时间来确定正确与否,依我所见有足够的成功希望.众所周知,费马猜想是说对于自然数 $n > 2$,不存在满足

$$x^n + y^n = z^n \qquad\qquad *$$

的自然数 x, y, z,上面所说的'理论上',意思是指对于充分大的(自然)数 $n > N$ 可以证明,而且这个 N 在理论上是可以计算的.该 N 可以用某个自守函数与数论不变量表示,但实际的数值计算似乎相当麻烦,并且还不知道是否对一切 n 确实都已解决.不过如果他的结果被确认是正确的,人们就会同时集中改善 N 的估计,有必要就动用计算机,那么最终解决也就为期不远了.但是,姑且不论宣传报道,对于我们纯数学工作者来说,本质是理论上的解决.

宫冈先生从去年下半年起对这个问题感兴趣并一直持续地进行研究,特别是今年在巴黎与梅热等讨论以后,他的研究工作迅速取得进展.偶尔在饭桌上听到他研究工作的进展情况,就是作为旁观者也感到心情激动,能成为这一历史事件的见证人我深感荣幸,何况宫冈

先生还是我最亲密而尊敬的朋友之一,其喜悦之情又添一分."

在其证明方法中,阿兰基洛夫 — 法尔廷斯(Arake-Faltings)的算术曲面理论起着中心的作用.

要说明什么是算术曲面是很难的,这就是在代数整数环(例如有理整数环 Z)上的代数曲线中,进一步考虑了曲线上的"距离".代数整数环在代数几何中说是一维的(曼宁称数论维数),整体当然是二维(曲面).从图上看,整数环成星状结构,例如在 Z 上就是只是该(数论)曲线在"0"处开着"孔",不具有紧流形那种好的性质.通过引入"距离"将其"紧化"后就是算术曲面.这一理论受韦尔批判的影响,本质上超越了格罗登迪克的概型理论.这回的结果如果正确,那么就是继法尔廷斯证明了莫德尔猜想之后,表明了这一理论在本质上的重要性.

实际上,官冈的理论给出了比莫德尔猜想本身,包括估计解的个数更强的形式,以及更自然的证明,他的结果的最大重要性正在于此.费马定理不过是一个应用例子(的确是个漂亮的应用).法尔廷斯在莫德尔猜想的证明之处展开了算术曲面理论,我们推测他恐怕是指望用后者证明莫德尔猜想.官冈的结果正是实现了法尔廷斯的这一目标.

他的理论包含了重要的新概念,今后必须详细加以研究.这一理论若能确立,将会给不定方程理论领域带来革命性的变化.它把黎曼曲面上的函数论与数论联系了起来,遗憾的是我们代数几何工作者看来似乎很难登台表演.

要对证明作详细介绍实在是无能为力,就按进展的情况来说说大概.首先由莫德尔猜想知道方程 *(当 n 确定时)的解的个数(除整数倍外)是有限多个.帕希恩利用巧妙的手法表明,类似于由官冈自己在 10 年前证明的一般复曲面的 Chern 数的不等式(Bogomolov – 官冈 – Yau 不等式)若在算术曲面成立,那么就可以证明较强形式的莫德尔猜想,进而利用弗雷的椭圆曲线这种特殊的算术曲面,就可以证明费马猜想(对于充分大的 n).

但是,在帕希恩的笔记中成问题的是,若按算术曲面中 Chern 数的定义类似地去做,就很容易作出不等式不成立的反例,一时间就怀疑帕希思的思想是否成立.但是梅热却想出摆脱这点的好方法,官冈进一步推进了这条路线.就是主张引入只依赖于特征 0 上纤维(本质上是有限个黎曼面)的别的不变量,使得利用它不等式就能成立.证明则是重新寻找复曲面的不等式(令人吃惊的是不只定理,甚至连证明方法都非常类似),此刻最大的障碍是没有关于向量丛的阿兰基洛夫 – 法尔廷斯理论,他援引了德利哥尼 – 比斯莫特(Bismut) – 基列斯莱等关于奎伦距离(解析挠理论)这种高度的解析手法的最新成果克服了这一困难(还应注意这一理论与物理的弦模型理论有着深刻的联系).

这一宏大理论的全貌涉及整个数论、几何、分析,它综合了许多人得到的深刻结果,宛如一座 Köln 大教堂. 恐怕可以这样说,宫冈作为这一建筑的明星,他把圆顶中央的最后一块石子镶嵌到了顶棚之上.

但宫冈的推论交叉着如此壮大的一般理论与包含相当技巧的精细讨论,就连要验证都很不容易,对他始终不渝的探索、最终找到这复杂迷宫出口的才能,浪川幸彦钦佩至极. 他出类拔萃的记忆为人称道,有人曾赠他"Walking encyclopedia"(活百科全书)的雅号,并对他灵活运用他那个丰富数据库似的才能惊讶万分.

在 3 月 29 日浪川幸彦的信中又说,此信虽是准备作为发往日本的特讯,但到底还是宣传报道机构的嗅觉灵敏,在此信到达以前日本早已轰动,就像在全世界捅了马蜂窝似的. 而且仅这方面的奇妙报道就不少,为此浪川幸彦想对事情经过作一简短报告,以正视听.

事情的发端是,2 月 26 日在研究所举行的讨论班上宫冈发表了算术曲面中类似的宫冈不等式看来可以证明的想法. 这时的笔记复印件由扎格(D. Zagier)(报纸上有各种读法)送给欧美的一部分专家,引起了振动.

因特网是 IBM 计算机的国际通信线路,可以很方便地与全世界通信联络,这回就是通过它把宫冈的消息迅速传遍数学界的. 因此其震源扎格那里从 3 月上旬起电话就多得吓人,铃声不断.

但是,具有讽刺意味的是 IBM 计算机当时在日本还没有普及,因特网在日本几乎没有使用,因此宫冈的消息除少数人知晓外,在日本还鲜为人知.

正当其时,3 月 9 日 UPI 通讯(合众国际社)以"宫冈解决了费马猜想吗?"为题作了报道,日本包括数学界在内不啻晴天霹雳,上下大为轰动.

但感到震惊的不仅日本,而且波及整个世界,此后宫冈处的电话铃声不绝,他不得不切断电话,暂时中止一切活动.

从效果上看,这一报道是过早了.UPI 电讯稿发出之时,正当宫冈将其想法写成(手写)的第一稿刚刚完成之际. 在数学界,将这种论文草稿(预印本)复印送给若干名专家,得到他们的评论后再确定在专业杂志发表的最终稿,这种做法司空见惯(不少还要按审稿者的要求再作修改). 像费马问题这样的大问题,出现错误的可能性相应的也要大些,因此必须慎之又慎. 在目前阶段还不能说绝对没有最终毫无结果的可能性.宫冈先生面临着巨大的不利条件,在一片吵闹声中送走了很重要的修改时期.

257

正如人们所预料的,实际上第一稿中确实包含了若干不充分之处.

官冈预定3月22日在波恩召开的代数几何研究集会上详细公布其结果.但经过与前一天刚刚从巴黎赶来的梅热反复讨论,到半夜时分就明白了还存在相当深刻的问题.为此次日的讲演就改为仅止于解说性的.

与此前后,还收到了法尔廷斯、德利哥尼等指出的问题(前者提的本质上与马祖尔相同).

后来才清楚,他的主要思想,即具有奎伦距离的讨论是好的(仅此就是独立的优秀成果).但紧接着的算术代数几何部分的讨论有问题,依照那样推导不出莫德尔型的定理.

这段时间大概是官冈最苦恼的时期了.事情已经闹大,退路也没有了. 不过这一周的研究集会中,欧洲各位同行老朋友来此聚会本身就大大搭救了他.大家都充分体会研究的甘苦,所以并不把费马作为直接话题,在无拘束的交谈之中使他重新振奋起了精神.

尽管如此,对于在如此状况下继续进行研究的官冈的顽强精神,浪川幸彦说他只能表示敬服.在大约一周之内,他改变了主要定理的一部分说法,修正了证明的过程,由此出现了克服最大问题点的前景.在浪川写这篇稿时,他已开始订正其他不齐备与错误之处,进行修改稿的完成工作.

因此,虽然一切还都处于未确定的阶段,但很难设想如此漂亮的理论最终会化为乌有,也许还可能修正一部分过程,但即使是官冈先生这种修正过程的技巧也是有定论的.

谷山、志村与官冈洋一以及望月新一的出现并非偶然,有着深刻的历史背景与现实原因,我们有必要探究一番.日本的数学发展较晚,与中国古代的数学成就相比稍显逊色,但交流是存在的.伴随律令制度的建立,中国的实用数学也很早就在日本传播开来.除了天文和历法的需要之外,班田制的实施、复杂的征税活动以及大规模的城市建设,都必须掌握实用的计算、测量技巧.早在7世纪初,来自百济的僧人观勒已经在日本致力于普及中国的算术知识.在大化革新(645)之后,日本仿照中国的学制设立了大学(671).当时算术是大学中的必修科目之一.在大宝元年(701)制定的大宝律令中,明确地把经、音、书、算作为大学的四门学科,在算学科中设有算博士1人、算生30人.在奈良时代(710—793),《周髀》《九章》《孙子》等著名算经已经成为在大学中培养官吏的标准教材.

我们从日本最古老的歌集《万叶集》(759)中可以见到九九口诀的一些习惯用法.例如把81称作"九和",把16称作"四四",这说明九九口诀在奈良时代

已相当流行.①

古代日本和中国一样,也是用算筹进行记数和运算.中国元朝末期发明的珠算,大约在15～16世纪的室町时代传入日本.在日本称算盘为"十露盘"(そツぼツ,Soroban).这个词的语源至今不明,但在1559年出版的一部日语辞典(天草版)中,已经收入了"そろぼ"这个词.除了从中国引进的"十露盘"之外,在日本的和算中还有一种称作"算盘"(さんぼん,Sanban)的计算器具,是在布、厚纸或木盘上画出棋盘状的方格,借助于大约6厘米长的算筹在格中进行运算.这两种不同的计算器具其汉字都可写作"算盘",但是发音不同,含义也不一样.

17世纪,日本人在中国传统数学的基础上创造了具有民族特色的数学体系——和算.和算的创始人是关孝和(1642—1708).

在关孝和以前,日本的数学和天文、历算一样,在很长一个时期(大约9～16世纪)处于裹足不前的状态.16世纪下半叶,织田信长和丰臣秀吉致力于统一全国,当时出于中央集权政治的需要,数学重新受到重视.以此为历史背景,明万历年间程大位所著《算法统宗》(1592)一书,出版不久即传入日本.江户早期的著名数学家毛利重能著《割算书》②(1622)一书,推广了《算法统宗》中采用的珠算法,而他的学生吉田光由(1598—1672)则以《算法统宗》为蓝本著《尘劫记》(1627)一书,用适合于日本人口味的体裁,把中国的实用算术普及到广大民间.

在日本影响较大的另一部算书是元朝朱世杰的《算学启蒙》(1299).此书出版不久即传至朝鲜,而在中国却一度失传,后由朝鲜返传回中国.日本流行的《算学启蒙》一书,据说是根据丰臣秀吉出征朝鲜之际带回的版本复刻而成(1658).

通过《算法统宗》和《算学启蒙》,日本人掌握了中国的算术和代数(即"天元术").关孝和就是在中国传统数学的影响下,青出于蓝而胜于蓝,在代数学中创造性地发展了有文字系数的笔算方法.他的《发微算法》(1674)为和算的发展奠定了基础.

这期间稍后的一位比较著名的数学家是会田安明(Aida Ammei,

---

① 从20世纪敦煌等地出土的木简可知,中国在很古老的时候已经形成了九九口诀.《战国策》中称,有人曾以九九之术赴齐桓公门下请求为士.

② 日文中的"割算"即除法.

1747—1817).会田安明生于山形(Yamagata),卒于江户(现在的东京).15 岁开始从师学习数学,22 岁到江户谋生,曾管理过河道改造和水利工程.业余时间刻苦自学数学,经常参加当时的学术争论.1788 年,他弃去公职,专门从事数学研究和讲学,逐渐扩大了在日本数学界的影响,他所建立的学派称为宅间派.会田安明的工作包括几何、代数、数论等几个方面.他总结了日本传统数学中的各种几何问题,深入研究了椭圆理论,指出怎样决定椭圆、球面、圆、正多边形的有关公式.探讨了代数表达式和方程的构造理论,提出用展开 $x_1^2 + x_2^2 + \cdots + x_n^2 = y^2$ 的方法,求 $k_1 x_1^2 + k_2 x_2^2 + \cdots + k_n x_n^2 = y^2$ 的整数解.利用连分数来讨论近似分数.还编制出以 2 为底的对数表.在他的著作中,大量地使用了新的简化的数学符号.会田安明非常勤奋,一年撰写的论文有五六十篇,一生的著作不少于 2 000 种.

日本数学的复兴是与对数学教育的重视分不开的.

日本从明治时代就非常重视各类学校的数学教育.数学界的元老菊池大麓、藤泽利喜太郎等人曾亲自编写各种数学教科书,在全国推广使用.因此,日本的数学教育在 20 世纪初就已经达到了国际水平.从大正时代开始,著名数学家层出不穷.特别是在纯数学领域,藤泽利喜太郎(东京大学)和他门下的三杰(高木贞治、林鹤一、吉江琢儿)发表了一系列有国际水平的研究成果.其中最著名的是高木贞治(1875—1960)关于群论的研究.在高木门下又出现了末纲恕一、弥永昌吉、正田健二郎三位新秀,他们以东京大学为基地,推动了数学基础理论的研究.

大约与此同时,在新建的东北大学形成了以林鹤一为中心的另一个重要的研究集团,其成员主要有藤原松二郎、洼田忠彦、挂谷宗一等人.日本著名数学教育家、数学史家小仓金之助也是这个集团的重要成员之一.林鹤一在 1911 年创办了日本最早的一个国际性专业数学刊物《东北数学杂志》,使日本的数学成就在世界上享有盛名.

进入 20 世纪 30 年代之后,沿着《东北数学杂志》的传统,在东北大学涌现了淡中忠郎、河田龙夫、角谷静夫、佐佐木重夫、深宫政范、远山启等著名数学家.此外,在大阪大学清水辰次郎(东京大学毕业)周围又形成了一个新兴的研究中心,其主要成员有正田健次郎(抽象代数)、三村征雄(近代解析)、吉田耕作(马尔可夫过程)等人.在东京大学,除了末恕纲一、弥永昌吉在整数论方面的卓越成就之外,更值得注意的是,在弥永昌吉门下出现了许多有才华的数学家,其中有小平邦彦(调和积分论)、河田敬义(整数论)、伊藤清(概率论)、古屋茂(函数方程)、安部亮(位相解析)、岩泽健吉(整数论)等人.到战后,以弥永的学生清水达雄为中心,展开了类似法国布尔巴基学派的新数学运动.

战时京都大学的数学研究似乎比较沉默,但也还是出现了一位引人注目的数学家冈洁.他在1942年发表了关于多复变函数论的研究,于1951年获日本学士院奖.到战后,围绕代数几何学的研究,形成了以秋月良夫为中心的京都学派.

可以看出,日本的纯数学研究从明治时代开始,到20世纪三四十年代,已经形成了一支实力相当雄厚的理论队伍.在战时动员时期,数学作为"象牙塔中的科学"仍然保持其稳步前进的势头,并取得了不少创造性成就.

日本数学与中国数学相比,虽然开始中国数学居于前列,并且从某种意义上充当了老师的角色,但随后日本数学后来居上.两国渐有差距,是什么原因促使这一变化的呢?关键在于对洋算的态度,及对和算的废止.

据华东师大张奠宙教授比较研究指出:

> "1859年,当李善兰翻译《代微积拾级》之时,日本数学还停留在和算时期.日本的和算,源于中国古算,后经关孝和(1642—1708)等大家的发展,和算有许多独到之处.行列式的雏形,可在和算著作中找到.19世纪以来,日本学术界,当然也尊崇本国的和算,对欧美的洋算,采取观望态度.1857年,柳河春三著《洋算用法》,1863年,神田孝平最初在开城所讲授西洋数学,翻译和传播西算的时间均较中国稍晚."

但是明治维新(1868)之后,日本数学发展极快.经过30年,中国竟向日本派遣留学生研习数学,是什么原因导致这一逆转?

日本的数学教育政策起了关键的作用.

这一差距显示了中日两国在科学文化方面的政策有很大不同.抚今追昔,恐怕会有许多经验值得我们吸取.

中国从1872年起,由陈兰彬、容闳等人带领儿童赴美留学,但至今不知有何人学习数学,也不知有何人回国后传播先进的西方算学.数学水平一直停留在李善兰时期的水平上.可是,日本的菊池大麓留学英国,从1877年起任东京大学理学部数学教授,推广西算.特别是1898年,日本的高木贞治远渡重洋,到德国的哥廷根大学(当时的世界数学中心)跟随希尔伯特(当时最负盛名的大数学家)学习代数数论(一门正在兴起的新数学学科),显示了日本向西方数学进军的强烈愿望.高木贞治潜心学习,独立钻研,终于创立了类域论,成为国际上的一流数学家,这是 1920 年的事.可是中国留学生专习数学的竟无一人.熊庆来先生曾提到一件轶事.1916 年,法国著名数学家波莱尔(E. Borel) 来华,曾提及他在巴黎求学时有一位中国同学,名叫康宁,数学学得很好,经查,康宁返国后在京汉铁路上任职,一次喝酒时与某比利时人发生冲突,竟遭枪杀.除此之外,中国到西洋学数学而有所成就者,至今未知一人.

1894 年,甲午战争失败后,中国向日本派遣留学生.1898 年,中日政府签订派遣留学生的决定.中国青年赴日本学数学的渐增,冯祖荀就是其中一位,他生于 1880 年,浙江杭县人,先在日本第一高等学校(高中) 就读,然后进入京都帝国大学学习数学,返国后任北京大学(1912) 数学教授.1918 年成立数学系时为系主任.

当然,尽管日本数学发展迅速,超过中国,但 20 世纪初的日本数学毕竟离欧洲诸国的水平很远,中国向日本学习数学,水平自然更为低下.第二次世界大战之后,随着日本经济实力的膨胀,日本的数学水平也在迅速提升.当今的世界数学发展格局是"俄美继续领先,西欧紧随其后,日本正迎头赶上,中国则还是未知数."中、日两国的数学水平,在 20 世纪 50 年代,曾经相差甚远,但目前又有继续扩大的趋势.

比较一下中日中小学数学教育的发展过程也是有益的.

1868 年,日本开始了"明治维新"的历史时期.明治 5 年,即 1872 年 8 月 3 日,日本颁布学制令.其中第 27 章是关于小学教科书的,在"算术"这一栏中明确规定"九九数位加减乘除唯用洋法".1873 年 4 月,文部省公布第 37 号文,指出"小学教则中算术规定使用洋算,但可兼用日本珠算",同年 5 月的 76 号文则称"算术以洋法为主".

一百多年后的今天,返观这项数学教育决策,确实称得上是明智之举,它对日本数学的发展、教育的振兴,起到了不可估量的作用.

最初在日本造此舆论的当推柳河春三.他在 1857 年出版的《洋算用法》序中说"唯我神州,俗美性慧,冠于万邦,而我技巧让西人者,算术其最也.……故今之时务,以习其术发其蒙,为急之尤急者."

明治以后,1871 年建立文部省.当时的文部大臣是大木乔任,他属改革派中的保守派,本人并不崇尚洋学,可是他愿意推行教育改革,相信"专家"的决策.当时,全国有一个"学制调查委员会",其中的多数人是著名的洋学家.例如,启蒙主义者箕作麟祥(曾在神田孝平处学过洋算),瓜生寅是专门研究美国的(曾写过《测地略》,用过洋算),内田正雄是荷兰学家(曾学过微积分),研究法国法律和教育的河津佑之是著名数学教授之弟,其余的委员全是西医学、西

262

洋法学等学家.在这个班子里,尽管没有一个洋算家,却也没有一个和算家,其偏于洋算的倾向,当然也就可以理解了.

在日本的数学发展过程中,国家的干预起了决定性作用,江户时代发展起来的和算,随着幕末西方近代数学的传入而日趋没落.从和算本身的演变来看,自18世纪松永良弼确立了"关派数学"传统之后,曾涌现出许多有造诣的和算家,使和算的学术水平遥遥领先于天文、历法、博物等传统科学部门.但另一方面,和算脱离科学技术的倾向也日益严重.这是因为和算有两个明显的弱点:第一,和算虽有卓越的归纳推理和机智的直观颖悟能力,却缺乏严密的逻辑证明精神,因而逐渐背离理论思维,陷于趣味性的智能游戏;第二,江户时代的封建制度使和算们的活动带有基尔特(guild)式的秘传特征,不同的流派各自垄断数学的传授,因而使和算陷于保守、僵化,没有能力应付近代数学的挑战.

由于存在上述弱点,和算注定是要走向衰落的.然而这些弱点并不妨碍和算能够在相当长一个时期独善其身地向前发展.事实上,直到明治初期,统治着日本数学的仍然是和算,而不是朝气蓬勃的西方近代数学①.如果没有国家的干预,和算是不会轻易让出自己的领地的.

明治五年(1872),新政府采纳洋学家的意见公布了新学制,其中明令宣布,在一切学校教育中均废止和算,改用洋算,这对和算是个致命的打击.在这之后,再也没有出现新的年轻和算家,老的和算家则意气消沉,不再有所作为.自从荻原信芳写成《圆理算要》(1878)之后,再也没有见到和算的著作问世.1877年创立东京数学会社时,在会员人数中虽然仍是和算家居多,但领导权却把持在中川将行、柳楢悦等海军系统的洋算家手中.这些洋算家抛弃了和算时代数学的秘传性,通过《东京数学会社杂志》把数学研究成果公诸于世.1882年,一位海军教授在《东京数学会社杂志》第52号上发表论文,严厉谴责了和算的迂腐,强调要把数学和当代科学技术结合起来.这是鞭挞和算的一篇檄文,小仓金之助称它为"和算的葬词".

此后不久,以大学出身的菊池大麓为首,在1884年发动了一次"数学政变",把一大批和算家驱逐出东京数学会社,吸收了一批新型的物理学家(如村冈范为驰、山川健次郎等)、天文学家(如寺尾寿等)入会,并把东京数学会社改称为东京数学物理学会.这次大改组,彻底破坏了和算家的阵容,至此结束了和算在日本的历史.

---

① 明治六年时,东京的和算塾102所,洋算塾40所,前者仍居于优势.

但凡一门艰深的学问要在一国扎根,生长点是至关重要的,高木贞治对于日本数论来说是一个高峰也是一个关键人物,是值得大书特书的.

　　高木贞治先生于 1875 年(明治八年)4 月 21 日出生在日本岐阜县巢郡的一色村.他还不满 5 岁就在汉学的私塾里学着朗读《论语》等书籍.童年时期,他还经常跟着母亲去寺庙参拜,时间一长,不知不觉地就能跟随着僧徒们背诵相当长的经文.

　　1880 年(明治十三年)6 月,高木开始进入公立的一色小学读书.因为他的学习成绩优异,不久就开始学习高等小学的科目.1886 年 6 月,年仅 11 岁的高木就考入了岐阜县的寻常中学.在这所中学里,他的英语老师是斋藤秀三郎先生,数学老师是桦正董先生.1891 年 4 月,高木以全校第一名的优异成绩毕业.经过学校的推荐,高木于同年 9 月进入了第三高级中学预科一类班学习.在那里,教他数学的是河合十太郎先生,河合先生对高木以后的发展有着重大的影响.在高中时期,高木的学友有同年级的吉江琢儿和上一年级的林鹤一等.1894 年 7 月,高木在第三高级中学毕业后就考入了东京帝国大学的理科大学数学系.在那里受到了著名数学家菊池大麓和藤泽利喜太郎等人的教导.在三年的讨论班中,高木在藤泽先生的直接指导下作了题为"关于阿贝尔方程"的报告.这篇报告已被收入《藤泽教授讨论班演习录》第二册中(1897).

　　1897 年 7 月,高木大学毕业后就直接考入了研究生院.当时也许是根据藤泽先生的建议,高木在读研究生时一边学习代数学和整数论,一边撰写《新编算术》(1898) 和《新编代数学》(1898).

　　1898 年 8 月,高木作为日本文部省派出的留学生去德国留学 3 年.当时柏林大学数学系的教授有许瓦兹、费舍、弗罗比尼乌斯等人.但许瓦兹、费舍二人因年迈,教学方面缺乏精彩性,而弗罗比尼乌斯当年 49 岁,并且在自己的研究领域(群指标理论)中有较大的突破,在教学方面也充满活力,另外他对学生们的指导也非常热情.当高木遇到某些问题向他请教时,他总是说:"你提出的问题很有趣,请你自己认真思考一下."并借给他和问题有关的各种资料.每当高木回想起这句"请你自己认真思考一下",总觉得是有生以来最重要的教导.

　　从第三高级中学到东京大学一直和高木要好的学友吉江比高木晚一年到德国留学.他于 1899 年夏季到了柏林之后就立即前往哥廷根.高木也于第二年春去了哥廷根.在高木的回忆录文章中记载着:"我于 1900 年到了哥廷根大学,当时在哥廷根大学有克莱茵、希尔伯特二人的讲座,后来又聘请了闵可夫斯基,共有三个专题讲座.使我感到惊奇的是,这里和柏林的情况不大一样,当时在哥廷根大学每周都有一次'谈话会',参加会议的人不仅是从德国,而且是从世界各国的大学选拔出的少壮派数学名家,可以说那里是当时的世界数学的中心.在那里我痛感到,尽管我已经25岁了,但所学的知识要比数学现状落后50年.当时,在学校除了数学系的定编人员之外,还有副教授辛弗利斯(Sinflies)、费希尔(Fischer)、西林格(Sylinger)、我以及讲师策梅罗(Zermelo)、亚伯拉罕(A'braham)等人."

高木从克莱茵那里学到了许多知识,特别是学会了用统一的观点来观察处理数学的各个分支的方法.而作为自己的专业研究方向,高木选择了代数学的整数论.这大概是希尔伯特的《数论报告》对他有很强的吸引力吧!特别是他对于被称之为"克罗内克的青春之梦"的椭圆函数的虚数乘法理论具有很浓的兴趣.在哥廷根时期,高木成功地解决了基础域在高斯数域情况下的一些问题(他回国后作为论文发表,也就是他的学位论文).

1901年9月底,高木离开了哥廷根,并在巴黎、伦敦等地作了短暂的停留之后,于12月初回到了日本,当时年仅26岁零7个月.由于1900年6月,高木还在留学期间就被东京大学聘为副教授,所以他回国后马上就组织了数学第三(科目)讲座,并和藤泽及坂井英太郎等人共同构成了数学系的班底.1903年,高木的学位论文发表后就获得了理学博士学位,并于第二年晋升为教授.

1914年夏季,第一次世界大战爆发后,德国的一些书刊、杂志等无法再进入日本.在此期间,高木只能潜心研究,"高木的类域理论"就是在这一时期诞生的.关于"相对阿贝尔域的类域"这一结果对于高木来说是个意外的研究成果.他曾反复验证这一结果的正确性,并以它为基础去构筑类域理论的壮丽建筑.而且关于"克罗内克的青春之梦"的猜想问题他也作为类域理论的一个应用作出了一般性的解决.并把这一结果整理成133页的长篇德语论文发表在1920年度(大正九年)的《东京帝国大学理科大学纪要》杂志上.同年9月,在斯特拉斯堡(Strasbourg)召开了第6届国际数学家大会.高木参加了这次会议并于9月25日在斯特拉斯堡大学宣读了这结果的摘要.然而,遗憾的是在会场上没有什么反响.这主要是因为第一次世界大战刚刚结束不久,德国的数学家没有被邀请参加这次会议,而当时数论的研究中心又在德国,因此,在参加会议的其他国家的数学家之中,能听懂的甚少.

1922年,高木发表了关于互反律的第二篇论文(前面所述的论文为第一篇论文).他运用自己的类域理论巧妙而又简单地推导出弗厄特万格勒(Futwängler)的互反律,并且对于后来的阿丁一般互反律的产生给出了富有启发性的定式化方法.

1922年,德国的西格尔把高木送来的第一篇论文拿给青年数学家阿丁阅读,阿丁以很大的兴趣读了这篇论文,并且又以更大的兴趣读完了高木的第二篇论文.在此基础上,阿丁于1923年提出了"一般互反律"的猜想,并把高木的论文介绍给汉斯(Hasse).汉斯对这篇论文也产生了强烈的兴趣,并在1925年举行的德国数学家协会年会上介绍了高木的研究成果.汉斯在第二年经过自己的整理后,把附有详细证明的报告发表在德国数学家协会的年刊上,从而向全世界的数学界人士介绍了高木的类域理论.另一方面,阿丁也于1927年完成了一般互反律的证明.这是对高木理论的最重要的补充.至此,高木–阿丁的类域理论完成了.

从此以后,高木的业绩开始在国际上享有盛誉.1929年(昭和四年),挪威的奥斯陆大学授予高木名誉博士称号.1932年在瑞士北部的苏黎世举行的国际数学家大会上,高木当选为副会长,并当选为由这次会议确定的菲尔兹奖评选委员会委员.

在国内,高木于1923年(大正十一年)6月当选为学术委员会委员.1925年6月,又当选为帝国学士院委员等职.1936年(昭和十一年)3月,他在东京大学离职退休.1940年秋季,在日本第二次授勋大会上荣获文化勋章.1951年获全日本"文化劳动者"称号.1955年在东京和日光举行的国际代数整数论研讨会上,高木当选为名誉会长.1960年2月28日,84岁零10个月的高木贞治先生因患脑出血和脑软化的合并症不幸逝世.

高木贞治先生用外文写的论文共有26篇,全部收集在《The Collected Papers of Teiji Takagi》(岩波书店,1973)中.他的著作除了前面提到的《新编算术》《新编代数学》以及《新式算术讲义》之外,还有《代数学讲义》(1920)、《初等整数论讲义》(1931)、《数学杂谈》(1935)、《过渡时期的数学》(1935)、《解析概论》(1938)、《近代数学史谈》(1942)、《数学小景》(1943)、《代数整数论》(1948)、《数学的自由性》(1949)、《数的概念》(1949)等.另外,高木先生还撰写了多册有关学校教育方面的教科书.

高木与菊池、藤泽等著名数学家完全不同,他从来不参加社会活动或政治活动,就连大学的校长、系主任或什么评议委员之类的工作也一次没有做过,而是作为一名纯粹的学者渡过了自己的一生.从高木的第一部著作《新编算术》到他的后期作品《数的概念》可以看出他对数学基础教育的关心.他的《解析概论》一书被长期、广泛地使用,使得日本的一般数学的素养得到了显著的提高.许多青年读了他的《近代数学史谈》之后都决心潜心研究数学,作出成果.在日本的数学家中,有许多人不仅受到了他独自开创整数论精神的鼓舞,而且还受到了他的这些著作的恩惠.在日本,得到高木先生直接指导的数学家有末纲恕

一、正田建次郎、管原正夫、荒又秀夫、黑田成腾、三村征雄、弥永昌吉、守屋美贺雄、中山正等人.

可以说在日本数学界的最近一百年的时间里,首先做出世界性业绩的是菊池先生,其次就是藤泽先生,第三位就是高木先生①.

宫冈洋一关于费马定理的证明尽管有漏洞,但他的证明的整体规模宏大、旁征博引,具有非凡的知识广度及娴熟的代数几何技巧. 望月新一关于 ABC 猜想的证明尽管未完全被承认,但他创造的宇宙际几何令人望洋兴叹,这一切都给人留下了深刻印象. 有人说:"一夜可以挣出一个暴发户,但培养一个贵族至少需要几十年." 宫冈洋一和望月新一引起的轰动决非偶然,它与日本数学的深厚积淀与悠久的代数几何传统息息相关. 提到日本的代数几何人们自然会想到三巨头 —— 小平邦彦、广中平佑、森重文. 而日本的代数几何又直接得益于美国的查里斯基,所以必须先讲讲他们的老师查里斯基. 伯克霍夫说:"今天任何一位在代数几何方面想作严肃研究的人,将会把查里斯基和塞缪尔(P. Samuel) 写的交换代数的两卷专著当作标准的预备知识."

查里斯基是俄裔美籍数学家. 1899 年 4 月 24 日生于俄国的科布林. 由于他在代数几何上的突出成就,1981 年荣获沃尔夫数学奖,时年 82 岁.

查里斯基 1913 ~ 1920 年就读于基辅大学. 1921 年赴罗马大学深造. 1924 年获罗马大学博士学位. 1925 ~ 1927 年接受国际教育委员会资助作为研究生继续在意大利研究数学. 1927 年到美国霍普金斯大学任教,1932 年被升为教授. 1936 年加入美国国籍. 1945 年访问巴西圣保罗. 1946 ~ 1947 年他是伊利诺易大学的研究教授. 1947 ~ 1969 年他是哈佛大学教授. 1969 年成为哈佛大学的名誉教授. 查里斯基 1943 年当选为美国国家科学院院士. 1951 年被选为美国哲学学会会员. 1965 年荣获由美国总统亲自颁发的美国国家科学奖章.

查里斯基对代数几何做出了重大贡献. 代数几何是现代数学的一个重要分支学科,与数学的许多分支学科有着广泛的联系,它研究关于高维空间中由若干个代数方程的公共零点所确定的点集,以及这些点集通过一定的构造方式导出的对象即代数簇. 从观点上说,它是多变量代数函数域的几何理论,也与从一般复流形来刻画代数簇有关. 进而它通过自守函数、不定方程等和数论紧密地结合起来. 从方法上说,则和交换环论及同调代数有着密切的联系.

查里斯基早年在基辅大学学习时,对代数和数论很感兴趣,在意大利深造

---

① 《理科数学》(日本科学史会编) 第一法规(1969) 第 7 章"高本 の 类体论".

期间,他深受意大利代数几何学派的三位数学家卡斯泰尔诺沃(G. Castelnuovo,1865—1952)、恩里克斯(F. Enriques,1871—1946)、塞维里(Severi,1879—1961)在古典代数几何领域的深刻影响.意大利几何学者们的研究方法本质上很富有"综合性",他们几乎只是根据几何直观和论据,因而他们的证明中往往缺少数学上的严密性.查里斯基的研究明显带有代数的倾向,他的博士论文就与纯代数学有密切联系,精确地说是与伽罗瓦理论有密切联系.他的博士论文主要是把所有形如 $f(x) + t \cdot g(x) = 0$ 的方程分类,这里 f 和 g 是多项式,x 可以解为线性参数 t 的根式表达式.查里斯基说明这种方程可分为 5 类,它们是三角或椭圆方程.取得博士学位后,他在罗马的研究工作仍然主要是与伽罗瓦理论有密切联系的代数几何问题.到美国后,他受莱夫谢茨(S. Lefschetz)的影响,致力于研究代数几何的拓扑问题.1927 ~ 1937 年间,查里斯基给出了关于曲线 C 的经典的黎曼 — 罗赫定理的拓扑证明,在这个证明中他引进了曲线 C 的 n 重对称积 $C(n)$ 来研究 C 上度数为 n 的除子的线性系统.

1937 年,查里斯基的研究发生了重要的变化,其特点是变得更代数化了.他所使用的研究方法和他所研究的问题都更具有代数的味道(这些问题当然仍带有代数几何的根源和背景).查里斯基对意大利几何学者的证明感到不满意,他确信几何学的全部结构可以用纯代数的方法重新建立.在 1935 年左右,现代化数学已经开始兴盛起来,最典型的例子是诺特与范·德·瓦尔登有关论著的发表.实际上代数几何的问题也就是交换环的理想的问题.范·德·瓦尔登从这个观点出发把代数几何抽象化,但是只取得了一部分成就,而查里斯基却获得了巨大成功.在 20 世纪 30 年代,查里斯基把克鲁尔(W. Krull)的广义赋值论应用到代数几何,特别是双有理变换上,他从这方面来奠定代数几何的基础,并且做出了实质性的贡献.查里斯基和其他的数学家在这方面的工作,大大扩展了代数几何的领域.

查里斯基对极小模型理论也做出了贡献.他在古典代数几何的曲面理论方面的重要成果之一,是曲面的极小模型的存在定理(1958).它给出了曲面的情况下代数 — 几何间的等价性.这就是说,代数函数域一经给定,就存在非奇异曲面(极小模型)作为其对应的"好的模型",而且射影直线如果不带有参数就是唯一正确的.因此要进行曲面分类,可考虑极小模型,这成了曲面分类理论的基础.

查里斯基的工作为代数几何学打下了坚实的基础.他不但对于现代代数几何的贡献极大,而且在美国哈佛大学培养起了一代新人,哈佛大学以他为中心形成了一个代数几何学的研究集体.1970 年度的菲尔兹奖获得者广中平佑

（Hironaka Heisuke，1931—　）和1974年度的菲尔兹奖获得者曼福德都出自他的这个研究集体．从某种意义上讲，广中平佑的工作可以说是直接继承和发展了查里斯基的成果．

查里斯基的主要论文有90多篇，收集在《查里斯基文集》中，共四卷．查里斯基的代表作有《交换代数》（共两卷，与P·塞缪尔合著，1958～1960）、《代数曲面》（1971）、《拓扑学》等．

查里斯基的关于代数簇的四篇论文于1944年荣获由美国数学会颁发的科尔代数奖．由于他在代数几何方面的成就，特别是在这个领域的代数基础方面的奠基性贡献，使他荣获美国数学会1981年颁发的斯蒂尔奖．他对日本代数几何的贡献是培养了几位大师，第一位贡献突出者是日本的小平邦彦．

小平邦彦（Kunihiko Kodaira，1915—1997）是第一个获菲尔兹奖的日本数学家，也是日本代数几何的推动者．

小平邦彦，1915年3月16日出生于东京．他小时候对数就显示出特别的兴趣，总爱反复数豆子玩．中学二年级以后，他对平面几何非常感兴趣，特别对那些需要添加辅助线来解答的问题十分着迷，以致老师说他是"辅助线的爱好者"．从中学三年级起，他就和一位同班同学一起，花了半年时间，把中学的数学课全部自修完毕，并把习题从头到尾演算了一遍．学完中学数学，他心里还是痒痒的，进行更深层次地学习．看见图书馆的《高等微积分学》厚厚一大本，想必很难，没敢问津，于是从书店买了两本《代数学》，因为代数在中学还是听说过的，虽然这两本1 300页的大书里还包含现在大学才讲的伽罗瓦理论，可是他啃起来却津津有味．

虽然他把主要精力放在数学上，却不知道世界上还有专门搞数学这一行的人，他只想将来当个工程师．于是他考相当于专科的高等学校时，就选了理科，为升大学做准备．理科的学校重视数学和外文，更促使他努力学习数学．他连当时刚出版的抽象代数学第一本著作范·德·瓦尔登的《近世代数学》都买来看．从小接受当时最新的思想对他以后的成长很有好处，在老师的指引下，他走上了数学的道路．

他于1932年考入第一高等学校理科学习．1935年考入东京大学理学院数学系学习．1938年在数学系毕业后，又到该校物理系学习三年，1941年毕业．1941年任东京文理科大学副教授．1949年获理学博士学位，同年赴美国在普林斯顿高等研究所工作．1955年任普林斯顿大学教授．此后，历任约翰大学、霍普金斯大学、哈佛大学、斯坦福大学的教授．1967年回到日本任东京大学教授．1954年荣获菲尔兹奖．1965年当选为日本学士会员．1975年任学习院大学教授. 他还被选为美国国家科学院和哥廷根科学院国外院士.

小平邦彦在大学二年级时,就写了一篇关于抽象代数学方面的论文,大学三年级时他醉心于拓扑学,不久写出了拓扑学方面的论文.1938 年他从数学系毕业后,又到物理系学习,物理系的数学色彩很浓,他主要是搞数学物理学,这对他真是如鱼得水.他读了冯·诺依曼(Von Neumann)的《量子力学的数学基础》,范·德·瓦尔登的《群论和量子力学》以及外尔的《空间、时间与物质》等书后,深刻认识到数学和物理学之间的密切联系.当时日本正出现研究泛函分析的热潮,他积极参加到这一门学科的研究中去,于 1937 ~ 1940 年大学学习期间共撰写了 8 篇数学论文.

正当小平邦彦踌躇满志,准备在数学上大展宏图的时候,战争爆发了.日本偷袭珍珠港,揭开了太平洋战争的序幕.日本与美国成了敌对国,大批日本在美人员被遣返.这当中有著名数学家角谷静夫.角谷在普林斯顿高等研究院工作时曾提出一些问题,这时小平邦彦马上想到可以用自己以前的结果来加以解决,他们一道进行研究,最终解决了一些问题.

随着日本在军事上的逐步失利,美军对日本的轰炸越来越猛烈,东京开始疏散.小平邦彦在 1944 年撤到乡间,可是乡下的粮食供应比东京还困难,他经历的那几年缺吃挨饿的凄惨生活,使他长期难以忘怀.但是,在这种艰苦环境下,他的研究工作不但没有松懈,反而有了新的起色.这时,他开始研究外尔战前的工作,并且有所创新.在战争环境中,他在一没有交流,二没有国外杂志的情况下,独立地完成了有关调和积分的三篇文章,这是他去美国之前最重要的工作,也是使他获得东京大学博士的论文的基础.但是直到 1949 年去美国之前,他在国际数学界还是默默无闻的.

战后的日本处在美国军队的占领之下,学术方面的交流仍然很少.角谷静夫在美国占领军当中有个老相识,于是托他把小平邦彦的关于调和积分的论文带到美国.1948 年 3 月,这篇文章到了《数学纪事》的编辑部,并被编辑们送到外尔的桌子上.

在这篇文章中小平对多变量正则函数的调和性质的关系给出极好的结果.著名数学家外尔看到后大加赞赏,称之为"伟大的工作".于是,外尔正式邀请小平邦彦到普林斯顿高等研究院来.

从 1933 年普林斯顿高等研究院成立之日起,聘请过许多著名数学家、物理学家.第二次世界大战之后,几乎每位重要的数学家都在普林斯顿待过一段.对于小平邦彦来讲,这不能不说是一种特殊的荣誉与极好的机会,他正是在这个优越的环境中迅速取得非凡成就的.

在外尔等人鼓励下,他以只争朝夕的精神,刻苦努力地研究,5年之间发表了20多篇高水平的论文,获得了许多重要结果.其中引人注目的结果之一是他将古典的单变量代数函数论的中心结果,代数几何的一条中心定理:黎曼——罗赫定理,由曲线推广到曲面.黎曼——罗赫定理是黎曼曲面理论的基本定理,概括地说,它是研究在闭黎曼曲面上有多少线性无关的亚纯函数(在给定的零点和极点上,其重数满足一定条件).所谓闭黎曼曲面,就是紧的一维复流形.在拓扑上,它相当于球面上连接了若干个柄.柄的个数 g 是曲面的拓扑不变量,称为亏格.黎曼——罗赫定理可以表述为,对任意给定的除子 D,在闭黎曼曲面 M 上存在多少个线性无关的亚纯函数 f,使 f 的除子 (f) 满足 $(f) \geq D$.如果把这样的线性无关的亚纯函数的个数记作 $l(D)$,同时记 $i(D)$ 为 M 上线性无关的亚纯微分 ω 的个数,它们满足 $(\omega) - D \leq 0$.那么,黎曼——罗赫定理就可表述为: $l(D) - i(D) = d(D) - g + 1$. $d(D) = \sum n_i$ 称为除子的阶数.由于这个定理将复结构与拓扑结构沟通起来的深刻性,如何推广这一定理到高维的紧复流形自然成为数学家们长期追求的目标.小平邦彦经过潜心研究,用调和积分理论将黎曼——罗赫定理由曲线推广到曲面.不久德国数学家希策布鲁赫(F. E. P. Hirzebruch)又用层的语言和拓扑成果把它成功地推广到高维复流形上.

小平邦彦对复流形进行了卓有成效的研究.复流形是这样的拓扑空间,其每点的局部可看作和 C^n 中的开集相同.几何上最常见而简单的复流形是被称为紧凯勒流形的一类.紧凯勒流形的几何和拓扑性质一直是数学家们关注的一个重要问题,特别是利用它的几何性质(由曲率表征)来获取其拓扑信息(由同调群表征).小平邦彦经过深入的研究得到了这方面的基本结果,即所谓小平消灭定理.例如,其中一个典型结果是,对紧凯勒流形 M,如果其凯勒度量下的里奇曲率为正,则对任何正整数 q,都有 $H^{(0,q)}(M,C)=0$,这里 $H^{(0,q)}(M,C)$ 是 M 上取值于 $(0,q)$ 形式芽层的上同调群.小平邦彦还得到所谓小平嵌入定理:紧复流形如果具有一正的线丛,那么它就可以嵌入复射影空间而成为代数流形,即由有限个多项式零点所组成.小平嵌入定理是关于紧复流形的一个重要结果.

由于小平邦彦的上述出色成就,1954年他荣获了菲尔兹奖.在颁奖大会上,著名数学家外尔对小平邦彦和另一位获奖者 J. P. 塞尔给予了高度评价,他说:"所达到的高度是自己未曾梦想到的.""自己从未见过这样的明星在数学天空中灿烂地升起.""数学界为你们所做的工作感到骄傲,它表明数学这棵长满节瘤的老树仍然充满着勃勃生机.你们是怎样开始的,就怎样继续吧!"

小平邦彦获得菲尔兹奖之后,各种荣誉接踵而来.1957年他获得日本学士院的奖赏,同年获得文化勋章,这是日本表彰科学技术、文化艺术等方面的最高荣誉.小平邦彦是继高木贞治之后第二位获文化勋章的数学家.

有的数学家在获得荣誉之后,往往开始走下坡路,再也作不出出色的工作了.对于小平邦彦这样年过40的人,似乎也难再有数学创造的黄金时代了.可是,小平邦彦并非如此,40岁后十几年间,他又写出30多篇论文,篇幅占他三卷文集的一半以上,而且开拓了两个重要的新领域.1956年起,小平邦彦同斯宾塞研究复结构的变形理论,建立起一套系统理论,在代数几何学、复解析几何学乃至理论物理学方面都有重要应用.60年代他转向另一个大领域:紧致复解析曲面的结构和分类.自从黎曼对代数曲线进行分类以后,意大利数学家对于代数曲面进行过研究,但是证明不完全严格.小平邦彦利用新的拓扑、代数工具,对曲面进行分类,他先用某个不变量把曲面分为有理曲面、椭圆曲面、$K3$ 曲面等等,然后再加以细致分类.这个不变量后来被日本新一代的代数几何学家称为小平维数.对于每种曲面,他都建立一个所谓极小模型,而同类曲面都能由极小曲面经过重复应用二次变换而得到.于是,他把分类归结为极小曲面的分类.

他彻底弄清了椭圆曲面的分类和性质.1960年,他得出每个一维贝蒂数为偶数的曲面都是一个代数曲面的变形.1968年,他得到当且仅当 S 不是直纹曲面时,S 具有极小模型.可以说,在代数曲面的现代化过程中,小平邦彦是最有贡献的数学家之一.对于解析纤维丛的分类只能对于某些限定的空间,也是由小平邦彦等人得出的.小平邦彦这些成就,有力地推动了20世纪60年代以来代数几何学和复流形等分支的发展.从1966年起,几乎每一届菲尔兹奖获得者都有因代数几何学的工作而获奖的.

在微分算子理论中,由小平邦彦和梯奇马什(Titchmarsh)给出了密度矩阵的具体公式而完成了外尔—斯通—小平—梯奇马什理论.

小平邦彦对数学有不少精辟的见解.他认为:"数学乃是按照严密的逻辑而构成的清晰明确的学问."他说:"数学被广泛应用于物理学、天文学等自然科学,简直起了难以想象的作用,而且有许多情况说明,自然科学理论中需要的数学远在发现该理论以前就由数学家预先准备好了,这是难以想象的现象.""看到数学在自然科学中起着如此难以想象的作用,自然想到在自然界的背后确确实实存在着数学现象的世界.物理学是研究自然现象的学问.同样,数学则是研究数学现象的学问.""数学就是研究自然现象中数学现象的科学.因此,理解数学就要'观察'数学现象.这里说的'观察',不是用眼睛去看,而是根据某种感觉去体会.这种感觉虽然有些难以言传,但显然是不同于逻辑推理能力之类的纯粹感觉,我认为更接近于视觉,也可称之为直觉.为了强调纯粹是感觉,不妨称此感觉为'数觉'……要理解数学,不靠数觉便一事无成.没有数觉的人不懂数学就像五音不全的人不懂音乐一样.数学家自己并不觉得例如在证明定理时主要是具备了数觉,所以就认为是逻辑上作了严密的证明,实际并非如此,如果把证明全部形式逻辑记号写下看看就明白了……谈及数学的感受,而作为数学感受基础的感觉,可以说就是数觉.数学家因为有敏锐的数觉,自己反倒不觉得了."对于数学定理,他说:"数学现象与物理现象同样是无可争辩实际存在的,这明确表现在当数学家证明新定理时,不是说'发明了'定理,而是说'发现了'定理.我也证明过一些新定理,但绝不是觉得是自己想出来的.只不过感到偶尔被我发现了早就存在的定理.""数学的证明不只是论证,还有思考实验的意思.所谓理解证明,也不是确认论证中没有错误,而是自己尝试重新修改思考实验.理解也可以说是自身的体验."对于公理系统他认为:"现代数学的理论体系,一般是从公理系出发,依次证明定理.公理系仅仅是假定,只要不包含矛盾,怎么都行.数学家当然具有选取任何公理系的自由.但在实际上,公理系如果不能以丰富的理论体系为出发点,便毫无用处.公理系不仅是无矛盾的,而且必须是丰富的.考虑到这点,公理系的选择自由是非常有限的……发现丰富的公理系是极其困难的."

关于数学的本质,他说:"数学虽说是人类精神的自由创造物,但绝不是人们随意杜撰出来的,数学乃是研究和描述实际存在的数学现象……数学是自然科学的背景.""为了研究数学现象,从开始起唯一明显的困难就是,首先必须对数学的主要领域有个全面的、大概的了解……为此就得花费大量的时间.没有能够写出数学的现代史我想也是由于同样的理由."

日本代数几何的第二位代表人物是广中平佑.

广中平佑是继小平邦彦之后日本的第二位菲尔兹奖获得者.他的工作主要是 1963 年发表的 218 页的长篇论文 "Resolution of singu - larities of an algebraic variety over a field of characteristic zero",在这篇论文中他圆满地解决了复代数簇的奇点解消问题.

1931 年广中平佑出生于日本山口县.当时正是日本对我国开始进行大规模侵略之际.他在小学受了 6 年军国主义教育,上中学时就赶上日本逐步走向失败的时候.当时,国民生活十分艰苦,又要经常躲空袭,因此他得不到正规学习的机会.中学二年级就进了工厂,幸好他还没到服兵役的年龄,否则就要被派到前线去充当炮灰.战争结束以后,他才上高中.他在 1950 年考入京都大学时,日本开始恢复同欧、美数学家的接触,大量新知识涌进日本.许多学者传抄

1946 年出版的韦尔的名著《代数几何学基础》,并组织讨论班进行学习,为日本后来代数几何学的兴旺发达打下了基础.1953 年,布尔巴基学派著名人物薛华荔到达日本,对日本数学界有直接影响.薛华荔介绍了 1950 年出版的施瓦兹的著作《广义函数论》.还没有毕业的广中立即学习了他的讲义,并写论文加以介绍.当时京都大学的老师学生都以非凡的热情来学习,这对广中有极大的鼓舞.他对数学如饥似渴的追求,使他早在 1954 年就开始自学代数几何学这门艰深的学科了.1954 年,他从京都大学毕业之后进入研究院,当时秋月康夫教授正组织年轻人攻克代数几何学.在这个集体中,后来培养出了井草准一、松阪辉久、永田雅宜、中野茂男、中井喜和等有国际声望的代数几何学专家,他们都是从那时开始他们的创造性活动的.在这种环境之中,早就以理解力和独创性出类拔萃的广中平佑更是如鱼得水,很迅速地成长起来.1955 年,在东京召开了第一次国际会议,代数几何学权威韦尔以及塞尔等人都顺便访问了京都.1956 年,前面提到的代数几何学权威查里斯基到日本,做了 14 次报告.这些大数学家的光临对于年轻的广中平佑来说真是难得的学习机会.他开始接触当时代数几何学最尖端的课题(比如双有理变换的理论),这对他的一生有决定性的影响,因为广中的工作可以说是直接继承和发展查里斯基的成果的.

广中平佑在家里是老大,下面弟妹不少,他在念研究生时,还不得不花费许多时间当家庭教师,干些零活挣钱养家糊口.尽管如此,他学习得仍旧很出色.

1957 年夏天在赤仓召开的日本代数几何学会议上,他表现十分活跃,他的演讲也得到大会一致好评.由于他的成绩突出,不久,他得以到美国哈佛大学学习,从此他同哈佛大学结下了不解之缘.当时代数几何学正进入一个突飞猛进的时期.第二次世界大战之后,查里斯基和韦尔已经给代数几何学打下了坚实的基础.10 年之后,塞尔又进一步发展了代数几何学.1964 年,格罗登迪克大大地推广了代数簇的概念,建立了一个庞大的体系,在代数几何学中引入了一场新革命.哈佛大学以查里斯基为中心形成了一个代数几何学的研究集体,几乎每年都请格罗登迪克来讲演,而听课的人当中就有后来代数几何学的新一代的代表人物——广中平佑、曼福德、小阿廷等人.在这样一个富有激励性的优越环境中,新的一代茁壮成长.1959 年,广中平佑取得博士学位,同年与一位日本留学生结婚.

这时,广中平佑处在世界代数几何学的中心,并没有被五光十色的新概念所压倒,他掌握新东西,但是不忘解决根本的问题.他要解决的是奇点解消问题,这已经是非常古老的问题了.

所谓代数簇是一个或一组代数方程的零点.一维代数簇就是代数曲线,二维代数簇就是代数曲面. 拿代数曲线来讲, 它上面的点一般来说大多数是常点, 个别的是奇点. 比如有的曲线(如双纽线) 自己与自己相交, 那么在这一交点处, 曲线就有两条不相同的切线, 这样的点就是普通的奇点; 有时, 这两条(甚至多条) 切线重合在一起(比如尖点), 表面上看起来好像同常点一样也只有一条切线, 而实际上是两条切线(或多条切线) 重合而成(好像代数方程的重根), 这样的点称为二重点(或多重点). 对于代数曲面来说, 奇点就更为复杂了. 奇点解消问题, 顾名思义就是把奇点分解或消去, 也就是说通过坐标变换的方法把奇点消去或者变成只有最简单的奇点. 这个问题的研究已有上百年的历史了. 而坐标变换当然是我们比较熟悉的尽可能简单的变换, 如多项式变换或有理式变换. 而行之有效最简单的变换是二次变换和双有理变换, 这一变换最早是由一位法国数学家提出的, 他名叫戎基埃尔(Jonguiéres, Ernest Jean Philippe Fauque de, 1820—1901), 生于法国卡庞特拉(Carpentras), 卒于格拉斯(Grasse) 附近. 1835 年进入布雷斯特(Brest) 海军学院学习, 毕业后, 在海军中服役达 36 年之久, 军衔至海军中将. 戎基埃尔在几何、代数、数论等几方面均有贡献, 而以几何学的成就最大. 他运用射影几何的方法研究初等几何, 探讨了当时流行的平面曲线、曲线束、代数曲线、代数曲面问题, 推广了曲线的射影生成理论, 发现了所谓双有理变换. 这种变换在非齐次坐标下有形式 $x' = x, y' = \dfrac{\alpha y + \beta}{\gamma y + \delta}$, 其中, α, β, γ 是 x 的函数, 且 $\alpha\delta - \beta\gamma \neq 0$. 1862 年, 戎基埃尔关于4 阶平面曲线的工作获得巴黎科学院奖金的三分之二. 1884 年, 他被选为法兰西研究院成员. 很早就已经证明, 代数曲线的奇点可以通过双有理变换予以解消. 从 19 世纪末起, 许多数学家就研究代数曲面的奇点解消问题, 但是论述都不能算很严格. 问题是通过变换以后, 某个奇点消去了, 是否还会有新奇点又生出来呢? 一直到 20 世纪 30 年代, 沃克和查里斯基才完全解决这个问题. 不久之后, 查里斯基于 1944 年用严格的代数方法解决了三维代数簇问题. 高维的情况就更加复杂了. 广中平佑运用许多新工具, 细致地分析了各种情况, 最后用多步归纳法才最终完全解决这个问题. 这简直是一项巨大的工程. 它不仅意味着一个问题圆满解决, 而且有着多方面的应用. 他在解决这个问题之后, 进一步把结果向一般的复流形推广, 对于一般奇点理论也做出了很重要的贡献.

广中平佑是一位精力非常充沛的人, 他的讲话充满了活力, 控制着整个讲堂. 他和学生的关系也很好, 每年总有几个博士出自他的门下. 在哈佛大学, 查里斯基退休之后, 他和曼福德仍然保持着哈佛大学代数几何学的光荣传统, 并推动其他数学学科向前迅速发展.

广中平佑 1975 年由日本政府颁授文化勋章(360 万日币终身年俸).

继广中平佑之后,将日本代数几何传统发扬光大的是森重文(Mori, shigefumi,1951—   ).森重文是日本名古屋大学理学部教授,他先是在 1988 年与东京大学理学部的川又雄二郎一起以"代数簇的极小模型理论"的出色工作获当年日本数学学会秋季奖.他们的工作属于 3 维以上代数几何.

代数簇是由多项式方程所定义的空间.它们的维数是标记一个点(的复数)的参数数目.曲线(在复数集合上的维数为 1,因而在实数上的维数为 2)的一个分类由亏格"g"给出,即由"孔穴"的数目来决定,这从 19 世纪以来已为人们所知.对一簇已知亏格的曲线的详细研究,是曼福德的主要工作,这使他于 1974 年获菲尔兹奖,同样的工作,使德利哥尼于 1978 年,法尔廷斯于 1986 年荣膺桂冠.他们把所开创并由格罗登迪克加以发展了的经典语言作了履行.一个曲面(复数上为 2 维,或者实数上为 4 维,因此很难描绘)的分类在 20 世纪初为意大利学派所尝试,他们的一些论证,被认为不太严格(这再次与上文所论情况相同),后被查里斯基及再后的小平邦彦重作并完成其结果.森重文的理论是非常广泛的,然而目前只限于 3 维范围.古典的工具是微分形式的纤维和流形上的曲线.森重文发现了另外一些变换,它们正好只存在于至少 3 维的情形,被称为"filp",更新了广中平佑对奇点的研究.

日本数学会理事长伊藤清三对上述获奖工作作了很通俗的评论:

"森重文、川又雄二郎两位最近在 3 维以上的高维代数几何学中,取得了世界领先的卓越成果,为高维代数几何今后的发展打下了基础.

这就是决定代数簇上正的 1 循环(one-cycle)构成的锥(cone)的形状的锥体定理;表示在一定的条件下在完备线性系中没有基点的无基点定理(base point free theorem);完全决定 3 维时关于收缩映射的基本形状的收缩定理;递变换的公式化与存在证明 —— 根据森、川又两位关于上述的各项基本研究,在 1987 年终于由森氏证明了,不是单有理的 3 维代数簇的极小模型存在.

这样,利用高维极小模型具有的漂亮性质与存在定理,一般高维代数簇的几何构造的基础也正在逐渐明了,可以期待对今后高维几何的世界性发展将做出显著的贡献.

森、川又两位的研究尽管互相独立,但在结果方面两者互相补充,从而取得了如此显著的成果,我认为授予日本数学会奖秋季奖是再合适不过的."

为了更多地了解森重文的工作,我们节选日本数学家饭高茂的通俗介绍.于此森重文工作可略见一斑.首先饭高茂指出:极小模型理论被选为日本数学会奖的对象,对于最近仍然发展显著的代数几何来说,是很光荣的,实在欣喜至极.

他先从双有理变换谈起:

代数几何学的起源是关于平面代数曲线的讨论,因此经常出现

$$x_1 = P(x,y), y_1 = Q(x,y)$$

型的变换,P, Q 是两变量的有理式.反过来若按两个有理式来解就成了二变量双有理变换的一个例子,特别地称为克雷莫纳(Cremona)变换.这是平面曲线论中最基本的变换.在双有理变换中,值不确定的点很多,这时可认为多个点对应于一个点.克雷莫纳变换若将线性情形除外,则在射影平面上一定存在没有定义的点,而以适当的有理曲线与该点对应.但是,当取平面曲线 C,按克雷莫纳变换 T 进行变换得到曲线 B 时,若取 C 与 B 的完备非奇异模型,则它们之间诱导的双有理变换就为处处都有定义的变换,即双正则变换.于是就成为作为代数簇的同构对应.

这样,由于 1 维时完备非奇异模型上双有理变换为同构,一切就简单了.但是即使在处理曲线时,只说非奇异的也不行.像有理函数、有理变换及双有理变换等都不是集合论中说的映射.因此里德(M. Reid)说道:"奉劝那些对于考虑值不唯一确定的对象感到难以接受的人立即放弃代数几何."

但 1 到 2 维,即使是完备非奇异模型,也会出现双有理变换却不是正则的情形,这就需要极小模型.查里斯基教授向日本年轻数学家说明极小模型的重要性时是 1956 年.查里斯基这一年在东京与京都举行了极小模型讲座,讲义已由日本数学会出版,讲义中对意大利学派的代数曲面极小模型理论被推广到特征为正的情形进行了说明.

查里斯基在远东讲授极小模型时,是否就已经预感到高维极小模型理论将在日本昌盛,并建立起巨大的理论呢?

适逢其时,他与年轻的广中平佑相遇,并促成广中到哈佛大学留学.以广中在该校的博士论文为基础,诞生了关于代数簇的正代数 1 循环构成的锥体的理论.广中建立的奇异点分解理论显然极为重要,是高维代数几何获得惊人发展的基础.

那么森重文的工作又该如何评价呢?

哈茨霍恩(Hartshorne) 的一个猜想说:具有丰富切丛的代数簇只有射影空间.森重文在肯定地解决该猜想上取得了成功,他在证明的过程中证明了:K 若不是 *nef*,则它与曲线的交恒为非负.若 K 是 *nef*,则 S 为极小模型.

已证明了一定存在有理曲线,并且存在特殊的有理曲线,而且重新对偶地抓住曲面时第一种例外曲线的本质,推广到高维,确立端射线的概念,从而明确把握了代数簇的正的 1 循环构成的锥体的构造,在非奇异的场合得到了锥体定理.以此为基础对 3 维时的收缩映射(contraction) 进行分类,所谓的森理论即由此诞生.它有效地给出了具体研究双有理变换的手段,确实成果卓著.

极小模型的存在一经确立,马上得到如下有趣的结果.

(1) 小平维数为负的 3 维簇是单直纹的

其逆显然,得到相当简明的结论,即 3 维单直纹性可用小平维数等于 $-\infty$ 来刻画,可以说这是 2 维时恩里克斯单直纹曲面判定法的 3 维版本,该判定法说,若 12 亏格是 0,则为直纹曲面.若按恩里克斯判定法,就立即得出下面耐人寻味的结果:直纹曲面经有理变换得到的曲面还是直纹曲面.但遗憾的是在 3 维版本中这样的应用不能进行.若不进一步进行单直纹簇的研究,恐怕就不能得到相当于代数曲面分类理论的深刻结果.

(2)3 维一般型簇的标准环是有限生成的分次环

这只要结合川又的无基点定理的结果便立即可得.与此相关,川又—松木确立的结果也令人回味无穷,即在一般型的场合极小模型只有有限个.

2 维时的双有理映射只要有限次合成收缩及其逆便可得到,这是该事实的推广.2 维时的证明用第一种例外曲线的数值判定便可立即明白,而 3 维时则远为困难.看看(1) 所完成的证明,似乎就明白了那些想要将 2 维时双有理映射的分解定理推广的众多朴素尝试终究归于失败的必然理由.

森在与科拉尔(Kollár) 的共同研究中,证明了即使在相对的情形下,也存在 3 维簇构成的簇极的小模型.利用此结果证明了 3 维时小平维数的形变不变性.多重亏格的形变不变性无法证明,是由于不能证明上述极小模型的典范除子是半丰富的.根据川又、官冈的基本贡献,当 $K^3 = 0, K^2$ 在数值上不为 0 时,知道只要小平维数为正即可.

如以上所见,极小模型理论是研究代数簇构造的关键,在高维代数簇中进行如此精密而深刻的研究,前不久连做梦都不敢想象.

1990 年 8 月 21 日至 29 日在日本东京举行了 1990 年国际(ICM - 90) 会议,在此次会上,森重文又喜获菲尔兹奖,并在大会上做了一小时报告.为了解森重文自己对其工作的评价,我们节选了其中一部分.

我们只讨论复数域 C 上的代数簇. 主要课题是 C 上函数域的分类.

设 X 与 Y 为 C 上的光滑射影簇, 我们称 X 双有理等价于 Y(记为 $X \sim Y$), 若它们的有理函数域 $C(X)$ 与 $C(Y)$ 是 C 的同构的扩域. 在我们的研究中, 典范线丛 K_x, 或全纯 n 形式的层 $\theta(K_X)$, $n = \dim X$, 起着关键作用. 换言之, 若 $X \sim Y$, 则有自然同构

$$H^0(X, \theta(vK_X)) \cong H^0(Y, \theta(vK_Y)), \forall v \geq 0$$

于是多亏格(plurigenera)

$$P_v(X) = \dim_C H^0(X, \theta(vK_X)), v > 0$$

是 X 的双有理不变量, 又小平维数 $k(X)$ 也是, 后者可用下式计算, 即

$$k(X) = \varlimsup_{v \to \infty} \frac{\lg P_v(X)}{\lg v}$$

这个由饭高(S. Iitaka)与 Moishezon 引进的 $k(X)$ 是代数簇双有理分类中最基本的双有理不变量. 它取 $\dim X + 2$ 个值: $-\infty, 0, \cdots, \dim X$, 而 $k(X) = \dim X, 0, -\infty$, 是对应于亏格大于等于 $2, 1, 0$ 的曲线的主要情况. 若 $k(X) = \dim X, X$ 被称为是一般型的.

从本维尼斯特(Benveniste)、川又(Y. kawamata)、科拉尔、森、里德与 Shokurov 在极小模型理论方面的最新结果, 可以得到关于 3 维簇的两个重要定理.

定理 1(本维尼斯特与川又的工作) 若 X 是一般型的 3 维簇, 则典范环

$$R(X) = \bigoplus_{p \geq 0} H^0(X, \theta(vK_X))$$

是有限生成的.

当 X 是具有 $k(X) < 3$ 的 3 维簇时, 藤田(Fujita)不用极小模型理论早就证明了 $R(X)$ 是有限生成的.

定理 2(宫冈的工作) 3 维簇 X 有 $k(X) = -\infty$(即 $P_v(X) = 0, \forall v > 0$)当且仅当 X 是单直纹的, 即存在曲面 Y 及从 $P^1 \times Y$ 到 X 的支配有理映射.

虽然在上列陈述中, 并未提到在与 X 双有理等价的簇中, 找一个"好"的模型 Y(极小模型)是至关重要的; 但选取正确的"好"模型的定义, 证明是个重要的起点.

定义(里德) 设 (P, X) 是正规簇芽. 我们称 (P, X) 是终端奇点, 若:

i 存在整数 $r > 0$ 使 rK_X 是个卡蒂埃(Cartier)除子(具有此性质的最小的 r 称为指标), 及

ii 设 $f: Y \to (P, X)$ 为任一消解, 并设 $E_1, \cdots, E_n$ 为全部例外除子, 则有

$$rK_Y = f^*(rK_X) + \sum a_i E_i, a_i > 0, \forall i$$

279

我们称代数簇 X 是个极小模型若 X 只有终端奇点且 K_X 为 nef(即对任一不可约曲线 C, 相交数 $(K_X \cdot C) \geqslant 0$), 我们称 X 只有 Q – 分解奇点若每个(整体积)韦尔除子是 Q – 卡蒂埃的.

此处的要点是尝试用双有理映射把 K_X 变为 nef(在维数大于 3 时仍是猜想), X 可能获得一些终端奇点, 它们是可以具体分类的. 下面是一个一般的例子.

设 a, m 是互素的整数, 令 $\mu_m = \{z \in C \mid z^m = 1\}$ 作用于 C^3 上, 有

$$\zeta(x, y, z) = (\zeta x, \zeta^{-1} y, \zeta^a z), \zeta \in \mu_m$$

则 $(P, X) = (0, C^3)/\mu_m$ 是个指标 m 的终端奇点.

极小模型理论认为:

定理 3 设 X 为任一光滑射影 3 维簇. 通过复合两种双有理映射(分别称为 flip 及除子式收缩) 若干次, X 变得双有理等价于一个只有 Q – 分解终端奇点的射影三维簇 Y 使:

i K_X 为 nef(极小模型情况), 或

ii Y 有到一个正规簇 Z 的映射, $\dim Z < \dim Y$ 而 $-K_Y$ 是在 Z 上相对丰富的.

暂时放开 flip 与除子式收缩的问题, 让我们看一下几个重要的推论.

在情况 ii 中, $k(X) = -\infty$, 而宫冈与森证明 X 是单直纹的. 在情况 i 中, 若继一般型, 则本维尼斯特与川又证明 (vK_X), 对某些 $v > 0$, 由整体截面所生成, 于是完成了定理 1. 宫冈证明情况 i 中 $k(X) \geqslant 0$, 于是完成了定理 2.

总结在一起, 我们有

定理 4 对光滑射影 3 维簇 X, 下列条件等价:

i $k(X) \geqslant 0$;

ii X 双有理等价于一个极小模型;

iii X 不是单直纹的.

用相对理论的框架, 3 维簇的双有理映射的粗略分解便得到了.

定理 5 设 $f: X \rightarrow Y$ 是在只有 Q – 分解奇点 3 维簇之间的映射, 则 f 可表达为 flip 与除子式收缩的复合.

在定理 3 与 5 中, 我们只从端射线(extremal rays) 所提供的信息去选 flip 与除子式收缩.

除子式收缩可视为曲面在一点吹开(blow up) 的 3 维类似. flip 是 3 维时的新现象, 它在原象与象的 1 维集以外为同构.

其实前面说了这么多就是想说明, 日本数学很行, 日本数论很行, 所以本书

很有价值,下面再介绍一下本书概况:

本书是"数论及其应用系列"丛书中的第六卷,丛书编辑为金光茂(日本近畿大学),编委成员有 V. N. Chubarikov(俄罗斯联邦莫斯科大学教授)、Christopher Deninger(德国明斯特大学教授)、贾朝华(中国科学院研究员)、刘建亚(中国山东大学副校长)、H. Niederreiter(新加坡国立大学教授)、M. Waldschmidt(法国巴黎第六大学教授),咨询委员会是 K. Ramachandra(印度塔塔基础研究院研究员,已退休)与 A. Schinzel(波兰科学院研究员).

本系列已出版书籍:

第一卷　算数几何和数论
第二卷　数论:在数论的海洋中航行
第四卷　实分析的问题与解决方法
第五卷　代数几何及其应用
第六卷　数论(梦幻之旅):第五届中日数论研论会演讲集
(第三卷为中国山东威海召开的一次数论会议的论文集)

本著作是 2008 年 8 月 27 ~ 31 日在日本东大阪市近畿大学举行的"数论(梦幻之旅):第五届中日数论研讨会"的演讲集,该会议的组织者是金光茂和刘建亚,以及当地的组织者青木崇教授.

这个会议名称听起来是浪漫且有异国情调的,闻名世界的诗人R. Browning 的亲戚 Tim. D. Browning 教授也是本次会议的参与者之一,对于会议的名称 R. Browning 表达了一个具有诗意的观点,即你可以梦想证明 RH 或是证明那些存在于梦中的最困难的东西. 但是主编选择这个名称是由于以下一些原因,大阪因为其最著名的大阪城堡而被人们熟知,大阪城堡的建造者是日本 16 世纪的英雄丰臣秀吉,他在临终时作了一首诗:"如晨露之坠地,如晨露之消失. 所有尘世盛行之物,亦不过梦中之梦." 原诗中 Naniwa 一词听起来与大阪的旧名字一样,并且他们中的许多人去过这个城市的中心难波公园(或许是为了德国音乐人 Stefan Honig 的作品 empty orchestra),难波是 Naniwa 的现代名称,这为命名本次研讨会提供了理由. 这个名字听起来虽然不诗意,逻辑性却很强. 会议由贾朝华教授作的一首诗结尾,再加上他们现在至少有参与者的四首诗,包括蔡天新(专业诗人)、贾朝华、刘建亚(曾在第四届中日研讨会中作诗一首)和 Tim D. Browning.

据本书主编在前言中所介绍:

"会议气氛如往常一样很愉快,我们相信每一个人都享受这丰富的五天时间. 我们组织了各种社会活动, 包括在喜来登酒店举行招待聚会, 感谢该酒店的好客与慷慨, 为我们提供了香槟酒瓶. 我们还去京都旅游(大巴车也是喜来登酒店安排的), 游览了当地的一些非去不可的地方, 一些外国参会者去了金阁封和清水寺多次. 晚上的活动也很有趣, 其中包括在不同地方举行的空管弦乐队活动. 我们发现, 不仅中国的参会者被认为是优秀的、热情的歌手, 而且大部分西方参会者也擅长唱歌. Trevor Wooley, Winfried Kohnen, Katsuya Miyake 和 Yumiko Hironaka 教授最具有娱乐精神. Jörg Brüdern 教授虽然没有唱歌, 但是许诺下次有机会的时候弹吉他. S. Kanemitsu 教授对于错过 Tim D. Browning 教授和 Koichi Kawada 教授参与的每晚举行的表演而遗憾."

现在我们来介绍一下这次研讨会的内容和这本演讲集. 这些演讲比较广泛地涉及了很多当代数论的内容, 就像从参会者的论文和本书的简介中看到的那样, 本著作的主题涵盖了解析数论(经典堆垒数论和现代堆垒数论的重点)、模形式理论、代数群和代数数论.

本演讲集不仅收集了参会者的论文, 而且还收集了没有参会的受邀者的论文, 比如 Andrzej Schinzel 教授、Igor Shparlinski 教授和 Ken Yamamura 教授.

中国的许多数学著作甚至名词都是从日本"引进"的, 中间可能会出现一点失误, 比如"无理数"这个词就有问题, 但大多数是好的, 希望这部书的出版再为我国的数论发展助一分力.

刘培志

2018 年 1 月 1 日
于哈工大

282

刘培杰数学工作室
已出版(即将出版)图书目录——高等数学

书　名	出版时间	定　价	编号
距离几何分析导引	2015-02	68.00	446
大学几何学	2017-01	78.00	688
关于曲面的一般研究	2016-11	48.00	690
近世纯粹几何学初论	2017-01	58.00	711
拓扑学与几何学基础讲义	2017-04	58.00	756
物理学中的几何方法	2017-06	88.00	767
几何学简史	2017-08	28.00	833
复变函数引论	2013-10	68.00	269
伸缩变换与抛物旋转	2015-01	38.00	449
无穷分析引论(上)	2013-04	88.00	247
无穷分析引论(下)	2013-04	98.00	245
数学分析	2014-04	28.00	338
数学分析中的一个新方法及其应用	2013-01	38.00	231
数学分析例选:通过范例学技巧	2013-01	88.00	243
高等代数例选:通过范例学技巧	2015-06	88.00	475
三角级数论(上册)(陈建功)	2013-01	38.00	232
三角级数论(下册)(陈建功)	2013-01	48.00	233
三角级数论(哈代)	2013-06	48.00	254
三角级数	2015-07	28.00	263
超越数	2011-03	18.00	109
三角和方法	2011-03	18.00	112
随机过程(Ⅰ)	2014-01	78.00	224
随机过程(Ⅱ)	2014-01	68.00	235
算术探索	2011-12	158.00	148
组合数学	2012-04	28.00	178
组合数学浅谈	2012-03	28.00	159
丢番图方程引论	2012-03	48.00	172
拉普拉斯变换及其应用	2015-02	38.00	447
高等代数.上	2016-01	38.00	548
高等代数.下	2016-01	38.00	549
高等代数教程	2016-01	58.00	579
数学解析教程.上卷.1	2016-01	58.00	546
数学解析教程.上卷.2	2016-01	38.00	553
数学解析教程.下卷.1	2017-04	48.00	781
数学解析教程.下卷.2	2017-06	48.00	782
函数构造论.上	2016-01	38.00	554
函数构造论.中	2017-06	48.00	555
函数构造论.下	2016-09	48.00	680
概周期函数	2016-01	48.00	572
变叙的项的极限分布律	2016-01	18.00	573
整函数	2012-08	18.00	161
近代拓扑学研究	2013-04	38.00	239
多项式和无理数	2008-01	68.00	22

刘培杰数学工作室
已出版(即将出版)图书目录——高等数学

书 名	出版时间	定 价	编号
模糊数据统计学	2008—03	48.00	31
模糊分析学与特殊泛函空间	2013—01	68.00	241
常微分方程	2016—01	58.00	586
平稳随机函数导论	2016—03	48.00	587
量子力学原理·上	2016—01	38.00	588
图与矩阵	2014—08	40.00	644
钢丝绳原理:第二版	2017—01	78.00	745
代数拓扑和微分拓扑简史	2017—06	68.00	791
受控理论与解析不等式	2012—05	78.00	165
不等式的分拆降维降幂方法与可读证明	2016—01	68.00	591
实变函数论	2012—06	78.00	181
复变函数论	2015—08	38.00	504
非光滑优化及其变分分析	2014—01	48.00	230
疏散的马尔科夫链	2014—01	58.00	266
马尔科夫过程论基础	2015—01	28.00	433
初等微分拓扑学	2012—07	18.00	182
方程式论	2011—03	38.00	105
Galois 理论	2011—03	18.00	107
古典数学难题与伽罗瓦理论	2012—11	58.00	223
伽罗华与群论	2014—01	28.00	290
代数方程的根式解及伽罗瓦理论	2011—03	28.00	108
代数方程的根式解及伽罗瓦理论(第二版)	2015—01	28.00	423
线性偏微分方程讲义	2011—03	18.00	110
几类微分方程数值方法的研究	2015—05	38.00	485
N 体问题的周期解	2011—03	28.00	111
代数方程式论	2011—05	18.00	121
线性代数与几何:英文	2016—06	58.00	578
动力系统的不变量与函数方程	2011—07	48.00	137
基于短语评价的翻译知识获取	2012—02	48.00	168
应用随机过程	2012—04	48.00	187
概率论导引	2012—04	18.00	179
矩阵论(上)	2013—06	58.00	250
矩阵论(下)	2013—06	48.00	251
对称锥互补问题的内点法:理论分析与算法实现	2014—08	68.00	368
抽象代数:方法导引	2013—06	38.00	257
集论	2016—01	48.00	576
多项式理论研究综述	2016—01	38.00	577
函数论	2014—11	78.00	395
反问题的计算方法及应用	2011—11	28.00	147
数阵及其应用	2012—02	28.00	164
绝对值方程—折边与组合图形的解析研究	2012—07	48.00	186
代数函数论(上)	2015—07	38.00	494
代数函数论(下)	2015—07	38.00	495

刘培杰数学工作室
已出版(即将出版)图书目录——高等数学

书 名	出版时间	定 价	编号
偏微分方程论:法文	2015—10	48.00	533
时标动力学方程的指数型二分性与周期解	2016—04	48.00	606
重刚体绕不动点运动方程的积分法	2016—05	68.00	608
水轮机水力稳定性	2016—05	48.00	620
Lévy 噪音驱动的传染病模型的动力学行为	2016—05	48.00	667
铣加工动力学系统稳定性研究的数学方法	2016—11	28.00	710
时滞系统:Lyapunov 泛函和矩阵	2017—05	68.00	784
粒子图像测速仪实用指南:第二版	2017—08	78.00	790
数域的上同调	2017—08	98.00	799
图的正交因子分解(英文)	2018—01	38.00	881
吴振奎高等数学解题真经(概率统计卷)	2012—01	38.00	149
吴振奎高等数学解题真经(微积分卷)	2012—01	68.00	150
吴振奎高等数学解题真经(线性代数卷)	2012—01	58.00	151
高等数学解题全攻略(上卷)	2013—06	58.00	252
高等数学解题全攻略(下卷)	2013—06	58.00	253
高等数学复习纲要	2014—01	18.00	384
超越吉米多维奇.数列的极限	2009—11	48.00	58
超越普里瓦洛夫.留数卷	2015—01	28.00	437
超越普里瓦洛夫.无穷乘积与它解析函数的应用卷	2015—05	28.00	477
超越普里瓦洛夫.积分卷	2015—06	18.00	481
超越普里瓦洛夫.基础知识卷	2015—06	28.00	482
超越普里瓦洛夫.数项级数卷	2015—07	38.00	489
超越普里瓦洛夫.微分、解析函数、导数卷	2018—01	48.00	852
统计学专业英语	2007—03	28.00	16
统计学专业英语(第二版)	2012—07	48.00	176
统计学专业英语(第三版)	2015—04	68.00	465
代换分析:英文	2015—07	38.00	499
历届美国大学生数学竞赛试题集.第一卷(1938—1949)	2015—01	28.00	397
历届美国大学生数学竞赛试题集.第二卷(1950—1959)	2015—01	28.00	398
历届美国大学生数学竞赛试题集.第三卷(1960—1969)	2015—01	28.00	399
历届美国大学生数学竞赛试题集.第四卷(1970—1979)	2015—01	18.00	400
历届美国大学生数学竞赛试题集.第五卷(1980—1989)	2015—01	28.00	401
历届美国大学生数学竞赛试题集.第六卷(1990—1999)	2015—01	28.00	402
历届美国大学生数学竞赛试题集.第七卷(2000—2009)	2015—08	18.00	403
历届美国大学生数学竞赛试题集.第八卷(2010—2012)	2015—01	18.00	404
超越普特南试题:大学数学竞赛中的方法与技巧	2017—04	98.00	758
历届国际大学生数学竞赛试题集(1994—2010)	2012—01	28.00	143
全国大学生数学夏令营数学竞赛试题及解答	2007—03	28.00	15
全国大学生数学竞赛辅导教程	2012—07	28.00	189
全国大学生数学竞赛复习全书(第2版)	2017—05	58.00	787

刘培杰数学工作室
已出版（即将出版）图书目录——高等数学

书　名	出版时间	定　价	编号
历届美国大学生数学竞赛试题集	2009—03	88.00	43
前苏联大学生数学奥林匹克竞赛题解（上编）	2012—04	28.00	169
前苏联大学生数学奥林匹克竞赛题解（下编）	2012—04	38.00	170
大学生数学竞赛讲义	2014—09	28.00	371
普林斯顿大学数学竞赛	2016—06	38.00	669
初等数论难题集（第一卷）	2009—05	68.00	44
初等数论难题集（第二卷）（上、下）	2011—02	128.00	82,83
数论概貌	2011—03	18.00	93
代数数论（第二版）	2013—08	58.00	94
代数多项式	2014—06	38.00	289
初等数论的知识与问题	2011—02	28.00	95
超越数论基础	2011—03	28.00	96
数论初等教程	2011—03	28.00	97
数论基础	2011—03	18.00	98
数论基础与维诺格拉多夫	2014—03	18.00	292
解析数论基础	2012—08	28.00	216
解析数论基础（第二版）	2014—01	48.00	287
解析数论问题集（第二版）（原版引进）	2014—05	88.00	343
解析数论问题集（第二版）（中译本）	2016—04	88.00	607
解析数论基础（潘承洞，潘承彪著）	2016—07	98.00	673
解析数论导引	2016—07	58.00	674
数论入门	2011—03	38.00	99
代数数论入门	2015—03	38.00	448
数论开篇	2012—07	28.00	194
解析数论引论	2011—03	48.00	100
Barban Davenport Halberstam 均值和	2009—01	40.00	33
基础数论	2011—03	28.00	101
初等数论 100 例	2011—05	18.00	122
初等数论经典例题	2012—07	18.00	204
最新世界各国数学奥林匹克中的初等数论试题（上、下）	2012—01	138.00	144,145
初等数论（Ⅰ）	2012—01	18.00	156
初等数论（Ⅱ）	2012—01	18.00	157
初等数论（Ⅲ）	2012—01	28.00	158
平面几何与数论中未解决的新老问题	2013—01	68.00	229
代数数论简史	2014—11	28.00	408
代数数论	2015—09	88.00	532
代数、数论及分析习题集	2016—11	98.00	695
数论导引提要及习题解答	2016—01	48.00	559
素数定理的初等证明. 第2版	2016—09	48.00	686
数论中的模函数与狄利克雷级数（第二版）	2017—11	78.00	837
数论：数学导引	2018—01	68.00	849

刘培杰数学工作室
已出版（即将出版）图书目录——高等数学

书　名	出版时间	定　价	编号
新编640个世界著名数学智力趣题	2014—01	88.00	242
500个最新世界著名数学智力趣题	2008—06	48.00	3
400个最新世界著名数学最值问题	2008—09	48.00	36
500个世界著名数学征解问题	2009—06	48.00	52
400个中国最佳初等数学征解老问题	2010—01	48.00	60
500个俄罗斯数学经典老题	2011—01	28.00	81
1000个国外中学物理好题	2012—04	48.00	174
300个日本高考数学题	2012—05	38.00	142
700个早期日本高考数学试题	2017—02	88.00	752
500个前苏联早期高考数学试题及解答	2012—05	28.00	185
546个早期俄罗斯大学生数学竞赛题	2014—03	38.00	285
548个来自美苏的数学好问题	2014—11	28.00	396
20所苏联著名大学早期入学试题	2015—02	18.00	452
161道德国工科大学生必做的微分方程习题	2015—05	28.00	469
500个德国工科大学生必做的高数习题	2015—06	28.00	478
360个数学竞赛问题	2016—08	58.00	677
德国讲义日本考题.微积分卷	2015—04	48.00	456
德国讲义日本考题.微分方程卷	2015—04	38.00	457
二十世纪中叶中、英、美、日、法、俄高考数学试题精选	2017—06	38.00	783
博弈论精粹	2008—03	58.00	30
博弈论精粹.第二版(精装)	2015—01	88.00	461
数学 我爱你	2008—01	28.00	20
精神的圣徒　别样的人生——60位中国数学家成长的历程	2008—09	48.00	39
数学史概论	2009—06	78.00	50
数学史概论(精装)	2013—03	158.00	272
数学史选讲	2016—01	48.00	544
斐波那契数列	2010—02	28.00	65
数学拼盘和斐波那契魔方	2010—07	38.00	72
斐波那契数列欣赏	2011—01	28.00	160
数学的创造	2011—02	48.00	85
数学美与创造力	2016—01	48.00	595
数海拾贝	2016—01	48.00	590
数学中的美	2011—02	38.00	84
数论中的美学	2014—12	38.00	351
数学王者　科学巨人——高斯	2015—01	28.00	428
振兴祖国数学的圆梦之旅:中国初等数学研究史话	2015—06	98.00	490
二十世纪中国数学史料研究	2015—10	48.00	536
数字谜、数阵图与棋盘覆盖	2016—01	58.00	298
时间的形状	2016—01	38.00	556
数学发现的艺术:数学探索中的合情推理	2016—07	58.00	671
活跃在数学中的参数	2016—07	48.00	675

刘培杰数学工作室
已出版(即将出版)图书目录——高等数学

书 名	出版时间	定 价	编号
格点和面积	2012—07	18.00	191
射影几何趣谈	2012—04	28.00	175
斯潘纳尔引理——从一道加拿大数学奥林匹克试题谈起	2014—01	28.00	228
李普希兹条件——从几道近年高考数学试题谈起	2012—10	18.00	221
拉格朗日中值定理——从一道北京高考试题的解法谈起	2015—10	18.00	197
闵科夫斯基定理——从一道清华大学自主招生试题谈起	2014—01	28.00	198
哈尔测度——从一道冬令营试题的背景谈起	2012—08	28.00	202
切比雪夫逼近问题——从一道中国台北数学奥林匹克试题谈起	2013—04	38.00	238
伯恩斯坦多项式与贝齐尔曲面——从一道全国高中数学联赛试题谈起	2013—03	38.00	236
卡塔兰猜想——从一道普特南竞赛试题谈起	2013—06	18.00	256
麦卡锡函数和阿克曼函数——从一道前南斯拉夫数学奥林匹克试题谈起	2012—08	18.00	201
贝蒂定理与拉姆贝克莫瑟尔定理——从一个拣石子游戏谈起	2012—08	18.00	217
皮亚诺曲线和豪斯道夫分球定理——从无限集谈起	2012—08	18.00	211
平面凸图形与凸多面体	2012—10	28.00	218
斯坦因豪斯问题——从一道二十五省市自治区中学数学竞赛试题谈起	2012—07	18.00	196
纽结理论中的亚历山大多项式与琼斯多项式——从一道北京市高一数学竞赛试题谈起	2012—07	28.00	195
原则与策略——从波利亚"解题表"谈起	2013—04	38.00	244
转化与化归——从三大尺规作图不能问题谈起	2012—08	28.00	214
代数几何中的贝祖定理(第一版)——从一道IMO试题的解法谈起	2013—08	18.00	193
成功连贯理论与约当块理论——从一道比利时数学竞赛试题谈起	2012—04	18.00	180
素数判定与大数分解	2014—08	18.00	199
置换多项式及其应用	2012—10	18.00	220
椭圆函数与模函数——从一道美国加州大学洛杉矶分校(UCLA)博士资格考题谈起	2012—10	28.00	219
差分方程的拉格朗日方法——从一道2011年全国高考理科试题的解法谈起	2012—08	28.00	200
力学在几何中的一些应用	2013—01	38.00	240
高斯散度定理、斯托克斯定理和平面格林定理——从一道国际大学生数学竞赛试题谈起	即将出版		
康托洛维奇不等式——从一道全国高中联赛试题谈起	2013—03	28.00	337
西格尔引理——从一道第18届IMO试题的解法谈起	即将出版		
罗斯定理——从一道前苏联数学竞赛试题谈起	即将出版		
拉克斯定理和阿廷定理——从一道IMO试题的解法谈起	2014—01	58.00	246
毕卡大定理——从一道美国大学数学竞赛试题谈起	2014—07	18.00	350
贝齐尔曲线——从一道全国高中联赛试题谈起	即将出版		
拉格朗日乘子定理——从一道2005年全国高中联赛试题的高等数学解法谈起	2015—05	28.00	480
雅可比定理——从一道日本数学奥林匹克试题谈起	2013—04	48.00	249
李天岩—约克定理——从一道波兰数学竞赛试题谈起	2014—06	28.00	349
整系数多项式因式分解的一般方法——从克朗耐克算法谈起	即将出版		

刘培杰数学工作室
已出版（即将出版）图书目录——高等数学

书　名	出版时间	定　价	编号
布劳维不动点定理——从一道前苏联数学奥林匹克试题谈起	2014—01	38.00	273
伯恩赛德定理——从一道英国数学奥林匹克试题谈起	即将出版		
布查特—莫斯特定理——从一道上海市初中竞赛试题谈起	即将出版		
数论中的同余数问题——从一道普特南竞赛试题谈起	即将出版		
范·德蒙行列式——从一道美国数学奥林匹克试题谈起	即将出版		
中国剩余定理：总数法构建中国历史年表	2015—01	28.00	430
牛顿程序与方程求根——从一道全国高考试题解法谈起	即将出版		
库默尔定理——从一道IMO预选试题谈起	即将出版		
卢丁定理——从一道冬令营试题的解法谈起	即将出版		
沃斯滕霍姆定理——从一道IMO预选试题谈起	即将出版		
卡尔松不等式——从一道莫斯科数学奥林匹克试题谈起	即将出版		
信息论中的香农熵——从一道近年高考压轴题谈起	即将出版		
约当不等式——从一道希望杯竞赛试题谈起	即将出版		
拉比诺维奇定理	即将出版		
刘维尔定理——从一道《美国数学月刊》征解问题的解法谈起	即将出版		
卡塔兰恒等式与级数求和——从一道IMO试题的解法谈起	即将出版		
勒让德猜想与素数分布——从一道爱尔兰竞赛试题谈起	即将出版		
天平称重与信息论——从一道基辅市数学奥林匹克试题谈起	即将出版		
哈密尔顿—凯莱定理：从一道高中数学联赛试题的解法谈起	2014—09	18.00	376
艾思特曼定理——从一道CMO试题的解法谈起	即将出版		
一个爱尔特希问题——从一道西德数学奥林匹克试题谈起	即将出版		
有限群中的爱丁格尔问题——从一道北京市初中二年级数学竞赛试题谈起	即将出版		
贝克码与编码理论——从一道全国高中联赛试题谈起	即将出版		
帕斯卡三角形	2014—03	18.00	294
蒲丰投针问题——从2009年清华大学的一道自主招生试题谈起	2014—01	38.00	295
斯图姆定理——从一道"华约"自主招生试题的解法谈起	2014—01	18.00	296
许瓦兹引理——从一道加利福尼亚大学伯克利分校数学系博士生试题谈起	2014—08	18.00	297
拉姆塞定理——从王诗宬院士的一个问题谈起	2016—04	48.00	299
坐标法	2013—12	28.00	332
数论三角形	2014—04	38.00	341
毕克定理	2014—07	18.00	352
数林掠影	2014—09	48.00	389
我们周围的概率	2014—10	38.00	390
凸函数最值定理：从一道华约自主招生题的解法谈起	2014—10	28.00	391
易学与数学奥林匹克	2014—10	38.00	392
生物数学趣谈	2015—01	18.00	409
反演	2015—01	28.00	420
因式分解与圆锥曲线	2015—01	18.00	426
轨迹	2015—01	28.00	427
面积原理：从常庚哲命的一道CMO试题的积分解法谈起	2015—01	48.00	431
形形色色的不动点定理：从一道28届IMO试题谈起	2015—01	38.00	439
柯西函数方程：从一道上海交大自主招生的试题谈起	2015—02	28.00	440

刘培杰数学工作室
已出版(即将出版)图书目录——高等数学

书　名	出版时间	定　价	编号
三角恒等式	2015—02	28.00	442
无理性判定:从一道2014年"北约"自主招生试题谈起	2015—01	38.00	443
数学归纳法	2015—03	18.00	451
极端原理与解题	2015—04	28.00	464
法雷级数	2014—08	18.00	367
摆线族	2015—01	38.00	438
函数方程及其解法	2015—05	38.00	470
含参数的方程和不等式	2012—09	28.00	213
希尔伯特第十问题	2016—01	38.00	543
无穷小量的求和	2016—01	28.00	545
切比雪夫多项式:从一道清华大学金秋营试题谈起	2016—01	38.00	583
泽肯多夫定理	2016—03	38.00	599
代数等式证题法	2016—01	28.00	600
三角等式证题法	2016—01	28.00	601
吴大任教授藏书中的一个因式分解公式:从一道美国数学邀请赛试题的解法谈起	2016—06	28.00	656
易卦——类万物的数学模型	2017—08	68.00	838
"不可思议"的数与数系可持续发展	2018—01	38.00	878
最短线	2018—01	38.00	879
从毕达哥拉斯到怀尔斯	2007—10	48.00	9
从迪利克雷到维斯卡尔迪	2008—01	48.00	21
从哥德巴赫到陈景润	2008—05	98.00	35
从庞加莱到佩雷尔曼	2011—08	138.00	136
从费马到怀尔斯——费马大定理的历史	2013—10	198.00	I
从庞加莱到佩雷尔曼——庞加莱猜想的历史	2013—10	298.00	II
从切比雪夫到爱尔特希(上)——素数定理的初等证明	2013—07	48.00	III
从切比雪夫到爱尔特希(下)——素数定理100年	2012—12	98.00	III
从高斯到盖尔方特——二次域的高斯猜想	2013—10	198.00	IV
从库默尔到朗兰兹——朗兰兹猜想的历史	2014—01	98.00	V
从比勒巴赫到德布朗斯——比勒巴赫猜想的历史	2014—02	298.00	VI
从麦比乌斯到陈省身——麦比乌斯变换与麦比乌斯带	2014—02	298.00	VII
从布尔到豪斯道夫——布尔方程与格论漫谈	2013—10	198.00	VIII
从开普勒到阿诺德——三体问题的历史	2014—05	298.00	IX
从华林到华罗庚——华林问题的历史	2013—10	298.00	X
数学物理大百科全书.第1卷	2016—01	418.00	508
数学物理大百科全书.第2卷	2016—01	408.00	509
数学物理大百科全书.第3卷	2016—01	396.00	510
数学物理大百科全书.第4卷	2016—01	408.00	511
数学物理大百科全书.第5卷	2016—01	368.00	512
朱德祥代数与几何讲义.第1卷	2017—01	38.00	697
朱德祥代数与几何讲义.第2卷	2017—01	28.00	698
朱德祥代数与几何讲义.第3卷	2017—01	28.00	699

刘培杰数学工作室
已出版(即将出版)图书目录——高等数学

书 名	出版时间	定 价	编号
闵嗣鹤文集	2011—03	98.00	102
吴从炘数学活动三十年(1951~1980)	2010—07	99.00	32
吴从炘数学活动又三十年(1981~2010)	2015—07	98.00	491
斯米尔诺夫高等数学.第一卷	2017—02	88.00	770
斯米尔诺夫高等数学.第二卷.第一分册	2017—02	68.00	771
斯米尔诺夫高等数学.第二卷.第二分册	2017—02	68.00	772
斯米尔诺夫高等数学.第二卷.第三分册	2017—02	48.00	773
斯米尔诺夫高等数学.第三卷.第一分册	2017—06	48.00	774
斯米尔诺夫高等数学.第三卷.第二分册	2017—02	58.00	775
斯米尔诺夫高等数学.第三卷.第三分册	2017—02	68.00	776
斯米尔诺夫高等数学.第四卷.第一分册	2017—02	48.00	777
斯米尔诺夫高等数学.第四卷.第二分册	2017—02	88.00	778
斯米尔诺夫高等数学.第五卷.第一分册	2017—04	58.00	779
斯米尔诺夫高等数学.第五卷.第二分册	2017—02	68.00	780
zeta 函数,q-zeta 函数,相伴级数与积分	2015—08	88.00	513
微分形式:理论与练习	2015—08	58.00	514
离散与微分包含的逼近和优化	2015—08	58.00	515
艾伦·图灵:他的工作与影响	2016—01	98.00	560
测度理论概率导论,第 2 版	2016—01	88.00	561
带有潜在故障恢复系统的半马尔柯夫模型控制	2016—01	98.00	562
数学分析原理	2016—01	88.00	563
随机偏微分方程的有效动力学	2016—01	88.00	564
图的谱半径	2016—01	58.00	565
量子机器学习中数据挖掘的量子计算方法	2016—01	98.00	566
量子物理的非常规方法	2016—01	118.00	567
运输过程的统一非局部理论:广义波尔兹曼物理动力学,第 2 版	2016—01	198.00	568
量子力学与经典力学之间的联系在原子、分子及电动力学系统建模中的应用	2016—01	58.00	569
算术域:第 3 版	2017—08	158.00	820
算术域	2018—01	158.00	821
高等数学竞赛:1962—1991 年的米洛克斯·史怀哲竞赛	2018—01	128.00	822
用数学奥林匹克精神解决数论问题	2018—01	108.00	823
代数几何(德语)	即将出版		824
丢番图近似值	2018—01	78.00	825
代数几何学基础教程	2018—01	98.00	826
解析数论入门课程	2018—01	78.00	827
中正大学数论教程	即将出版		828
数论中的丢番图问题	2018—01	78.00	829
数论(梦幻之旅):第五届中日数论研讨会演讲集	2018—01	68.00	830
数论新应用	2018—01	68.00	831
数论	2018—01	78.00	832

刘培杰数学工作室
已出版(即将出版)图书目录——高等数学

书　名	出版时间	定　价	编号
数学王子——高斯	2018—01	48.00	858
坎坷奇星——阿贝尔	2018—01	48.00	859
闪烁奇星——伽罗瓦	2018—01	58.00	860
无穷统帅——康托尔	2018—01	48.00	861
科学公主——柯瓦列夫斯卡娅	2018—01	48.00	862
抽象代数之母——埃米·诺特	2018—01	48.00	863
电脑先驱——图灵	2018—01	58.00	864
昔日神童——维纳	2018—01	48.00	865
数坛怪侠——爱尔特希	2018—01	68.00	866

联系地址:哈尔滨市南岗区复华四道街 10 号　哈尔滨工业大学出版社刘培杰数学工作室
网　　址:http://lpj.hit.edu.cn/
邮　　编:150006
联系电话:0451—86281378　　13904613167
E-mail:lpj1378@163.com